National Safe Tractor and Machinery Operation Program

Student Manual

General Class Information

Instructor(s)	Instructor 1	Instructor 2	Instructor 3
Name:			
Address:			
Phone:			
Email:			

Class Schedule:

Location:

Date	Day	Time	Topics	Task Sheets

Written Test Date: ___________ Location:____

Skills and Driving Test Date: ___________ Location:___________________________________

Introduction

The United States Department of Labor (USDOL) Fair Labor Standards Act of 1938 identified several hazardous occupations where individuals employed in these occupations had to be 18 years of age or older. The identified occupations included manufacturing and mining. In 1968, the Fair Labor Standards Act of 1938 was amended to include the Hazardous Occupations Order for Agriculture (AgHO). This order established the age for employment in agriculture as 16 years of age or 14 years of age with special training. The order also identified hazardous operations and farm tasks, such as operating a tractor over 20 horsepower and operating other specific farm machinery, including, but not limited to, a corn picker, combine, hay baler, feed grinder, forklift or power post-driver. Exempt from this law are (1) minors who are employed on a farm owned and operated by their parents or guardians and (2) minors ages 14 and 15 who have received training and certification from an approved tractor and farm machinery safety certification program. Each state is responsible for providing the approved safety training that allows minors ages 14 or 15 to legally be employed to operate a tractor or other specified machinery. Youth can receive a USDOL Certificate of Training for tractor driving by completing 4 hours of orientation to on-farm hazards and general safety and that have participated in a 10-hour tractor safety course. With an additional 10-hour machinery safety course, youth receive a certificate of training for machinery operation. Individuals 16 years of age and older can be employed in agriculture without this certification.

In the late 1960s and early 1970s, several tractor and machinery programs were developed based on the 1968 AgHO. Over the years, these programs deteriorated with a loss of interest and implementation. Little is known about the extent that tractor and machinery safety certification programs reach young tractor and machinery operators. Great variations exist in the type of teaching materials, the number of hours, the forms of instruction, the testing procedures, and the skills assessment in these programs. The need for current and better quality training materials was cited by both certification program instructors and coordinators.

In recognition of these shortcomings, the United States Department of Agriculture (USDA) funded a major project with Penn State University, Ohio State University and the National Safety Council to develop a National Safe Tractor and Machinery Operation Program (NSTMOP). This project was funded in 2001 under USDA Agreement No. 2001-41521-01263, "Establishing a 'National Safe Tractor and Machinery Operation Certification' Program" and continuing in 2010 under USDA Agreement No. 2010-41521-20839.

If you would like to become a **student or instructor** in the NSTMOP, find out more at:

www.nstmop.psu.edu

or

National Safe Tractor and Machinery Operation Program
The Pennsylvania State University
Agricultural & Biological Engineering Department
246 Agricultural Engineering Building
University Park, PA 16802

Phone: 814-893-8124
Fax: 814-863-1031
Email: nstmop@psu.edu

Regulations

The United States Department of Labor (USDOL) regulations that relate to occupations in agriculture particularly hazardous for the employment of children younger than age 16 are found in the Code of Federal Regulations, Title 29, Part 570, Subpart E-1 (29 CFR 570 Subpart E-1), officially referred to as AgHO. The HOSTA-NSTMOP Task Sheet 1.2.1, *Hazardous Occupations Order in Agriculture,* summarizes what youth younger than age 16 and their parents and employers need to know about the regulations. The official regulations, 29 CFR 570 Subpart E-1, can be found at the USDOL website (www.dol.gov).

Training Programs

Sufficient meeting time to provide a minimum of ***24 hours of instruction*** to teach the Minimum Core Content Areas (MCCA) based curriculum is needed. This instruction may take place over several weeks and can include group discussions, demonstrations, field trips to farms or equipment dealers, and hands-on activities. In addition to classroom instruction, you should be expected to complete study or field assignments, which may be used to meet the 24-hour requirement.

Curriculum Materials
The NSTMOP Task Sheets are the primary curriculum resource for the NSTMOP program and may be used alone or supplemented with additional information and instructor knowledge. An example of an additional resource: Deere & Company's *Farm and Ranch Safety Management* book, other written texts, other instructional task sheets, student worksheets, tractor and machine operator and service manuals, demonstrations, vendor tractor and equipment safety videos, and guest speakers.

Instructional Methods
The NSTMOP materials are designed to be used in a variety of instructional settings. They can be used in:

- a traditional classroom setting (e.g. a formal high school agricultural classroom setting),
- an extension/4-H evening or Saturday meeting format,
- an independent study format, or
- some combination.

Testing Procedures

1. **You must be at least 14 years of age on the test date to be allowed to take the Skills and Driving Tests**. If you are under the age of 14, you are permitted to attend educational classes and take the Written Test but are not allowed to operate tractors or other equipment if practice sessions are held.
2. The Written, Skills, and Driving Tests must be passed with a minimum score of 70% before you can be awarded a USDOL Certificate of Training. The locations where these tests are given hereafter are referenced as Test Stations.
3. The Written Test will consist of a combination of 50 multiple-choice and true/false questions from a question bank based on NSTMOP Task Sheets that cover the MCCA. The Written Test must be passed with a 70% or better before you can move to the Skills Test.
4. You must pass first the Written Test and then the Skills Test before you can take the Driving Test.

5. A personal, laminated Certificate of Completion wallet card (see below) will be issued by the NSTMOP Office if you successfully pass all of the tests and the $5 fee is submitted. This wallet card does not replace the *USDOL Certificate of Training* that must be kept on file by the trainer, employer and the employee.

Front

National Safe Tractor and Machinery Operation Program
Certificate of Completion

Name
«Name»

Date of Birth
«DOB»

Address
«Address1»
«Address2» «Zip»

Certificate Number
«Cert Number»

Back

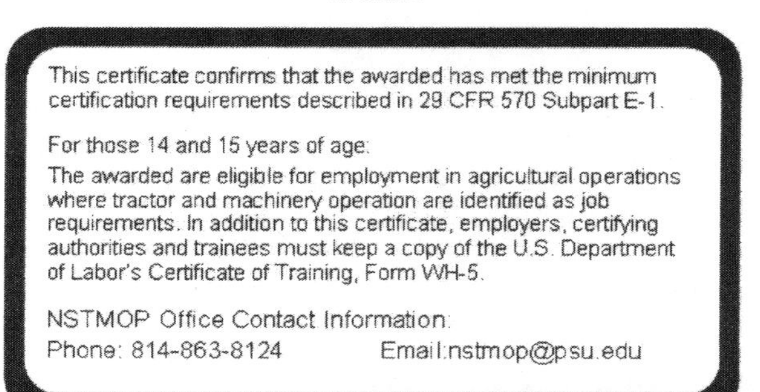

This certificate confirms that the awarded has met the minimum certification requirements described in 29 CFR 570 Subpart E-1.

For those 14 and 15 years of age:
The awarded are eligible for employment in agricultural operations where tractor and machinery operation are identified as job requirements. In addition to this certificate, employers, certifying authorities and trainees must keep a copy of the U.S. Department of Labor's Certificate of Training, Form WH-5.

NSTMOP Office Contact Information:
Phone: 814-863-8124 Email:nstmop@psu.edu

Written Test

The goal of the Written Test is to evaluate your knowledge of the agricultural safety and health topics that are listed in the Minimum Core Content Areas (MCCA) for the National Safe Tractor and Machinery Operation Program. The Written Test will consist of a combination of 50 multiple-choice and true/false questions from a question bank based on those NSTMOP Task Sheets that cover the MCCA. The Written Test must be passed before you can move to the Skills Test. The Written Test is passed with a score of 70% or higher. If you fail, you will not be allowed to retake the Written Test without additional study.

Skills and Driving Test

The goal of the Skills Test is to evaluate your ability to safely and efficiently start a tractor and hitch to a wheeled or 3-point implement. The goal of the Driving Test is to test your ability to safely and efficiently drive a tractor pulling a two-wheel implement through a specified course with spaces and borders. Neither the Skills nor Driving test activity is a competitive event; actual barrier measurements and times are not recorded. The Skills and Driving Tests must each be passed with a minimum score of 70%, and zero "automatic failure" violations. Experienced tractor operators may complete each activity within 8-10 minutes while less experienced operators may require a few minutes more. *Any operator who cannot complete each activity within 15 minutes automatically fails the test.*

You should be appropriately dressed for the Skills and Driving Tests. Snug-fitting clothes in good repair, long pants, and solid shoes with slip-resistant treads are recommended. Inappropriate dress includes baggy pants, shorts, sandals or open-toed footwear, and jewelry, including rings, watches, necklaces and dangling earrings. You may be forbidden to take either test if you are not appropriately dressed in the opinion of the Community Lead Instructor.

Suggested skills and driving courses and evaluation forms are provided on the next few pages. Different courses are acceptable as long as the layouts include similar maneuvers.

Skills and Driving Test Layout Map

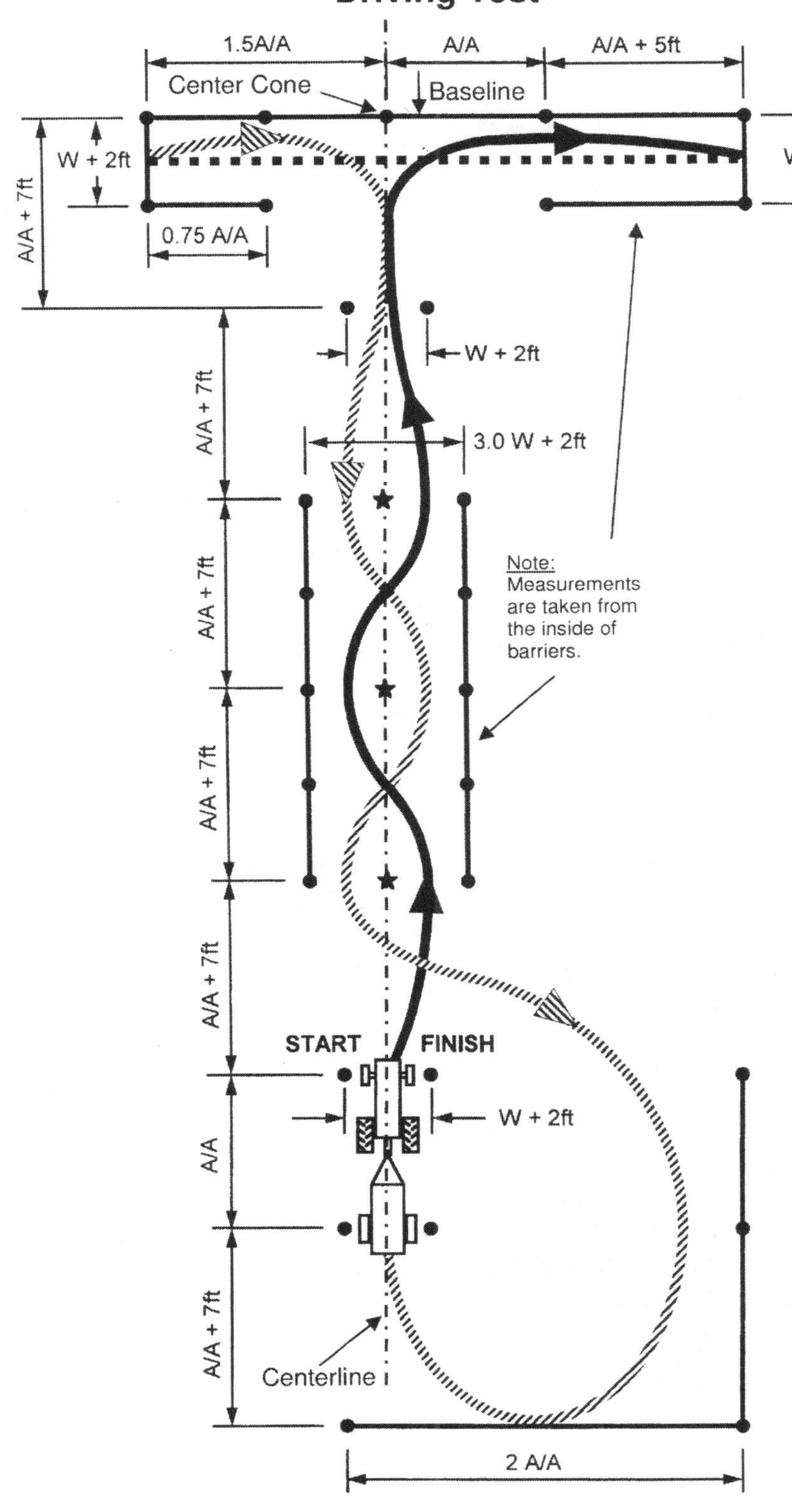

Calculations

A/A = ________

2 A/A = ________

1.5 A/A = ________

.75 A/A = ________

A/A + 5 = ________

A/A + 7 = ________

W = ________

W + 2 = ________

3 W + 2 = ________

Length 7 A/A + 42 = ________

Width 3.5 A/A + 5 = ________

★ — Use traffic cones, stakes, buckets, etc. to identify serpentine path

● — Use traffic cones, stakes, straw or hay, etc. as markers.

| — Use rope, baler twine, straw or hay bales to form a continuous line.

Start path

Back-up path

Return path

A/A — Means axle to axle. The distance between center of front axle of tractor and center of axle of towed equipment.

W — Width in feet of the tractor or two-wheel towed equipment, whichever is wider.

Skills Test

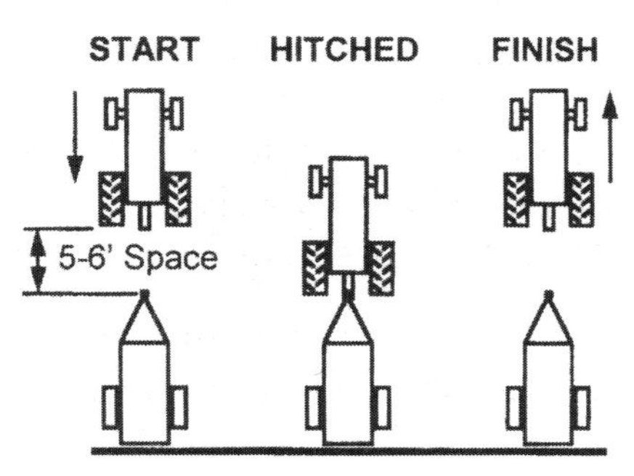

Note: Raise or lower implement 3 or 4 inches before starting.

Skills Test Evaluation Form

Student Name: ____________ Skills Examiner: ____________

Test Date: ____________ Test Location: ____________

Skill Activities (15 Minutes Maximum):

☐ Student is dressed appropriately.

	S	U	NA	
				Tractor entry and start-up
1				Can demonstrate universal hand signal for ____________ (Instructor Randomly Picks)
2				Use handholds and steps to mount tractor
3				Adjust seat, steering wheel (if necessary)
4				Buckle seat belt
5				Check major controls (PTO, hydraulics, gear shift stick) for the neutral (or PARK) position.
6				Adjust throttle to one-third open, push clutch in, move gear selection to START position if so equipped.
7				Can demonstrate and verbalize how to safely stop the tractor.
				DISCUSS UNSATISFACTORY PERFORMANCE HERE
				Tractor Back-up and Implement Hitching
8				With clutch pushed in, start tractor, idle down throttle
9				Select slow/low reverse gear.
10				Slowly and smoothly back tractor to within a few inches of implement tongue.
11				Stop tractor, shift to neutral and set brakes or place in PARK, reduce throttle (if moved from start-up), shut tractor off **(Not shutting off tractor is automatic Skills Test failure)**, unbuckle seat belt.
12				Dismount tractor by facing tractor and using handholds and steps.
				DISCUSS UNSATISFACTORY PERFORMANCE HERE
13				Use handholds and steps to mount tractor. Buckle seat belt, adjust throttle to one-third open, push clutch in, move gear selection to START position if so equipped, start tractor.
14				Back tractor to align drawbar and implement tongue holes, stop tractor, place gear in neutral and set brakes or place gear in PARK, reduce throttle (if moved from start-up), shut tractor off **(Not shutting off tractor is automatic Skills Test failure)**, dismount tractor by facing tractor and using handholds and steps. Student may need to repeat steps 13-14.

	S	U	NA	
15				Place tractor pin in hole to connect the implement, using the tractor pin safety-locking device; connect safety chains (if present) to tractor.
16				Correctly connect the implement PTO, hydraulic hoses, or both.
17				Wind-up jack, swing to storage position
				INSTRUCTOR INSPECTS ALL CONNECTIONS. CORRECT MISTAKES
				Tractor Unhitching and Shut-off
18				Lower jack stand, adjust to take weight off hitch points
19				Disconnect the implement PTO, hydraulic hoses, or both.
20				Remove tractor pin
21				Use handholds and steps to mount tractor
22				Buckle seat belt, adjust throttle to one-third open, push clutch in, move gear selection to START position if so equipped, start tractor
23				Move tractor forward 5 to 6 feet **(Not completely disconnecting the implement from the tractor or to move the jack stand to storage position is automatic Skills Test failure)**.
24				Stops tractor, lowers throttle (if adjusted upward), places gear in neutral and sets brakes or puts gear in PARK, shuts tractor off, unbuckles seat belt. **(Not shutting off tractor is automatic Skills Test failure)**
25				Dismount tractor by facing tractor and using handholds and steps.
				Not completing the Skills Test in 15 minutes is automatic failure.
				DISCUSS UNSATISFACTORY PERFORMANCE HERE
Automatic Failure – Please note the reason for the automatic failure.				

Driving Test Evaluation Form

Student Name: ____________ Skills Examiner: ____________

Test Date: ____________ Test Location: ____________

Driving Activity (15 minutes maximum): Students must explain each activity as performed unless otherwise noted. Instructors should correct student errors where indicated.

☐ Student is dressed appropriately.

	S	U	NA	Driving Activities
				Pre-operation inspection: student identifies and explains what they are inspecting.
1				Check Fluid Levels (Oil reservoir dipstick, Fuel tank cap, Water/antifreeze fill location, Hydraulic fluid dipstick, etc.)
2				Battery condition
3				Tire condition (Tractor and machine)
4				Guards and Shields (Tractor and machine)
5				Hitch and related connections
6				Walk around Tractor and Machine and look for hazards
				DISCUSS UNSATISFACTORY PERFORMANCE HERE
				Tractor entry
7				Use handholds and steps to mount tractor
8				Buckle seat belt
9				Adjust seat, steering wheel
10				Check major controls (PTO, hydraulics, gear shift position) for the neutral or PARK position
11				Adjust throttle, push clutch in
12				Demonstrate and verbalize how to safely stop the tractor
				DISCUSS UNSATISFACTORY PERFORMANCE HERE
				Tractor start-up and Driving: If too high a gear and/or speed is used, or if driver appears to not have complete control or awareness of tractor and implement positioning, the driver should be stopped immediately, and an automatic failure should be recorded.
13				Starts tractor movement smoothly, has selected low gear, using low throttle setting

	S	U	NA	Driving Activities
14				Drives through serpentine (or similar) course without running over or crushing barriers (a light touch or brushing is acceptable) or having to back up
15				Drives into pull-in stall without running over or crushing any barriers (a light touch or brushing is acceptable) and stops within a foot of end barriers without running over or crushing them (a light touch or brushing is acceptable)
16				Backs into back-in stall within two passes, stops within a foot of end barriers without running over or crushing them (a light touch or brushing is acceptable). 1 pull-up is acceptable
17				Drives out of back-in stall without running over or crushing any barriers (a light touch or brushing is acceptable)
18				Drives back through serpentine course (or similar) course without running over or crushing any barriers (a light touch or brushing is acceptable)
19				Drives tractor back into middle of START/STOP position without running over or crushing any barriers (a light touch or brushing is acceptable)
				Tractor shut-off and exit
20				Stops tractor, lowers throttle (if adjusted upward), places gear in neutral, and set brakes or puts in PARK, shuts tractor off **(Not shutting off tractor is automatic Driving Test failure)**
21				Dismount tractor by facing tractor and using handholds and steps
				Not completing the Driving Test in 15 minutes is automatic failure.
				DISCUSS UNSATISFACTORY PERFORMANCE HERE

Automatic Failure-Please note the reason for automatic failure.

Minimum Core Content Areas (MCCA) and Task Sheets

In response to a request by the Hazardous Occupations Safety Training in Agriculture (HOSTA) National Steering Committee, an expert panel of agricultural safety and health professionals developed a set of Minimum Core Content Areas (MCCA) for training youth younger than age 16 who wish to work in an agricultural occupation and operate tractors and machinery (see *MCCA – Training Content*). The MCCA are based on the current United States Department of Labor Regulations in 29 CFR 570 Subpart E-1.

There are 92 Task Sheets provided with this curriculum. Not all of these Task Sheets are needed to address the required training prescribed by the USDOL AgHO. If you have limited familiarity and experience with production agriculture, work hazards and/or agricultural safety issues, the additional task sheets will help you understand hazards associated with agricultural operations.

The 48 Task Sheets identified as *core* with an (**C**) cover the MCCA topics and should be used to prepare for the NSTMOP Written Test. Written test questions come from these Task Sheets. To meet current requirements of the USDOL AgHO exemption, ***at least 24 hours should be devoted to these topics.***

1. **INTRODUCTION**
   - 1.1. The Work Environment **C**
   - 1.2. Safety and Health Regulations **C**
     - *1.2.1. Hazardous Occupations Order in Agriculture* **C**
     - *1.2.2. Occupational Safety and Health Act (OSHA)*
     - *1.2.3. Worker's Compensation Laws*
     - *1.2.4. Worker Protection Standard*
     - *1.2.5. State Vehicle Codes*
   - 1.3. Environmental Regulations
   - 1.4 State Agricultural Safety Professionals
   - 1.5 National Agricultural Safety and Health Organizations
2. **SAFETY BASICS**
   - 2.1. Injuries Involving Youth **C**
   - 2.2. Risk Perception
   - 2.3. Reaction Time
   - 2.4. Age-Appropriate Tasks **C**
   - 2.5. Severe Weather
     - *2.5.1. Heat and Sun*
     - *2.5.2. Cold*
     - *2.5.3. Lightning, Tornadoes, and Rain*
   - 2.6. Housekeeping
   - 2.7. Personal Dress **C**
   - 2.8. Hazard Warning Signs **C**
   - 2.9. Hand Signals **C**
   - 2.10. Personal Protective Equipment **C**
   - 2.11. First Aid and Rescue **C**
3. **AGRICULTURAL HAZARDS**
   - 3.1. Mechanical Hazards **C**
   - 3.2. Noise Hazards and Hearing Protection
   - 3.3. Respiratory Protection
     - *3.3.1. Using Respiratory Protection*
   - 3.4. Working with Livestock **C**
   - 3.5. Agricultural Pesticides **C**
   - 3.6. Electrical Hazards **C**

6. MATERIALS HANDLING

6.1. Skid Steers

6.1.1 Starting and Stopping a Skid Steer

6.1.2 Skid Steer- Ground Movement

6.1.3 Skid Steer- Attaching Accessories

6.1.4 Skid Steer- Using Hydraulic Attachments

6.2. ATVs and Utility Vehicles

6.3. Telehandlers

6.4. Using a Tractor Front-End Loader

6.5. Dump Trucks and Trailers- Farm Use Only

6.6. Loading and Towing Equipment on a Trailer

6.8 Augers and Elevators

6.9 Silage Defacers

6.10 Silage Bagging and Bale Wrapping Equipment

Minimum Core Content Areas – Training Content

The following Minimum Core Content Areas (MCCA) identify the essential subject matter to be included in any training program designed to prepare youth to perform the following tasks as prescribed by the "Hazardous Occupations Order in Agriculture" for employment of children younger than age 16 (USDOL, effective date 1970).

1. Operating a tractor over 20 PTO horsepower, or connecting or disconnecting an implement or any of its parts to or from such a tractor.
2. Operating or assisting to operate (including starting, stopping, adjusting, feeding, or any other activity involving physical contact associated with the operation) any of the following machines:
   a. Corn picker, cotton picker, grain combines, hay mower, forage harvester, hay baler, potato digger or mobile pea viner;
   b. Feed grinder; crop dryer; forage blower, auger conveyor, or the unloading mechanism of a nongravity-type self-loading wagon or trailer; or
   c. Power post-hole digger, power post-driver, or nonwalking rotary tiller.

NOTE: The order and organization in which the Minimum Core Content Areas (MCCA) are taught is left to the discretion of the instructor. The items in parentheses are suggested topics and not inclusive of those which can be taught under any respective MCCA.

Programs that address GENERAL agricultural safety shall incorporate the following MCCA:	**Programs that address safe TRACTOR operation shall incorporate the following MCCA:**	**Programs that address MACHINERY safety shall incorporate the following MCCA:**
G1. Applicable state and federal regulations (OSHA, USDOL, WPS …) **G2.** Characteristics of the safe farm worker (dress, attitude, behavior, developmentally appropriate ...) **G3.** Mechanical hazards (pinch points, pull-ins, wrap points ...) **G4.** Chemical hazards (labeling, pesticide restrictions, petroleum products …) **G5.** Electrical hazards (Grounding/GFCIs, overhead powerlines, underground power …) **G6.** Livestock hazards (working prohibitions, zoonotic diseases, animal behavioral characteristics …) **G7.** Structural hazards (working prohibitions, falls, grain bins, silos …) **G8.** Understanding and use of universal hand signals **G9.** Environmental hazards (weather, insects, noise, terrain …) **G10.** ATVs and utility vehicles **G11.** Personal protective equipment (PPE) **G12.** First aid & emergency response	**T1.** Tractor controls, instruments & gauges **T2.** Tractor operational systems (fuel, electrical, hydraulics ...) **T3.** Tractor safety systems & features (seatbelt, ROPS, master shield, safe starting systems, enclosed cabs …) **T4.** Tractor pre-operational check (fuel, tires, air, water, oil …) **T5.** Operator's manuals (tractor specifications, maintenance, warnings, controls …) **T6.** Tractor operating procedures (start, move forward/backward, stop, mounting/dismounting …) **T7.** Tractor operational hazards (stability, fires, runovers, slips/falls, extra riders, PTO …) **T8.** Public road use (operator age, lighting & marking, equipment width & length, highway regulations …)	**M1.** Operator's manuals (specific machines) **M2.** Implement power transfer (PTOs, hydraulic, mechanical, electrical …) **M3.** Implement hitching (3-point, front mounted, hydraulics, drawbar ...) **M4.** Machine operating hazards (corn picker, cotton picker, grain combines, hay mower, forage harvester, hay baler, potato digger or mobile pea viner, feed grinder; crop dryer; forage blower, auger conveyor, or the unloading mechanism of a nongravity-type self-loading wagon or trailer, power post-hole digger, power post-driver, or nonwalking rotary tiller …) **M5.** Materials handling (large round bales, flowing grain, safe lifting …)

Minimum Core Content Areas – Testing Procedures

	Written Exam	Skills		
		Pre-operation	Operating Skills	Driving Skills
General	X A			
Tractor & Machinery	X B	X C	X D	X E

A&B: The written exam shall be a 50 question multiple choice/T or F test based, minimally, upon the MCCA.

C: Pre-operational check (tractor & machinery)
The pre-operational activities may include:
- Checking of oil, fuel, cooling fluids, etc.
- Visual inspection of tires, shields, guards, fasteners, leaks, etc.
- Demonstrate universal hand signals (ASAE)
- Safe mounting and dismounting, adjustment of the operator's station, location of controls, etc.

D: Operating Skills
The operating skills component will include:
- Starting tractor
- Backing up tractor
- Connecting tractor to a machine (PTO, hydraulics, electrical, hitch, etc.)
- Disconnecting machines
- Stopping and parking tractor

E: Driving Skills
The driving skills component will include:
- Operate tractor & machinery through a serpentine driving course, forward & reverse operation, back in & pull out of prescribed space.

Testing parameters:

- The testing is done in a sequential fashion. That is, the youth must pass each phase of the examination process in the following order: written, pre-operational, operating and driving. Students shall not be allowed to move to the next examination phase until the previous one has been successfully passed.
- A minimum score of 70% must be obtained on each of the phases (written, pre-operational, operating and driving) to pass.
- Students may retake any failed portion of the exam process. The student should not be allowed to retake any failed portion of the exam process without additional study or practice on his or her own time. If the student fails the written exam portion of the evaluation process, a different exam will be used for the retesting.
- State and local exam procedures may vary from the suggested examples provided as long as the minimum key components of each exam (written, pre-operation, operating and driving) are met.

TESTING PROCEDURES:

Linear process ↓

Step 1: Written exam
Step 2: Pre-operational
Step 3: Operating Skills (drive forward/backward and simple machine hookup)
Step 4: Driving Skills (drive tractor and 2-wheel implement through a course)

THE WORK ENVIRONMENT

HOSTA Task Sheet 1.1 **Core**

NATIONAL SAFE TRACTOR AND MACHINERY OPERATION PROGRAM

Introduction

Farm work is different from most other jobs. Many of these differences increase the chance that you will get hurt.

This task sheet talks about why and how the farm work environment can be different from other jobs, and why these differences increase hazards and risk of injury.

Farm Differences and Safety

A farm is defined by the U.S. Dept. of Agriculture as any place from which $1,000 or more of agricultural products was produced. A farm can be as small as 10 acres or as large as ten thousand acres. In 2007, 64% of all farms were 180 acres or less, and just over 80% of all farms sold less than $100,000 of agricultural products.

Farming is characterized by a big variety of products. For example, in addition to farms with beef, dairy, hog, sheep and poultry, there are farms with mules, llamas, buffalo, mink, fish and bees. Common farm crops are corn, soybean, wheat, and hay, but less common ones are grass and tree, bush and flower plants. Added to these types of farms are orchard, nut, fruit, and vegetable farms.

Figure 1.1.a. Agricultural work is done under changing conditions. Often the work is done hurriedly to beat the approaching storm. Safety must still be a major priority.

Farming takes place on land that is flat for miles around, but also on land that can be hilly and mountainous. Some farming takes place where there are buildings that protect against the cold, wind and rain, but most farming takes place where there is little or no shelter from the sun, rain, cold and wind.

All these differences make it hard to find easy ways to reduce your chance of injury. However, constantly thinking about what can go wrong will help you avoid getting hurt.

Farming is more diverse than most people think.

Learning Goals

- To understand the variability of agriculture and how this relates to farm safety and health
- To identify factors and situations that contribute to agricultural hazards and risks

Related Task Sheets:

Safety and Health Regulations	**1.2**
Hazardous Occupations Order in	**1.2.1**

Cooperation provided by The Ohio State University and National Safety Council.

Farm work includes many characteristics and factors that increase risk of injury.

Unique Characteristics

There are many ways to organize the information that describes why the farm work environment is different from other types of work environments. One of the simplest ways is to list the four main characteristics of farming that makes it different from other types of work environments.

1. A lack of uniformity and control of workplaces and work activities
2. An overlap of home and work sites
3. Most farms are operated by family members using labor without age related restrictions
4. Little government regulation of work hazards and risk (except with pesticides)

The combined effect of these four characteristics helps make farming one of the most hazardous occupations.

Factors That Make It Difficult To Improve Safety

The four main characteristics are simple to learn, but hide a better understanding of why farming has many hazards and risks and why they are hard to eliminate or control.

Another way to look at this issue is to think of "factors that influence farm work and risk of injury". This method results in factor or areas:

- Environmental
- Personal
- Work Activity
- Social, Economic and Political

These four factors are explained in greater detail in the following tables. Review these tables for a good understanding of why so many hazards and risks remain a part of farming and why there are so many injuries.

Environmental factors that influence farm work and risk of injury

Weather	Farm work must often be completed regardless of weather extremes.
Work sites	Commonly overlap with residence.
Emergency services	Not readily available; often involves a delayed response due to isolation of work site.
Isolation of work	Co-workers often not in eyesight or hearing distance when trouble occurs.
Personal hygiene	Often required and made available in other occupations. Up to individual workers in agriculture.
Environmental hazards (noise, vibration, lighting, dusts, etc.)	Hazards and exposures are not monitored or regulated in agriculture as they are in most hazardous industries.

Cooperation provided by The Ohio State University and National Safety Council.

Personal factors that influence farm work and risk of injury

Young workers	Children younger than 16 years old, and as young as five, are commonly exposed to and interact with work hazards and environments that are beyond their normal physical, mental and/or emotional abilities to respond to safely.
Senior workers	There is no standard retirement age in agriculture. This results in farmers with significant physical limitations and slow reaction times continuing to work in high-risk situations.
Minimal physical limits	Initial physical exams or minimum performance requirements are often required to begin work or to continue work in other hazardous occupations.
Physical exams	Routine medical surveillance is not common.
Special care for physical or mental conditions	Special care is not available or only by self-imposed restrictions. These issues are tightly controlled in other hazardous occupations.
Transfers to light duty	Transfer of workers to light duty is usually not usually an option in agriculture.
Dispersion of workforce	It is difficult to provide health and safety services because of geographic dispersion and mobility of the workforce.
Farm operators	The farm population ranges from those with advanced college degrees to those with a high school education or less; from farming full time and working significant hours off the farm; to working full time off the farm and farming for supplemental income; to farming only as a hobby or lifestyle statement.

Work Activity factors that influence farm work and risk of injury

Work hours	60 to 80 hour work weeks are common hours of labor in agriculture
Labor and management functions	Usually these jobs represent separate functions in other hazardous occupations, but not in farming.
Work pace	The work pace can be highly erratic rather than steady, and is frequently affected by weather situations and machinery breakdowns.
Work routine	The work routine can be highly irregular with many tasks being seasonal or done once or twice per season or year.
Specialization	Specialization is not normally possible; the phrase, "jack-of-all-trades" often applies.
Instructions	Farmers often learn their trade by observation and experience.
Holidays and vacations	Days off are normal for most occupations, but not for the farm worker.
Labor demands	Farmers frequently make use of any temporarily available labor: migrant, spouse, children, friends, visitors, new acquaintances, and off-the-street employees.
Uncertainty	Farming is characterized by an uncertain future. Weather, fast spreading plant and animal disease, broad economic policy, and unexpected world events can result in financial hardship for the farmer.
Agriculture production	There are great differences in size and type of farms, and the technology used. This makes grouping the types of modern agriculture difficult.

Social, Economic and Political factors that influence farm work and risk of injury.

Lifestyle vs.	Farming is commonly viewed as a "way of life" rather than as an occupation.
Agrarianism	This is a term applied to agriculture that says farmers are owed a social debt by society because they suffer so that a democratic society can prosper.
Day care	Often not available, practical, or affordable in rural areas. Results in parents baby-sitting infants, toddlers, preschoolers and other children during farm work.
Occupational safety and health legislation	New standards and regulations often exempt production agriculture because of a combination of lack of practicality to farming, lack of ability to enforce the standards or regulations, and the burden on farmers to comply.
Cultural beliefs about farm safety and health	There is a cultural belief that farming is a hazardous and unpredictable occupation. This contributes to the belief by farm workers that little can be done about farm safety and health except to be careful.
Market forces	Farmers do not set their own prices for products produced. They cannot add the costs of safety and health to products to recoup costs.
Self-reliance for safety	Farmers primarily rely on their own knowledge and awareness of hazards to work safely, and often accept blame when an injury occurs, especially when they commit an unsafe behavior that directly results in an injury.
Enculturation	Children are taught values, responsibility, good work ethics, decision making, and about life and death. Strong bonds among children, parents, grandparents, neighbors and communities are developed and nourished from the shared experiences of farming.

Safety Activities

1. How many of these characteristics and factors are present on the farms where you live or work. Discuss these with your parents, instructor, or mentor.
2. Show the factors tables to one or two area farmers and have them identify how many factors may have contributed to a farm work injury to themselves or someone on their farm.
3. How many of these factors might be present in non-farm work environments? Are there any occupations with high numbers of serious injuries that have as many of these factors as farming? Discuss these with your parents, instructor, or mentor.

References

1. Aherin RA, Murphy DJ, Westaby JD (1992). Reducing Farm Injuries: Issues and Methods. St. Joseph, MI: American Society of Agricultural Engineers. 58 pp.
2. Area farmers.

Contact Information

National Safe Tractor and Machinery Operation Program
The Pennsylvania State University
Agricultural and Biological Engineering Department
246 Agricultural Engineering Building
University Park, PA 16802
Phone: 814-865-7685
Fax: 814-863-1031
Email: NSTMOP@psu.edu

Credits

Developed by WC Harshman, AM Yoder, JW Hilton and D J Murphy, The Pennsylvania State University. Reviewed by TL Bean and D Jepsen, The Ohio State University and S Steel, National Safety Council. Revised 3//2013

This material is based upon work supported by the National Institute of Food and Agriculture, U.S. Department of Agriculture, under Agreement Nos. 2001-41521-01263 and 2010-41521-20839. Any opinions, findings, conclusions, or recommendations expressed in this publication are those of the author(s) and do not necessarily reflect the view of the U.S. Department of Agriculture.

SAFETY AND HEALTH REGULATIONS

HOSTA Task Sheet 1.2 **Core**

NATIONAL SAFE TRACTOR AND MACHINERY OPERATION PROGRAM

Introduction

Safety and health regulations affecting agricultural workers have existed for many years. Not as many safety regulations are applied to agriculture as there are in other hazardous occupations. The ones that do exist are important because they help keep you and co-workers from being hurt or killed. Fines and imprisonment may also be given to employers who do not follow these regulations.

This task sheet explains safety and health regulations important to youth who plan to work in the field of agriculture.

Figure 1.2.a. Safety and health regulations govern not only agricultural equipment in traffic situations, but also the protection of young people as they work in the field of agriculture. Safety and health regulations are designed to protect you, not to prevent you from learning new skills or earning money.

Hazardous Occupations Order in Agriculture

Since 1969, the U.S. Department of Labor has declared many agricultural tasks to be hazardous for youth younger than age 16. With certain exemptions, employment of youth under age 16 for these tasks is illegal. The law does not apply to youth younger than age 16 who are employed, either with or without compensation, by their parents or legal guardian.

As part of this declaration, a procedure was established by the Department of Labor so that youths 14 and 15 years of age could be exempted from certain portions of the law. This exemption applies to agricultural tractors and specific types of farm machinery. This exemption is explained in more detail in Task Sheet 1.2.1.

Penalties for subjecting youth to hazardous occupations are relatively strict. Youth are not penalized for the infractions, but the employer can be. The penalty to the employer for the first offense can be up to a $10,000 fine for a willful violation. For a second offense, up to a $10,000 fine and/ or imprisonment for not more than six months can be assessed.

HOOA

Hazardous Occupations Order in Agriculture

Learning Goal

- To become aware of the regulations that affect agricultural workers

Related Task Sheets:

Hazardous Occupations Order in Agriculture	1.2.1
OSHA Act	1.2.2
Worker's Compensation Laws	1.2.3
Worker Protection Standards	1.2.4
State Vehicle Codes	1.2.5
Environmental Regulations	1.3
Operating the Tractor on Public Roads	4.14

Cooperation provided by The Ohio State University and National Safety Council.

OSHA

Figure 1.2.b. Check the www.osha.gov website to search for the rules that govern our work in agriculture.

OSHA
Occupational Safety and Health Act

WPS

Figure 1.2. c. The Environmental Protection Agency oversees the pesticide laws of this country, including the regulations that assure workers are not exposed to the hazards of agricultural pesticides. See the Worker Protection Standard section.

Occupational Safety and Health Act (OSHA)

Important points about OSHA regulations are sometimes confusing or misunderstood. An employer/employee relationship has to exist for OSHA to apply to a business or operation. This means that if a farm operator uses only his or her own labor, or uses only family labor, OSHA has no jurisdiction in that operation.

OSHA became effective in 1971 but has had little direct influence upon most agricultural operations since October 1976. That is when Congress restricted OSHA from expending any funding to enforce rules on farms employing fewer than 10 employees. This restriction, known as the "small farm exemption," has been in effect since 1976.

This does not mean that farms with 10 or fewer employees are exempt from OSHA's requirements, only that OSHA cannot inspect those farms for compliance. Although these two statements appear to be similar, the differences could be significant in a court of law.

Worker's Compensation Laws

Businesses, including farm operations, must pay into a worker's compensation insurance fund to cover medical and rehabilitation costs of injuries to workers injured on the job. Since the cost of this program is high, employers must be sure that employees are trained in safety and have a safe work attitude.

Concerns about injury and worker's compensation costs may cause your employer to be especially concerned about your safety behavior. Do not feel that you are being singled out as not being able to work safely.

Worker Protection Standard

EPA's Worker Protection Standard (WPS) is a regulation aimed at reducing the risk of pesticide poisonings and injuries among agricultural workers and pesticide handlers. The WPS offers protection to over 3.5 million people who work with pesticides at over 560,000 workplaces. The WPS contains:

- requirements for pesticide safety training
- notification of pesticide applications
- use of personal protective equipment
- restricted entry intervals following pesticide application
- decontamination procedures and supplies
- emergency medical assistance recommendations

The Worker Protection Standard regulates the workplace and promotes guidelines for the safety and health of workers.

Insurance Company

Some insurance companies have rules and regulations which they insist be followed by their customers. Usually this is based upon studies they make of customers' claims (actuarial studies). Since agriculture is known to be a particularly hazardous occupation, some insurance companies may view the employment of young workers as a liability risk. Some farmers have been notified not to use young people for certain jobs because of the possibility of increased insurance premiums. Additionally, if you are going to work for a farmer, you or your parents may want to ask if the farmer has insurance coverage in case of an injury.

Vehicle Codes

Most state vehicle codes will contain provisions that apply to the movement of agricultural equipment on public roadways. The rules and regulations vary greatly from state to state. Check your state vehicle code for information regarding the following points:

- Definition of "public road or highway." Your state may define highway as," the entire width between the boundary lines of every highway publicly maintained when any part is open to the use of the public for purposes of vehicle travel." Any road open to the public is referred to as a highway, including shoulders and berms.
- Your state's vehicle code may have a statement that requires all persons who operate motor vehicles upon a highway to have a license unless specifically exempted elsewhere in the code. Exemptions to the licensing requirement may show some language similar to the following: "Persons 14 or 15 years of age are restricted to the operation of implements of husbandry on one- and two-lane highways which bisect or immediately adjoin the premises upon which such person resides."
  In other words, 14- and 15-year-old youths can operate farm tractors only on public roadways that bisect or adjoin their place of residence.

Consider these points also.

- Use of the SMV emblem laws are fairly constant nationwide. The SMV emblem must be used properly. See Task Sheet 4.14.
- Lighting and marking regulations can be found in most state vehicle codes.
- Load restrictions for width, length, weight, number of towed implements, and safety chains use can be found in vehicle codes.
- State vehicle codes may also address trucks licensed for farm use only, riders as passengers on the bed of a truck, and farm use of ATVs.

Check with your local Highway Police and Department of Transportation to learn what the traffic laws are in your state.

Figure 1.2.d. The vehicle code of your state must be followed. Your state highway patrol can be a source of information for you. Contact your local state police headquarters.

Be sure you check the motor vehicle code of your state!

Figure 1.2.e. You must follow the rules of the road when operating farm equipment on public roads. Know the rules of the road! Are there any rules being violated in this picture?

Safety Activities

1. Use the Internet to search for information about the federal safety regulations mentioned in this task sheet Find additional details on how the regulations may affect agricultural workers.

2. Talk to your parents' insurance agent, and ask about injury and liability concerns that he or she may have regarding your employment in agriculture.

3. Complete the following chart with a list of tasks you have done, mark whether or not the task was covered by federal safety regulations, and note what hazard you encountered. Are there any jobs/tasks you have done which may be prohibited for youth your age?

Tasks I Have Completed	Is the Task Covered By Federal Safety Regulations?	Safety Hazard of the Task

References

1. www.osha.gov
2. www.epa.gov/oppfead1/safety/workers/workers.htm
3. Your state's motor vehicle code. The code may be on the Internet or a printed copy may be available in your community library.
4. Local Highway Patrol troopers

Contact Information

National Safe Tractor and Machinery Operation Program
The Pennsylvania State University
Agricultural and Biological Engineering Department
246 Agricultural Engineering Building
University Park, PA 16802
Phone: 814-865-7685
Fax: 814-863-1031
Email: NSTMOP@psu.edu

Credits

Developed by WC Harshman, AM Yoder, JW Hilton and D J Murphy, The Pennsylvania State University. Reviewed by TL Bean and D Jepsen, The Ohio State University and S Steel, National Safety Council. Revised 3/2013

This material is based upon work supported by the National Institute of Food and Agriculture U.S. Department of Agriculture, under Agreement Nos. 2001-41521-01263 and 2010-41521-20839. Any opinions, findings, conclusions, or recommendations expressed in this publication are those of the author(s) and do not necessarily reflect the view of the U.S. Department of Agriculture

HAZARDOUS OCCUPATIONS ORDER IN AGRICULTURE

HOSTA Task Sheet 1.2.1 **Core**

NATIONAL SAFE TRACTOR AND MACHINERY OPERATION PROGRAM

Introduction

Since 1969, the U.S. Department of Labor (DOL) has declared many agricultural tasks to be hazardous for youth younger than age 16. With certain exemptions, employment of youth under 16 for these tasks is illegal. However, the regulation does not apply to youth younger than age 16 who are employed, either with or without compensation, by their parents or legal guardian.

The Exemption

As part of the DOL's Fair Labor Standards Act, a declaration known as the Hazardous Occupations Order in Agriculture (HOOA) established a procedure whereby youth 14 and 15 years of age could be exempted from certain portions of the regulation. This exemption has to do with the operation of agricultural tractors and specific types of farm machinery.

Specifically, the exemption states that with successful completion of a 10-hour training program, 14- and 15-year-old youth can be employed to: "operate a tractor of over 20 PTO horsepower, or connect or disconnect an implement or any of its parts to or from such a tractor."
Additionally, with successful completion of a 20-hour training program, these youth can be employed to: "operate or assist to operate (including starting, stopping, adjusting, feeding, or any other activity involving physical contact associated with the operation) any of the following machines:

(i) corn picker, cotton picker, grain combine, hay mower, forage harvester, hay baler, potato digger or mobile pea viner;

(ii) feed grinder, crop dryer, forage blower, auger conveyor, or the unloading mechanism of a non-gravity type self-unloading wagon or trailer;

(iii)power post-hole digger, power post driver, or non-walking rotary tiller.

With the 10-hour training program, youth are allowed only to operate a tractor with no powered equipment attached. To do field work of any kind, youths need to complete the 20-hour training program.

The law defines "agriculture" as: "farming in all its branches including: preparation for market, delivery to market, delivery to storage, or to carriers for transportation to market."
This statement allows a properly trained youth to haul produce and other products to markets, between farms, etc. Provisions in your state vehicle code may preclude this activity by 14- and 15-year-olds.

Not all jobs are considered hazardous for young people. There are many tasks on farms that are not considered hazardous

-continued on page 2-

Prohibited Work

HOOA prohibits all 14 and 15-year-olds from these tasks (no exemptions):

- Handling animal sires or sows and cows with newborns within a pen or corral
- Working more than 20 feet above the ground
- Working with Category I and II agricultural chemicals
- Handling and using explosives and anhydrous ammonia

HOOA
Hazardous Occupations Order in Agriculture

Learning Goals

- To understand the Fair Labor Standards Act and HOOA
- To understand the reason for Hazardous Occupations Safety Training in Agriculture

Related Task Sheets:

The Work Environment	**1.1**
Safety and Health Regulations	**1.2**
Worker Protection Standards	**1.2.4**

Cooperation provided by The Ohio State University and National Safety Council.

The Exemption *(from page 1)*

by the DOL and are permitted under the Fair Labor Standards Act. Some of these include:

- Loading and unloading trucks
- Operating small tractors (under 20 horsepower)
- Picking vegetables and berries
- Placing vegetables and fruits on conveyors or into boxes
- Clearing brush and harvesting trees up to 6 inches in butt diameter
- Working with animals on the farm or at fairs and shows (except for specified breeding stock in confined areas, such as cows with newborn calves in closed box stalls, bulls, or sows with newborn piglets)
- Raising and caring for poultry
- Milking cows
- Cleaning barns, equipment, and storage buildings
- Mowing lawns
- Riding, driving, or exercising horses
- Picking cotton
- Handling irrigation pipes
- Riding on transplanters

Penalties for subjecting youth to hazardous occupations are relatively strict. Youth are not penalized for the infractions; the employer is held accountable. First offense—up to a $10,000 fine for willful violation. Second offense—up to a $10,000 fine and/ or imprisonment for not more than six months.

Workers younger than age 14

HOOA regulations do not permit youth younger than age 14 to complete the exemption training. This means youth younger than age 14 cannot be hired by an agricultural employer to operate tractors or machinery.

When a federal regulation and state regulation are in conflict, the most restrictive regulation is enforced. Check with state authorities when in doubt.

Safety Activities

1. Make a list of jobs or tasks you have done on the farm. How many of them are included in the list of activities prohibited by the Hazardous Occupations Order in Agriculture for youth younger than age 14?
2. Discuss with your classmates or interested friends why you think some tasks have been included in the Hazardous Occupations Order in Agriculture list and why other tasks have not.

References

1. USDA publications. These publications are available from many state farm safety specialists located at land grant universities.
2. www.dol.gov, website of the US Department of Labor.

Contact Information

National Safe Tractor and Machinery Operation Program
The Pennsylvania State University
Agricultural and Biological Engineering Department
246 Agricultural Engineering Building
University Park, PA 16802
Phone: 814-865-7685
Fax: 814-863-1031
Email: NSTMOP@psu.edu

Credits

Developed by WC Harshman, AM Yoder, JW Hilton and D J Murphy, The Pennsylvania State University. Reviewed by TL Bean and D Jepsen, The Ohio State University and S Steel, National Safety Council. Revised 3/2013

This material is based upon work supported by the National Institute of Food and Agriculture, U.S. Department of Agriculture, under Agreement Nos. 2001-41521-01263 and 2010-41521-20839. Any opinions, findings, conclusions, or recommendations expressed in this publication are those of the author(s) and do not necessarily reflect the view of the U.S. Department of Agriculture.

OCCUPATIONAL SAFETY AND HEALTH ACT

HOSTA Task Sheet 1.2.2
NATIONAL SAFE TRACTOR AND MACHINERY

Introduction

The 1971 Occupational Safety and Health Administration (OSHA) regulations were created to save lives, prevent injuries, and to protect the health of all American workers.

Since 1971 workplace fatalities have been decreased by 50%. Workplace injury and illness numbers have been decreased by 40%. This has happened despite the fact that workforce numbers and job sites doubled in numbers.

This task sheet examines how OSHA affects agricultural work places. The entire law cannot be presented here.

OSHA's Jurisdiction

An employer/employee relationship has to exist in order for OSHA to apply to a business or operation. If a farm operator uses only his or her own labor, or uses only family labor, OSHA does not apply. Since 1976 Congress has restricted OSHA from expending any administrative funds to enforce rules and regulations on any farm with 10 or fewer employees.

The 10 or fewer employees restriction in agriculture is known as the "small farm exemption." Small farms, however, are not actually exempt from OSHA regulations. Legally OSHA covers all farms, even though OSHA cannot inspect farms with 10 or fewer employees. One important reason for understanding that small farms still fall under OSHA is that, in a court of law, OSHA rules and regulations may be used to identify safe and unsafe conditions on the farm.

General OSHA Rules

A general rule of OSHA requires employers to provide employees a place of employment that is free from recognized hazards that have caused or are likely to cause death or serious injury. A second part of this rule states that employers must comply with OSHA safety and health standards. These two rules apply to small farms as well as larger farms. This could also be important in a court of law if an employee is killed or injured from farm work.

OSHA also requires that each employee comply with safety and health rules, such as shutting off power to equipment before working on any machine; wearing personal protective equipment; and informing employers of hazards. An employee who is injured or causes injury to another worker by deliberately acting in an unsafe way may find themselves in legal difficulty due to the OSHA standards.

www.osha.gov

Figure 1.2.2.a. To study the law in detail, check out the Internet website shown above.

OSHA Occupational Safety and Health Act

Learning Goals

- To become aware of OSHA regulations affecting agricultural work

Related Task Sheets:

Safety and Health Regulations	1.2
Mechanical Hazards	3.1
Electrical Hazards	3.6
Confined Spaces	3.8
Tractor Stability	4.12
Operating the Tractor on Public Roads	4.14
Lighting and Marking	4.14.1

Cooperation provided by The Ohio State University and National Safety Council.

Farming has long been recognized as one of the most hazardous occupations.

Fig. 1.2.2.b. The business of farming presents daily exposure to hazards. Livestock, machinery, and the environment present safety concerns. OSHA regulations may limit inspections of farms, but safe work habits and the workplace conditions must be made a daily concern.

Agriculture and OSHA

There are just a few OSHA regulations that are specific to farming. Some of the rules are not related to the type of farm work that 14- and 15-year-olds are allowed to do under the Hazardous Occupations Order in Agriculture (see Task Sheet 2.1). The OSHA agricultural standards most important to tractor and machinery operators are the Tractor Rollover Protection, Machinery Guarding, and Accident Prevention Signs and Tags regulations. They are discussed in more detail in the following sections of this task sheet.

Tractor Rollover Protection (ROPS)

Rollover Protective Structure (ROPS) Requirements

ROPS have been required on all tractors operated by employees since 1976. In addition, OSHA regulations state that employers are also required to provide safe operating instructions to employees at initial assignment and on an annual basis thereafter. Employers are to insure that seatbelts are used by the employees on ROPS-equipped tractors. Exempted from the standards are low-profile tractors used in orchards, greenhouses, and other buildings.

Operating Instructions

The following instructions are to be provided to the employee at their initial assignment and at least once a year thereafter:

- If the tractor has a ROPS, use the seat belt.
- Avoid ditches, embankments, and holes.
- Reduce speed when turning, crossing slopes, and on rough, slick, or muddy surfaces.
- Avoid slopes too steep for safe operation.
- Exercise care at row ends, on roads, and around trees.
- Do not permit extra riders on the tractor.
- Operate the tractor smoothly with no jerky starts, turns, and stops.
- Hitch only to the drawbar and recommended hitch points.
- Set the brakes and use the park lock if available when the tractor is stopped.

Accident Prevention Signs and Tags

SMV Emblems

The OSHA accident prevention signs and tags regulation defines use of the SMV emblem. The SMV emblem must be displayed at the rear of the tractor and/or tractor implement combination to warn others that the farm vehicle is incapable of traveling at more than 25 mph. See Task Sheet 4.14.

Properly use the SMV emblem. Be sure it is clean and visible if you are required to operate farm tractors and equipment on public roads. In some states, it is illegal to improperly use SMV emblems as driveway and mailbox markers.

Machinery Guarding

Moving Parts Guarding and Instruction

Guarded machine parts prevent the worker from exposure to entanglement and dismemberment risks. OSHA Machine Guarding Standards require the following:

All farm field and farmstead equipment, regardless of date of manufacture, must be provided with PTO guarding.

All power transmission components on new field and farmstead equipment must be provided with nip point guarding. Nip points are pinch points on gears, belts, and pulleys. See Task Sheet 3.1.

Means must be provided to prevent accidental application of electrical power to farmstead equipment. Electrical power devices must be locked out (LO) or tagged out (TO) during maintenance and service of the equipment (See Task Sheet 3.6).

Employee education is part of this OSHA standard as well. The law states, "Employees must be instructed in the safe operation and servicing of all equipment which they operate or will operate." The following instructions must be given at the time of assignment and at least once a year:

Keep all guards in place when the machine is in operation.

Permit no riders on farm field equipment other than those necessary for instruction or assistance.

Stop the engine, disconnect the power source, and wait for all machine movement to stop before servicing, adjusting, cleaning, or unclogging the equipment except where the machine must be running to be serviced or maintained. If the machine must be running to do such tasks, then employees are to be instructed in all steps and procedures to safely do the service or maintenance.

Clear the machine area before starting the engine, engaging the power, or operating the machine.

Lock out electrical power before working on farmstead equipment.

You have a Right to a Safe and Healthy Workplace.

It's the Law!

Figure 1.2.2.c. This is the top portion of an OSHA workplace poster. Affected businesses must display the entire poster where employees can see it.

Ignorance is no excuse for violating safety laws.

1.2.2.d. OSHA regulations specifically cover ROPS, seat belts, and guarding of moving machinery parts. Be responsible and know what is required.

Confined Spaces

Although OSHA regulations for confined spaces do not apply to agriculture, the general duty clause expresses that hazards such as confined spaces (silos, manure pits, grain bins and elevators, and controlled atmosphere storages) must be explained to the employee. No worker should be exposed to risk of injury or death while working within a confined space. See Task Sheet 3.8.

For more information on OSHA Confined Space standards, go to www.osha.gov. Search the website for OSHA standard 1910.146 to learn more.

Figure 1.2.2.e. Workers must be instructed about the dangers of confined space work before they begin a task. Confined spaces may expose the worker to deadly gases, entrapment or suffocation.

Safety Activities

Use the Internet to access the OSHA website. Search the website for information regarding agricultural operations. Report on specific training and instructions employers must provide to employees about tractor and ROPS use, machine guarding, SMV emblems, and field sanitation.

What percentage of the farms in your community employ more than 10 employees? Hint: Do a survey of the total number of farmers in your community. This is the denominator. The numerator will be the total number of farms employing more than 10 employees. Divide the number of farms with more than 10 employees into the total number of farms. Make the calculation.

Form as many words as you can from the title "Occupational Safety and Health Act." If you can, include in your list words or phrases that are related to safety, risk, or injury. For instance, the words "safe" and "unsafe action" can be found. Make your list here or on a separate sheet of paper. Score yourself as an expert in recognizing safety if you get more than 10 words dealing with safety, risk, or injury.

References

www.osha.gov.

Murphy, D.J. ,1992, Safety and Health for Production Agriculture, St. Joseph, Michigan: American Society of Agricultural Engineers.

Contact Information

National Safe Tractor and Machinery Operation Program
The Pennsylvania State University
Agricultural and Biological Engineering Department
246 Agricultural Engineering Building
University Park, PA 16802
Phone: 814-865-7685
Fax: 814-863-1031
Email: NSTMOP@psu.edu

Credits

Developed by WC Harshman, AM Yoder, JW Hilton and D J Murphy, The Pennsylvania State University. Reviewed by TL Bean and D Jepsen, The Ohio State University and S Steel, National Safety Council. Revised 3/2013

This material is based upon work supported by the National Institute of Food and Agriculture, U.S. Department of Agriculture, under Agreement Nos. 2001-41521-01263 and 2010-41521-20839. Any opinions, findings, conclusions, or recommendations expressed in this publication are those of the author(s) and do not necessarily reflect the view of the U.S. Department of Agriculture.

WORKER'S COMPENSATION LAWS

HOSTA Task Sheet 1.2.3

NATIONAL SAFE TRACTOR AND MACHINERY OPERATION PROGRAM

Introduction

In the early years of the Industrial Revolution, laws protecting workers did not exist. To correct this problem worker's compensation laws were passed. Worker's compensation laws provide financial help to workers injured on the job no matter who is at fault.

This task sheet discusses worker's compensation laws. Each state's law may be worded a little differently. Federal Worker's Compensation laws apply only to federal employees.

The Law

Use the Internet to check your state's worker's compensation rules. For example, www.state.pa.us takes you to the Pennsylvania state website. Typing in the keyword, "worker's compensation" leads you to this information. Most states have a minimum level of hours worked or pay received before worker's compensation takes effect.

Regarding agriculture, the Pennsylvania law states, "Any employer employing persons in agricultural labor shall be required to provide worker's compensation coverage for such persons if such employer is covered by the law or if during the calendar year wages in excess of $1,200 are paid to one employee for agricultural labor, or employment to one employee in excess of 30 or more days is provided. If such conditions are met, then all employees are to be provided the workmen's compensation coverage."

How the Law Works:

The following information is important for an agricultural employee.

- Compensation for injury, disability, and death is provided as a benefit to employees and their surviving family members by law.
- Employers and employees pay into the Worker's Compensation fund of their home state according to the hours of employment provided.
- Claims are filed with the employer and medical attention is provided by approved providers.
- Depending upon the extent of the injury, compensation during recuperation is paid, but is limited to two-thirds of the statewide average weekly wage.
- Medical checkups may be required to determine the return-to-work date or how long the benefits will be paid.

Figure 1.2.3.a. The Industrial Revolution brought men, women, and children into the workplace with few rules to protect them. Injuries and deaths brought hardships to the worker and their families. Laws to assist the injured worker were later passed.

Worker's Compensation laws protect the worker.

Learning Goal

- To understand worker's rights to compensation when injury and disability and death have occurred at the workplace.

Related Task Sheet:

Safety and Health Regulations **1.2**

Cooperation provided by The Ohio State University and National Safety Council.

What You Should Expect

As a beginning worker, these points will help you understand how Worker's Compensation relates to your participation in a work environment.

- Notification of employees rights and filing of claims should be clearly posted for employees to see.
- The notice should state, "Remember, it is important to tell your employer about your injuries."
- Report all injuries no matter how small. For example, a deeply imbedded splinter can become infected. This could lead to blood poisoning resulting in emergency medical treatment and/or amputation.
- Injuries must be reported within 72 hours of the occurrence to be covered by compensation.
- If the worker has suffered some disability, he/she has the right to be transferred to a different job or a modified job when he/she returns to work.

These points do not represent legal opinions. This may alter the procedures you will encounter if you must file a claim.

No matter how big or how small an injury, notify your employer when you are injured!

Safety Activities

1. Locate a Worker's Compensation Notice at your place of employment and read the notice. If you are not employed, ask any employer to show you one of these documents.
2. Conduct a survey of farm employees or classmates employed by farmers to determine if any of them have received worker's compensation due to injuries in the workplace.
3. Visit your state government website to research the worker's compensation laws. The law may be several hundred pages long; therefore, do not print it.

References

1. Visit www.state.pa.us. Type "Worker's Compensation" in the keyword search. Scroll for information.
2. The website of your state government.

Contact Information

National Safe Tractor and Machinery Operation Program
The Pennsylvania State University
Agricultural and Biological Engineering Department
246 Agricultural Engineering Building
University Park, PA 16802
Phone: 814-865-7685
Fax: 814-863-1031
Email: NSTMOP@psu.edu

Credits

Developed by WC Harshman, AM Yoder, JW Hilton and D J Murphy, The Pennsylvania State University. Reviewed by TL Bean and D Jepsen, The Ohio State University and S Steel, National Safety Council. Revised 3/2013

This material is based upon work supported by National Institute of Food and Agriculture, U.S. Department of Agriculture, under Agreement Nos. 2001-41521-01263 and 2010-41521-20839. Any opinions, findings, conclusions, or recommendations expressed in this publication are those of the author(s) and do not necessarily reflect the view of the U.S. Department of Agriculture

WORKER PROTECTION STANDARD

HOSTA Task Sheet 1.2.4

NATIONAL SAFE TRACTOR AND MACHINERY OPERATION PROGRAM

Introduction

The Worker Protection Standard (WPS) regulations of the Environmental Protection Agency (EPA) require employers to take steps to reduce the risk of pesticide-related illness and injury to those persons who use or are potentially exposed to pesticides.

This task sheet discusses the WPS. *Youthful farm workers younger than age 16 years are prohibited from being involved with pesticide applications (see Task Sheet 1.2.1 for allowable work tasks)*. Youth farm workers may, however, come into contact with pesticide-treated areas in the course of their daily work. Understanding the WPS will explain the need for safety when exposure to agricultural chemicals exists.

The Standard

Workers who perform hand work in fields (farm and orchards), forests, nurseries, and greenhouses, as well as employees who handle (transport, mix, load, apply) pesticides in agricultural operations, must be provided information about the materials they are using.

There are no exemptions based on the size of the farming operation.

WPS regulations require information to be provided to workers.

Minimum standards include:

- Oral (verbal) or posted, written notice of a pesticide application and the restricted entry interval (See page 2, Figure 1.2.4.a.)
- Pesticide safety training
- Pesticide safety posters placed where all workers and handlers can access the information
- Informing workers of pesticide label safety information
- Centrally posted list of recently applied pesticides

WPS regulations also require employers to provide:

- Decontamination facilities nearby to work sites
- Periodic pesticide safety training and ongoing information availability
- Notice of pesticide application and pesticide information
- Clean and well-maintained personal protective equipment
- Location and contact information for emergency assistance

As an employee, you may see or be informed of pesticide safety information even though you are not eligible to apply the pesticides. If you are asked to apply pesticides, inform your employer that you are ineligible for the work.

Note: WPS's also require the employer to monitor and assist workers in avoiding heat stress.

You may have to water, prune, weed, or harvest a pesticide-treated area. Be aware of what the WPS means to you!

Learning Goals

- To become aware of the risks of exposure to agricultural pesticides
- To gain knowledge of Worker Protection Standard (WPS) designed to reduce personal exposure to agricultural pesticides

Related Task Sheets:

Safety and Health Regulations	1.2
Hazardous Occupations Order in Agriculture	1.2.1
Heat and Sun	2.5.1
Personal Protective Equipment	2.10
Agricultural Pesticides	3.5

Cooperation provided by the Ohio State University and National Safety Council

MSDS (SDS) Information

Material Safety Data Sheets (MSDS), recently renamed Safety Data Sheets (SDS), are provided to consumers for products ranging from paints and solvents to medicines and pesticides. These data sheets provide the consumer with much information regarding the product they have purchased.

MSDS's (SDS's) supplement the pesticide labels. MSDS data does not offset the need to keep pesticide labels on file to meet WPS record-keeping requirements.

Restricted Entry Interval (REI)

MSDS's (SDS's) should be kept on file.

WPS designed signs must be used at entrances to treated areas to warn workers and others that a pesticide treatment has been made.

These are the rules for posting signs.

- Post signs no more than 24 hours before the pesticide application.
- Keep signs posted during the REI period for 4-48 hours.
- Remove signs within 3 days of application.
- Keep workers out of the area

Figure 1.2.4.a. Restricted entry interval signs must be posted to alert workers and handlers of pesticide application. Pesticide label information may show 4 to 48 hours REI. No one should enter those areas until warnings are removed.

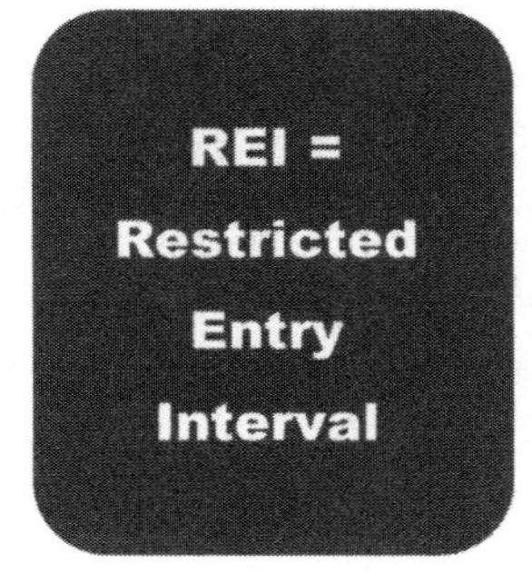

Safety Activities

1. Ask your employer or local agricultural chemical sales representative to show you a pesticide label from the pesticide files. Use the label and/or MSDS (SDS) to answer these questions. MSDS (SDS) information can also be found on the Internet.
   a) What are the health hazards of the product to humans?
   b) What personal protective equipment is required to use the product?
   c) What are the spill control procedures to use for the product?
   d) What is the REI of the product?
2. Use the Internet to search for specific WPS regulations for farm, greenhouse, nursery, and forest pesticide applications.
3. Review the Hazardous Occupations Order in Agriculture for the exact wording of the rule which prohibits workers younger than age 16 from working with pesticides.

References

1. www.usda.gov.
2. www.epa.gov.
3. Worker Protection Checklist, Penn State Pesticide Education Program, www.pested.psu.edu/resources/fApplicator topics/factsheet.
4. Agrichemical Fact Sheet 12, Penn State Cooperative Extension Service, EPA Worker Protection Standard for Agricultural Pesticides.

Contact Information

National Safe Tractor and Machinery Operation Program
The Pennsylvania State University
Agricultural and Biological Engineering Department
246 Agricultural Engineering Building
University Park, PA 16802
Phone: 814-865-7685
Fax: 814-863-1031
Email: NSTMOP@psu.edu

Credits

Developed by WC Harshman, AM Yoder, JW Hilton and D J Murphy, The Pennsylvania State University. Reviewed by TL Bean and D Jepsen, The Ohio State University and S Steel, National Safety Council. Revised 3/2013.

This material is based upon work supported by the National Institute for Food and Agriculture, U.S. Department of Agriculture, under Agreement Nos. 2001-41521-01263 and 2010-41521-20839. Any opinions, findings, conclusions, or recommendations expressed in this publication are those of the author(s) and do not necessarily reflect the view of the U.S. Department of Agriculture.

STATE VEHICLE CODES

HOSTA Task Sheet 1.2.5

NATIONAL SAFE TRACTOR AND MACHINERY OPERATION PROGRAM

Introduction

Each state's legislative body has passed laws that govern motor vehicle use in their state. Since farmers sometimes use the highways to transport farm equipment and products, special rules are included in the motor vehicle codes to assure agricultural producers use the roads safely.

This task sheet discusses State Vehicle Codes from the Pennsylvania viewpoint. *Inclusion of every state's interpretation or language regarding farm implements is not possible.*

See the Safety Activity Section for an assignment for your location.

The Pennsylvania Code

The Pennsylvania Vehicle Code includes several provisions that apply to the movement of agricultural equipment upon public roadways. The definitions for implements of husbandry and highway are of concern to agricultural employers and youthful tractor operators. References concerning licensing and exemptions from licensing are also noteworthy.

Implement of Husbandry Defined

"Implement of husbandry" is defined as "a vehicle designed or adapted and determined by the Department of Transportation to be used exclusively for agricultural operations and only infrequently operated or moved upon highways."

Highway Defined

A second definition of importance is that of "highway." Highways include the entire width between the boundary lines of every way publicly maintained when any part is open to the use of the public for purposes of vehicle travel." Any road open to the public is referred to as a highway, including shoulders and berms.

Licenses Required

Section 1501 of the PA Code has a general statement that requires all persons who operate motor vehicles upon a highway to have a license unless specifically exempted elsewhere in the Code. Section 1502 then goes on to explain exemptions to the licensing requirement. Part (5) says:

> "Persons 14 or 15 years of age are restricted to the operation of implements of husbandry on one and two-lane highways which bisect or immediately adjoin the premises upon which such person resides."

In other words, 14 and 15-year-old youths can operate farm tractors only on public roadways that bisect or adjoin their place of residence.

Many farm employers, parents and youth are probably unaware of this restriction.

Figure 1.2.5.a. Are there any rules for taking this farm tractor and mower-conditioner onto public roadways? Check the vehicle code for your state to find out. Write a one-page report for your club, class, or mentor about your findings.

Can 14- and 15-year-olds legally operate a tractor on the road?

Learning Goals

- To understand the state regulations that affect implements of husbandry used on public roads

Related Task Sheets:

Safety and Health Regulations	**1.2**
Operating the Tractor on Public Roads	**4.14**

Other Rules of the Road

Regulations, and the exemptions to those regulations, standardize the "rules of the road." Vehicle codes may exempt farm equipment from brake systems, bumpers, mirrors, horns, lights, and inspection.

Wide Loads and Passing

PA law states that a wide load (wider than a single lane) should be pulled entirely off the road at the first reasonable and safe location to allow following motorists to pass. Be sure to use the correct signals to show your intended actions. **Never wave the traffic around you as that makes you responsible for what the other driver does.**

Load Listing

Towed loads that deflect from the path of the drawing vehicle creates a hazard. PA law covers pulled loads that weave back and forth. Towed loads may have no more than 6 inches of deflection from the path of the drawing vehicle wheels.

Safety Chains

PA law does not provide an exemption to agriculture regarding safety chain use. Use safety chains to secure the load. See Task Sheet 4.14.

In some cases common courtesy must also be used. Promote agriculture's positive image by sharing the road safely and responsibly.

Figure 1.2.5.b. Promote the positive image of agriculture. Share the road safely and responsibly by obeying your state's vehicle code.

Did you answer the question about 14- and 15-year-old drivers?

Safety Activities

1. Use the Internet to access your state government website. Search for the vehicle code for your state.
2. Using any Internet search engine, type in "implement of husbandry" and "public roadways" to search for your state's vehicle code or information regarding this subject.
3. If you cannot find the information in Question 1 above, contact your local representative to the state House of Representatives, and ask for a copy of the state motor vehicle code. This is a long document. Use the Table of Contents and the index to locate the rules and exemptions your state makes for agriculturists using the public roadways. *The local public library may also be a good source for this document.*
4. Use a poster presentation with local farm groups to review the requirements of your state's vehicle code.

References

1. State Motor Vehicle Code of your State.

Contact Information

National Safe Tractor and Machinery Operation Program
The Pennsylvania State University
Agricultural and Biological Engineering Department
246 Agricultural Engineering Building
University Park, PA 16802
Phone: 814-865-7685
Fax: 814-863-1031
Email: NSTMOP@psu.edu

Credits

Developed by WC Harshman, AM Yoder, JW Hilton and D J Murphy, The Pennsylvania State University. Reviewed by TL Bean and D Jepsen, The Ohio State University and S Steel, National Safety Council. Revised 3/2013

This material is based upon work supported by the National Institute of Food and Agriculture, U.S. Department of Agriculture, under Agreement Nos. 2001-41521-01263 and 2010-41521-20839. Any opinions, findings, conclusions, or recommendations expressed in this publication are those of the author(s) and do not necessarily reflect the view of the U.S. Department of Agriculture.

ENVIRONMENTAL REGULATIONS

HOSTA Task Sheet 1.3

NATIONAL SAFE TRACTOR AND MACHINERY OPERATION PROGRAM

Introduction

Twelve-year-old Jesse was assigned to haul manure down the highway to the leased farm. Traffic was heavy that day, and he panicked in the sharp turn. The manure spreader upset into the road ditch spilling the load. Jesse ran the 1/2 mile to the barn afraid of what he had done. Someone stopped by the farm a short time later to tell the owner about the tractor and spreader sitting in the ditch.

There are several problems described in this short story. Can you identify them? This task sheet will discuss the environmental regulations that farm equipment operators must know.

Environmental Rules

Environmental laws are enforced by the Environmental Protection Agency (EPA). These laws include provisions for clean air, clean water, safe pesticide use, and safe drinking water standards.

These federal laws also have state and local counterparts and enforcement officials. States have Department of Environmental Resources (DER), and local governments have ordinances as well. Farmers and farm employees should have an understanding of all the regulations that are designed to protect our environment.

What Typical Laws Cover

Laws that regulate environmental hazards do not have agricultural exceptions. Typically farmers are held to high standards in protecting the environment. What do you know about these areas?

- Water pollution
- Air pollution
- Drinking water standards
- Pesticide rules and regulations
- Shifting load violations
- Used tire disposal
- Trash burning hours and rules
- Battery disposal
- Oil and fuel spills
- Used oil disposal
- Sink hole protection
- Manure spreading

Each of these subjects will have a federal, state, or local ordinance which affects each citizen. Penalties for violating the law can include fines for breaking the law and payment for property damages.

If you think a task you are assigned can pollute the air, water, or soil, ask your employer if you are causing a legal problem for him or her.

Figure 1.3.a. The Environmental Protection Agency (EPA) enforces the laws of the land in regard to pollution. State and local laws also support these efforts to maintain a healthy environment.

EPA
Environmental Protection Agency

Learning Goals

- To understand that farm equipment operators are responsible for environmental protection
- To understand what to do if a spill poses a hazard to the environment

Related Task Sheets:

Cooperation provided by The Ohio State University and National Safety Council.

Roadway spills require:

- **Control**
- **Containment**
- **Clean up**

Figure 1.3. b. Manure spills may not be as drastic as this one, but manure spills on highways must be kept from leaking into streams (controlled and contained) and cleaned up. The nearest Department of Environmental Resources must be notified. Your best action is to notify local police and fire companies to direct and control traffic during the cleanup.

Manure Handling and Spills

Manure and pesticide applications can pollute water if not done properly. This section will discuss manure loading, transporting, and application.

Manure Handling

Manure handling can take many forms. Solid, semi-solid, and liquid manure handling involves several types of equipment. Front-end loaders or gravity flow storages may be used. Gravity fill liquid manure tanks are more likely to pose environmental spill risks than a manure fork used to clean a calf pen. A stuck manure pit valve can cause immense problems due to spills.

Manure Transportation/Spills

Drivers of farm equipment who use the highway pose a risk to others using the same road. Hauling manure poses a greater threat to safety, since manure can take different forms and can be difficult to handle. Shifting load violations carry penalties under law. Pennsylvania regulations require farmers to use methods, equipment and facilities in such a way that do not pose a health or safety risk to the environment. Should a spill occur, the operator must take immediate steps to control, contain, and clean up the spill. In addition the Department of Environmental Resources must be notified. Penalties may be assessed. Notifying local police and fire officials is important if traffic is to be controlled and directed.

Manure Application

Manure application on farm fields should be done with water quality and nutrient management regulations in mind. Here are a few points to consider:

- Manure spread on frozen soil eventually finds its way into waterways.
- Manure spread close to streams, ponds, wells and springs contaminates these water resources.
- Manure contains nutrients such as nitrogen and phosphorus that feeds plants, but in excess can pollute underground water and streams.

Farms should have a plan in place to deal with manure leaks or spills. The plan should be posted and known by employees. Adequate equipment and supplies should be available, and phone numbers of local police and fire officials should be available too.

Pesticides

Handling of pesticides in any manner by workers younger than age 16 is forbidden by labor laws.

Burning Trash

Youthful farm workers may be assigned the task of burning trash from around the farm. While such a job seems easy, there may be some hidden environmental risks involved. Toxic materials may pose air pollution threats. Local burning laws may be violated.

Toxic materials

Pesticide containers, chemical cleaners, and tires have found their way to burning areas. The toxic fumes released from these materials may make you sick or cause severe health problems. Ask your employer what hazard is associated with what he or she has assigned you to burn.

Burning Ordinances

Local government laws may limit burning to certain items on certain days and at certain times of the day. Ask your employer about these local laws.

Fuel, Oil, Lubrication—Spills and Disposal

Laws exist to protect the environment, but farmers should also want to prevent their own properties from becoming polluted. Waste from equipment service and maintenance often becomes a source of pollution.

Sources of farm shop machinery, and buildings pollution include:

- Used oil
- Oil filters
- Antifreeze
- Paint and solvents
- Air-conditioner refrigerant
- Spilled or dumped fuel
- Fuel, oil and lubricant containers

Material spills happen. If fuel, oil, lubricants, or coolants are spilled, check the container label for the method of cleanup. Major spills require contacting local and state authorities.

Disposal information for hazardous materials can also be found on labels. Community collection points can be used to dispose of many materials. Contact your local recycling coordinator or Cooperative Extension Service for information on local recycling efforts.

Tire, Battery, and Garbage Disposal

Some materials are more difficult or costly to discard. Tires laying around become water-filled breeding grounds for mosquitoes. Batteries pile up in the corner. Some garbage should not be burned. What should be done?

Tire dealers and battery suppliers must accept these items from you. A disposal fee may be charged. Alternative uses for tires may be found as well.

Garbage that cannot be burned should be disposed of properly. Onfarm burial or use of landfills is possible. Read the labels on all materials to know the proper disposal methods.

Figure 1.3. c. Pesticide containers will specify the method of disposal on the label. Burning may be the correct method, but returning containers to the dealer, or burying them, may be indicated as well. Youths younger than age 16 should not be exposed to pesticides.

Contact EPA at 1-800-424-9346 to learn more about the disposal of hazardous materials.

Figure 1.3. d. Do not dispose of tires by dumping them over a hill.

Cooperation provided by The Ohio State University and National Safety Council

Safety Activities

1. Write a report concerning the problems you can identify as you read the introduction to this task sheet. In the report, name the problem and explain why there is a problem.

2. Word Find. Make as many words (three or more letters) as you can from the title "Environmental Protection Agency." Score 1 point for each word you find. To challenge yourself further, list only words that deal with clean air, clean water, safe pesticide use, and soil contamination.

3. Contact local municipal authorities (township supervisors) to request a copy of local burning ordinances.

4. 4. Ask a state highway officer to tell you about farm machinery accidents involving manure and pesticide spills. Ask them about shifting load violation penalties also.

5. Write a short essay about how to control, contain, and clean up a manure spill.

6. Research the subject of nutrient management to determine how much nitrogen, phosphorus, and potassium is needed by corn, alfalfa, and soybeans. Explain how nitrogen and phosphorus from manure can become a pollutant in our water supplies.

7. What problem does excess nitrogen and phosphorus cause in our waterways?

References

1. www.epa.gov (Environmental Protection Agency)
2. www.dep.state.pa.us (or Department of Environmental Protection for your state)
3. www.dot.state.pa.us (or Department of Transportation for your state)

Contact Information

National Safe Tractor and Machinery Operation Program
The Pennsylvania State University
Agricultural and Biological Engineering Department
246 Agricultural Engineering Building
University Park, PA 16802
Phone: 814-865-7685
Fax: 814-863-1031
Email: NSTMOP@psu.edu

Credits

Developed by WC Harshman, AM Yoder, JW Hilton and D J Murphy, The Pennsylvania State University. Reviewed by TL Bean and D Jepsen, The Ohio State University and S Steel, National Safety Council. Revised 3/2013.

This material is based upon work supported by the Natioal Institute of Food and Agriculture, U.S. Department of Agriculture, under Agreement Nos. 2001-41521-01263 and 2010-41521-20839. Any opinions, findings, conclusions, or recommendations expressed in this publication are those of the author(s) and do not necessarily reflect the view of the U.S. Department of Agriculture.

STATE AG SAFETY AND HEALTH RESOURCES

HOSTA Task Sheet 1.4

NATIONAL SAFE TRACTOR AND MACHINERY OPERATION PROGRAM

Introduction

There are "safety" experts found throughout the United States. Do you know where to find information about agricultural safety in your state? Safety professionals offer a wide variety of information, materials, demonstrations, and programs.

This task sheet discusses state agricultural safety resources who can help farm youth learn more about working safely and successfully on the farm. Learn who your state resources are and how to contact them.

State Level Resources

Some of these state safety resource programs employ specialists who provide safety training, may be able to travel to meet with you and your group, and can guide you to other resources to answer safety questions. Here are safety resources you may have in your state.

- College of Agriculture specialists in agricultural safety and health
- Cooperative Extension Service (offices in each county)
- Agriculture and Extension Education program specialists in 4-H and FFA at the state level (Contact your state 4-H office and Department of Education)
- State Farm Bureau Safety Leaders
- Colleges of Health/Nursing and University Medical Centers/ Hospitals
- Veterinary Medicine Colleges
- State Departments of Health
- State Fire Instructors

How can you contact these resources? The government section of the phone book and the Internet provide information, but may take time. Consider your state's Land Grant University as well.

A goal of your state's Land Grant University is to provide agricultural training as a means of improving agriculture. Agricultural safety is one area of this training. In Pennsylvania for instance, Penn State University started as a "Farmers High School." *Where is your land grant university located*?

Contact your State University **Ag Safety and Health Specialist** using the Internet to find the Land Grant University in your state. For example www.cas.psu.edu accesses the College of Ag Science at Penn State University. Search for "agricultural safety" sources. Next search "Cooperative Extension Service" sites to access your county Cooperative Extension Service location and contacts.

Figure 1.4.a. The Department of Agriculture and Biological Engineering at Penn State University is just one of many such departments in each state that offers a source of agricultural safety information. Take time to learn about the Land Grant University(ies) in your state.

Your state has ag safety resources. Learn who these experts are.

Learning Goal

- To become familiar with your state's ag safety professionals as a source of safety information

Related Task Sheet:

National Ag Safety and Health Resources **1.5**

Cooperation provided by The Ohio State University and National Safety Council.

Community Level Resources

There are local safety resources that provide safety information.

Public Sources

Public organizations are government-related and are tax-payer supported. Information may be free or inexpensive. Some of these resources include:

- State Police or Highway Patrol (traffic laws and road hazards)
- County Coroner (investigations into farm-related fatalities)
- Regional Departments of Agriculture (statewide and county data on the scope of agriculture, agricultural fairs and expositions, and grants for farm safety projects)
- Local Departments of Health (safety information)

Private Sources

Private sources are businesses that serve the agricultural industry. Several examples include:

- Electrical service suppliers and vendors (safety programs)
- Machinery and equipment dealers (Films on safety and equipment operation training materials)
- Veterinarians (animal health and animal handling safety)
- Local doctors and nurses (emergency medical help for farm accident victims and injury prevention ideas)
- Ag pesticide representatives (pesticide use and safety training seminars)
- Volunteer Fire Departments (fire prevention and agricultural rescue programs.)
- American Red Cross chapters (CPR and first aid training)

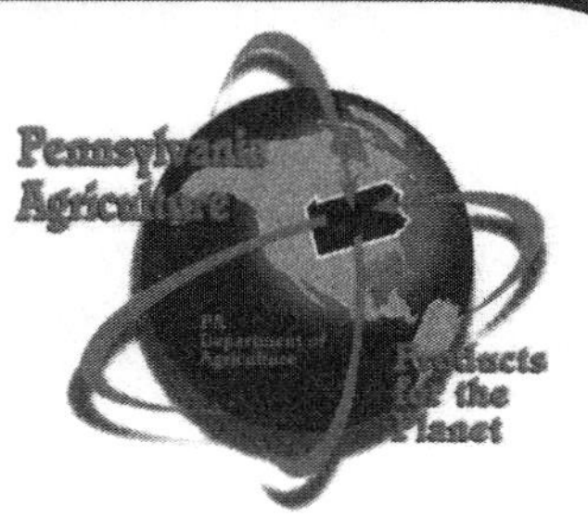

Figure 1.4.b. Check the Internet for the website of your state's Department of Agriculture. What is the logo that they use to promote agriculture in your state? What agricultural safety programs does your state offer?

- Emergency Medical Services (ambulance services and first aid classes)
- Safety Consultants (Private businesses in safety consultation charge a fee to assist in safe work practices)
- Insurance Companies (brochures on safety issues and presentations about safety)

Safety Activities

1. Use the Internet to visit the website of your Land Grant University(ies) to learn more about farm safety.
2. Use the Internet to visit the website of your state's Department of Agriculture to learn more about farm safety programs. Do they have a grant program for youth organizations to conduct safety activities? Do they have a Safety Quiz Bowl competition? Learn how you can participate.
3. Ask your local Extension Agent to sponsor and help train a Safety Quiz Bowl Team for competition.

References

1. Your State Land Grant University. For example, the website for the Pennsylvania State University is www.cas.psu.edu/
2. Your State Department of Agriculture
3. Your local county Cooperative Extension Service, or County Agent's Office
4. www.ffa.org (National FFA Organization website)
5. www.4-H.org (National 4-H Organization website)

Contact Information

National Safe Tractor and Machinery Operation Program
The Pennsylvania State University
Agricultural and Biological Engineering Department
246 Agricultural Engineering Building
University Park, PA 16802
Phone: 814-865-7685
Fax: 814-863-1031
Email: NSTMOP@psu.edu

Credits

Developed by WC Harshman, AM Yoder, JW Hilton and D J Murphy, The Pennsylvania State University. Reviewed by TL Bean and D Jepsen, The Ohio State University and S Steel, National Safety Council. Revised 3/2013

This material is based upon work supported by the National Institute of Food and Agriculture, U.S. Department of Agriculture, under Agreement Nos. 2001-41521-01253 and 2010-41521-20839. Any opinions, findings, conclusions, or recommendations expressed in this publication are those of the author(s) and do not necessarily reflect the view of the U.S. Department of Agriculture.

NATIONAL AG SAFETY AND HEALTH RESOURCES

HOSTA Task Sheet 1.5

NATIONAL SAFE TRACTOR AND MACHINERY OPERATION PROGRAM

Introduction

Agricultural safety issues do not rest in the hands of a few concerned people. There are many groups at the national level who understand the hazards of the agricultural industry. They are dedicated to protecting a vital part of the farm workforce—young people.

This task sheet discusses national sources of farm safety information. Contact them to learn how you can increase your safety knowledge.

Public/Governmental Agencies

Federal and state government departments are considered public agencies because they exist due to public funding through tax dollars. Many of these can be contacted through the Internet.

OSHA: The Occupational Safety and Health Administration is the safe workplace regulatory agency. See Task Sheet 1.2.2 to learn more about OSHA regulations relating to agriculture. Use www.osha.gov to access the website.

USDOL: The U.S. Department of Labor is the labor regulatory agency of the U.S. government. Child labor laws such as the Hazardous Occupation Order in Agriculture (Task Sheet 1.2.1) are enforced through this agency. Use www.dol.gov to learn more.

USDA: The United States Department of Agriculture serves rural America and the agricultural community through education, research, and regulation of food production and safety, conservation, and worldwide market development. Go to www.usda.gov.

CES: The Cooperative Extension Service of USDA brings safety information to the state and local level. A Cooperative Extension Service office can be found in your county or parish.

AgrAbility: Education, technical and financial information, and support systems to farmers with disabilities, is the function of AgrAbility programs in several states. AgrAbility works with non-profit disability service organizations (e.g. Easter Seals) to provide services to those farmers suffering disability. USDA and National Easter Seals sponsor this program.

NIOSH: The National Institute for Occupational Safety and Health (NIOSH) is a branch of the Centers for Disease Control and Prevention (CDC). NIOSH is responsible for conducting research and making recommendations for the prevention of work-related injury and illness, including agriculture. Check out www.cdc.gov/NIOSH

Consumer Product Safety Commission

Figure 1.5.a. Many government agencies can provide information about farm safety. What are these two organizations and what do they do?

Ag safety info can come from many national sources.

Learning Goal

- To become aware of the many national agricultural organizations available as resources for safety information

Related Task Sheet:

Safety and Health Regulations	1.2
Hazardous Occupations Order in Agriculture	1.2.1
Occupational Safety and Health Act	1.2.2
State Agricultural Safety and Health Resources	1.4

Cooperation provided by The Ohio State University and National Safety Council.

Careers are available in agricultural safety work.

Figure 1.5.b. Corporate manufacturing and supply vendors are sources of safety information about their products. Some of the information can be accessed via the Internet, while local dealers can provide training videos, brochures, field days, field trips, and demonstrations for you or your group. Read more about national level associations and organizations on page 3.

More Governmental Agencies

ASH Centers: Agriculture Safety and Health Centers are supported by NIOSH. Centers currently serve 10 areas of the United States. Regional ASH Center locations include:

- Pacific Northwest, Washington
- Western, California
- High Plains Intermountain, Colorado
- Southwest, Texas
- Great Plains, Iowa/Nebraska
- National Farm Medicine Center, Wisconsin/Minnesota
- Southeast, Kentucky
- Northeast, New York
- Great Lakes, Ohio
- Southern Coastal Area, North Carolina

These centers provide safety education programs specific to their geographic location. Use the Internet to locate each center through www.cdc.gov/niosh.

National Children's Center for Rural and Agricultural Health and Safety: This center, also sponsored by NIOSH, promotes farm safety for children. One program creating safe play areas on farms draws attention to helping small children grow up safely on farms. Explore their resources by contacting www.marshfieldclinic.org.

CPSC: The U.S. Consumer Product Safety Commission is a federal regulatory agency working with industry to develop and implement standards for safety in consumer products. This agency can recall unsafe products. Contact www.cpsc.gov.

EPA: The Environmental Protection Agency of the federal government is assigned the responsibility to protect the air, water, and natural resources of the U.S. Pesticide laws and air and water pollution regulations affect our farms. See www.epa.gov to learn more about this regulatory agency.

OVR: An Office of Vocational Rehabilitation can be found in each state as part of the state's Bureau of Labor and Industry (Pennsylvania designation). This agency assists citizens with disabilities to gain economic independence. Specialized services are available from OVR offices. Financial aid may be available to assist disabled farmers.

Corporate Sources

Many corporate groups are sources of information about agricultural safety. A few are listed here. You may discover more as you develop your safety awareness. Try finding them on the Internet.

Vendors:

- Gempler's Inc.
- NASCO

Equipment Manufacturer:

- Deere and Company
- Case IH
- New Holland
- AGCO
- Kubota

Chemical Company:

- Dow
- Monsanto
- DuPont
- Novartis

This listing is used as an example and does not represent endorsement of any specific vendor or manufacturer (Figure 1.4.2.b.).

Other Sources

Some organizations or associations exist as nonprofit groups. They work toward a common good for their industry or interests.

NSC: The National Safety Council is a federally chartered nonprofit, nongovernmental source of safety and health information. Education in safety, safety resources, and farm safety statistics are available from this group. Check out www.nsc.org.

AEM/FEMA: The Association of Equipment Manufacturers (AEM) and the Farm Equipment Manufacturers Association (FEMA) represent large and small companies. AEM is a trade and development resource. FEMA represents the common interests of hundreds of smaller companies. Find them at www.aem.org and www.farmequip.org.

ASAE: The American Society of Agricultural and Biological Engineers is a professional and technical organization dedicated to the advancement of engineering in agriculture, food, and biological systems. Find them at www.asae.org.

NLSI/NLPI: The National Lightning Safety Institute (NLSI) and the National Lightning Protection Institute (NLPI) are two similar associations. The NLSI promotes lightning safety for people and structures. The NLPI promotes high quality, safe design, and safe installation of lightning protection systems. Use www.lightningsafety.com or www.lightning.org to access their websites.

You may also try these sources:

- National 4-H Organization
- The National FFA
- National SAFE KIDS Campaign
- Farm Safety 4 Just Kids
- The National Center for Farm Worker Health

For example, the Farm Safety 4 Just Kids (FS4JK) program provides educational opportunities and resources to make the farm a safe and healthy environment for children. Contact them at www.fs4jk.org.

NASD: The National Agriculture Safety Database is the national central storehouse of agricultural health, safety, and injury prevention materials. Agricultural statistics on injury and death can be found there. Funding for this effort comes from USDA and NIOSH. Use www.cdc.gov/nasd to locate this source.

Would you consider a career in the field of agricultural safety?

Figure 1.5.c. These youth organizations serve agriculture. What safety programs do they offer? Use the Internet or local 4-H leaders and agriculture teachers to find out more.

Safety Activities

1. Organize a list of Internet websites that discuss agricultural safety. Hint: Try the Land Grant University in your state first. Then begin using any search engine on the Internet to look for those references discussed in this task sheet. You can expect to find dozens of sources.

2. Use the NAGCAT website (www.nagcat.org) to find out more about this resource. Find out how to produce a safety calendar. The website describes how to customize a safety calendar for your family or group. Perhaps you could make a farm safety calendar for your home, club, or school.

3. Call your local County Cooperative Extension Service and ask to have safety publications mailed to you.

4. Use a national chain store catalog (Sears, Gemplers, etc.) to make a list of their available safety materials. List the price tag as well. Safety is a large and important business.

5. Call your local Volunteer Fire Department to inquire as to whether they have a Junior Member eligibility. Perhaps you could join the group to learn more about fire safety and rescue techniques.

6. Volunteer at local Red Cross and/or Easter Seals chapters to help these groups help others in the community.

References

1. The Internet; use any search engine.
2. Local Cooperative Extension Service offices
3. Local Secondary Agricultural Education Instructors
4. State Land Grant Universities
5. Federal and State Government Agencies
6. Safety Associations and Corporations

Contact Information

National Safe Tractor and Machinery Operation Program
The Pennsylvania State University
Agricultural and Biological Engineering Department
246 Agricultural Engineering Building
University Park, PA 16802
Phone: 814-865-7685
Fax: 814-863-1031
Email: NSTMOP@psu.edu

Credits

Developed by WC Harshman, AM Yoder, JW Hilton and D J Murphy, The Pennsylvania State University. Reviewed by TL Bean and D Jepsen, The Ohio State University and S Steel, National Safety Council. Revised 3/2013

This material is based upon work supported by the National Institute of Food and Agriculture, U.S. Department of Agriculture, under Agreement Nos. 2001-41521-01263 and 2010-41521-20839. Any opinions, findings, conclusions, or recommendations expressed in this publication are those of the author(s) and do not necessarily reflect the view of the U.S. Department of Agriculture.

INJURIES INVOLVING YOUTH

HOSTA Task Sheet 2.1 **Core**

NATIONAL SAFE TRACTOR AND MACHINERY OPERATION PROGRAM

Introduction

"I'm always careful! I'll never suffer a work injury!" You may say this to yourself as you begin to read this task sheet. But this same thinking is what injures and kills hundreds of workers in farm accidents each year.

This task sheet looks at the numbers of fatalities and injuries that have caused great concern in farming and ranching.

The Situation

The work death rate per 100,000 workers regularly ranks agriculture among the most hazardous industries in the U.S. Youths are included in these injury numbers. Other industries that have many serious work hazards, like mining and construction, do not have a youth injury problem because youth younger than 16 do not usually work in these industries.

Youth Farm Injury Statistics

Accurate numbers of youth work fatalities and injuries are difficult to determine because youth do not work regularly enough or in large enough numbers to be counted in most official injury statistics. Special studies relying on voluntary cooperation by farmers are done to find out about youth farm work injury. As a result, the statistics that are developed are considered lower than what the actual numbers may be. The facts below are national data.

Fatality Facts

- Currently, estimates show that slightly more than 100 youth younger than age 20 are killed each year in farm work-related incidents.
- Between 1995 and 2002, 907 farm deaths among youth were documented (most between 16 and 19 years of age).
- 25% of the fatalities involved machinery, 17% motor vehicles (including ATVs) and 16% drowning.
- Males, age 20 and younger, accounted for most of the fatalities.

Injury Facts:

- In 2009 over 15,000 youth injuries occurred on farms.
- Falls accounted for 40% of the injuries.
- Hand, head, and leg injuries are typical of the injury.
- Livestock and dairy farms have more injuries than crop farms

State Data

Contact the safety specialist at your land grant university to learn of farm injury statistics for your state.

Figure 2.1.a. Youthful workers are called upon to complete many tasks that may present several hazards. For example, this tractor does not have a ROPS or a seat belt.

If you are studying this task sheet, you are part of the ag industry. Don't become part of the sad statistics.

Learning Goals

- To learn about the numbers and types of injuries associated with youth working in agriculture

Related Task Sheets:

Safety and Health Regulations	1.2
Hazardous Occupations Order in Agriculture	1.2.1
Age-Appropriate Tasks	2.4

Cooperation provided by The Ohio State University and National Safety Council.

How Can I Use This Information?

More than 2 million youths younger than age 20 are potentially exposed to agricultural hazards each year according to estimates by the National Institute for Occupational Safety and Health. Farm family workers, hired workers, children of seasonal and migrant workers, and farm visitors can all encounter a wide range of hazards. Machinery, livestock, farm storage structures, and farm ponds all present unique farm safety challenges.

Follow these safety suggestions to avoid becoming a farm injury or fatality statistic.

1. Identify agricultural hazards in the work area to which you are assigned.
2. Develop a plan to deal with the hazards you identified.
3. Use safety practices all of the time.
4. Think about the consequences of your actions before taking a chance.
5. Reinforce safe work habits by helping others to work safely.
6. Wear personal protective equipment suggested for the job.
7. Speak up for your safety on the job.

Being safe is largely a matter of choice.

Figure 2.1.b. Agriculture presents unique safety challenges. You are challenged to work safely in many conditions. Not all work is done with tractors and machinery.

What are the consequences of your unsafe actions?

Safety Activities

1. Review what you have read by completing this quiz:
   a. True or False? Most fatal injuries to farm youth occur to females.
   b. What are the three leading causes of injuries?
   c. True or False? Most farm injuries involve working with fruit trees.
   d. What percentage of farm fatalities involved machinery?
2. Using the Internet sites www.nsc.org (National Safety Council) and www.niosh.gov (National Institute for Occupational Safety and Health), locate information comparing the work fatality of agriculture with other industries. Use a computer to make a chart or graph to summarize the data. If you do not have access to a computer, make a full-size poster of the information to share with your group.

References

1. National Children's Center for Rural and Agricultural Health and Safety, 2013 Fact Sheet, Childhood Agricultural Injuries.
2. Farm and Ranch Safety Management, John Deere Publishing, 2009. Illustrations reproduced by permission. All rights reserved.

Contact Information

National Safe Tractor and Machinery Operation Program
The Pennsylvania State University
Agricultural and Biological Engineering Department
246 Agricultural Engineering Building
University Park, PA 16802
Phone: 814-865-7685
Fax: 814-863-1031
Email: NSTMOP@psu.edu

Credits

Developed by WC Harshman, AM Yoder, JW Hilton and D J Murphy, The Pennsylvania State University. Reviewed by TL Bean and D Jepsen, The Ohio State University and S Steel, National Safety Council. Revised 3/2013

This material is based upon work supported by the National Institute of Food and Agriculture, U.S. Department of Agriculture, under Agreement Nos. 2001-41521-01263 and 2010-41521-20839. Any opinions, findings, conclusions, or recommendations expressed in this publication are those of the author(s) and do not necessarily reflect the view of the U.S. Department of Agriculture

RISK PERCEPTION

HOSTA Task Sheet 2.2

NATIONAL SAFE TRACTOR AND MACHINERY OPERATION PROGRAM

Introduction

Why do people take risks? Has past experience taught you that taking risk is acceptable? Have you also learned that risk-taking increases your chances of injury?

Risk can be measured. The odds of injury and a prediction of the consequences of risk-taking have been studied by safety specialists. A person's risk perception (how we judge risk) about work risks comes from personal judgments made about a work situation.

This task sheet discusses risk-taking and the perceptions people have about risks. Risk-taking behavior is a topic that all workers must understand.

The Nature Of Risk

No one can deny that all people take risks. We risk our lives and health each day. Some risks are minor. We don't expect that everyone will smash their finger in the car door. Other risks are major. Driving too fast increases the risk of a crash and possible injury. We are exposed to risk each day.

Risk can be defined as "the chance you take of becoming injured by a hazard."

Risk measurement starts with probability (odds or chances). What are the odds or chances that we can be injured by a specific hazard? Most people do not judge the probability of risk very well. Odds of risk can be placed in categories. See page 2.

Risk measurement also includes how serious you can be injured by a hazard. Risks can be great (death) to negligible (splinter). Page 3 discusses the severity of the consequences of risk.

Risk perception is an important concept in safe work activity. Human perceptions of risk are not very accurate.

Our judgments about risks are based upon several things. One important factor is how familiar we are with a hazard. If we think we know a lot about a hazard because we are often exposed to the hazard, we often underestimate the degree of risk.

Another factor is whether or not we are voluntarily interacting with a hazard. When we voluntarily take a risk, we usually underestimate the chances of being hurt.

A third factor is how much attention a hazard brings if it hurts someone. We tend to think that there is a great risk in flying in an airplane (kills many people at one point in time and gets more attention). We underestimate the hazard of driving a car. An automobile crash may kill one or two persons at a time but receive only local attention.

A person must understand risk, the probability of danger, and the personal consequences which can result.

Figure 2.2.a. We view some jobs as more hazardous than others. When we understand that the consequences of taking a risk can lead to injury or death, we improve our safe work habits greatly.

To reduce risks, a person must understand the consequences of their own actions.

Learning Goals

- To understand how we think about risk

Related Task Sheets:

The Work Environment	**1.1**
Injuries Involving Youth	**2.1**
Reaction Time	**2.3**
Age-Appropriate Tasks	**2.4**

Cooperation provided by The Ohio State University and National Safety Council.

What are the odds you will be injured while working?

Figure 2.2.b. What are the odds that you will be hit by lightning? Using the rating system below, a measure of the probability of risk would be determined by how often you were exposed to the dangers of a lightning storm. Since most people take shelter during a lightning storm, the probability or remote.

Probability

Work, and all other activity, involves risks. Some risks are very small. Other risks are great. What are the odds (or chances) that you will be injured while engaged in an activity?

The subject of probability is a study of the odds or chances of a single event actually occurring out of the possible times it could occur. For example, if you roll a single die (dice), the odds of rolling a 1 is 1 out of 6.

Safety experts have rated the probability of exposure to risks in several ways. One rating system is discussed here.

This rating uses a time frequency that can be measured.

The frequency rating system includes these categories:

- Frequent exposure – Probability is likely/possible on a daily basis. As an example, daily use of a PTO-powered implement is a frequent exposure to this hazard.
- Probable exposure – Probability is likely/possible on a weekly or monthly basis. As an example, weekly or monthly inspections of the silo unloader gives a probable exposure to the hazards of a fall.
- Occasional exposure – Probability is likely/possible over a year or many year time period. As an example, a yearly skiing trip provides the occasional exposure to the risk of a ski injury.
- Remote exposure – Probability is not likely, but is possible over many years, even a lifetime. As an example, the painting of a barn roof is done only rarely by the owner of the barn; so the exposure to a fall injury is considered a remote probability. The barn roof painter, however, is frequently exposed.
- Improbable exposure – Probability is unlikely, but still possible. As an example, nuclear power radiation poses an improbable exposure.

From these probability ratings we can see that the less exposure to risk that we have, the less likely the odds of injury or death.

Select a work activity which you perform, and rate its probability for your exposure to risk.

Consequences of Risk Exposure

Just as risk exposure probabilities can be assigned measurement categories (page 2), the consequences of risk exposure can be assigned a measurement category. One method to rate the consequences of risk exposure for severity of the outcome is discussed here.

Categories of consequences of severity of risk exposure can include:

- Catastrophic severity – Injury or death is imminent (near), and there is potential for widespread loss. As an example, death from operating a non-ROPS tractor that rolls over poses a great risk.
- Critical severity – Severe or permanent injury, long-term illness, and temporary property loss is possible. As an example, trying to unplug a corn picker that is running can lead to entanglement and potential loss of an arm or leg.
- Marginal severity – Less serious risk exposure with shorter term losses. As as example, falling from a horse and breaking an arm is less severe than having an arm amputated due to a PTO entanglement.
- Negligible severity – Risk exposure event results in need for first aid, or property losses that are easily repaired. As an example, a splinter from plywood can be treated with basic first-aid supplies. If the splinter caused the plywood to be dropped, the loss is slight.

Select a work activity which you perform, and rate the severity of the risk.

The probability of risk exposure and the consequences of the risk can then be treated as an equation with a resulting answer (what to do to reduce risk). See if you can use the Risk Matrix Table (page 4) to answer the question, "What is the risk of climbing over a turning, unguarded PTO shaft every day?"

Can you rank all of your work activities with this matrix?

Reducing Risks

People take risks everyday. Some risks are seen as acceptable because of past experiences, our own notions and overconfidence of the risk situation, and our willingness to accept the risk.

The following points are important to consider in reducing the risk to which a young worker is exposed.

1. Recognize your own traits that increase risk. Are you impatient in getting work done?
2. Recognize when you need more training to do a job. Risk-taking behavior can be reduced with knowledge of hazards.
3. Remove hazards from the work place. The fewer hazards that exist in the work zone, the less risk of danger that exists.
4. Use safe technology correctly. Modern farm machines are engineered to reduce risks to the operator. The operator must use this technology safely.

Figure 2.2.c. Modern farm equipment has many more safety technologies than older machinery. Future advancements may include sensors that detect when a person enters a hazardous area. Then the tractor and machine may shut down to keep the operator or bystander from harm.

Risk can be equated with expected damage or injury consequences. (See page 4)

Cooperation provided by The Ohio State University and National Safety Council.

Applying A Risk Matrix Table to Reduce Risk Probability

Severity → / Frequency ↓	Catastrophic (1)	Critical (2)	Marginal (3)	Negligible (4)
Frequent (A)	Shut down immediately; correct problem	Shut down immediately; correct problem	Correct ASAP	Correct sometime
Probable (B)	Shut down immediately; correct problem	Correct ASAP	Correct soon	Correct sometime
Occasional (C)	Correct ASAP	Correct soon	Correct sometime	Correct sometime
Remote (D)	Correct sometime	Correct sometime	Correct sometime	Correct sometime
Improbable (E)	Correct with preventative maintenance	Correct with preventative maintenance	Correct with preventative maintenance	Correct with preventative maintenance

Table 2.2.a. The Risk Matrix Table provides a means of evaluating a risk and what to do to reduce the consequences of the risk exposure.

Safety Activities

1. Write a short essay about a time or event in which you took a risk.
2. From your essay, what were your feelings after you had time to look back on the risk you took. Write a few notes about your feelings.
3. Make a list of risk-taking situations that you have experienced. Place these examples into the appropriate risk category.
4. What do you recognize about yourself that might be an indicator that you are a risk-taker?
5. Take a safety tour of a farm area. List the hazards, and then list the chores that you find risky.

References

1. Safety and Health for Production Agriculture, Dr. Dennis J. Murphy, 1992, American Society of Agricultural Engineers, St. Joseph, Michigan.

Contact Information

National Safe Tractor and Machinery Operation Program
The Pennsylvania State University
Agricultural and Biological Engineering Department
246 Agricultural Engineering Building
University Park, PA 16802
Phone: 814-865-7685
Fax: 814-863-1031
Email: NSTMOP@psu.edu

Credits

Developed by WC Harshman, AM Yoder, JW Hilton and D J Murphy, The Pennsylvania State University. Reviewed by TL Bean and D Jepsen, The Ohio State University and S Steel, National Safety Council. Revised 3/2013

This material is based upon work supported by the National Institute of Food and Agriculture, U.S. Department of Agriculture, under Agreement Nos. 2001-41521-01263 and 2010-41521-20839. Any opinions, findings, conclusions, or recommendations expressed in this publication are those of the author(s) and do not necessarily reflect the view of the U.S. Department of Agriculture.

REACTION TIME

HOSTA Task Sheet 2.3

NATIONAL SAFE TRACTOR AND MACHINERY OPERATION PROGRAM

Introduction

How fast can you react? **Reaction time is defined as the time it takes for a person to react to an event or an emergency.** Emergencies occur without warning. Our past experience, along with our reaction time, determines how well we respond to an emergency event.

This task sheet discusses reaction time as it relates to you and the speed of the machines with which you work. Machines are much faster than a human's reaction time. There are no super heroes faster than a speeding machine.

Reactions Are More Complex Than You Think

Reacting to an emergency involves a complex sequence of events. Consider when an animal jumps in front of your car as you travel down a road. What happens next?

- Your eye gathers the information, "Animal in road," and sends a message to your brain.
- Your brain receives the information, processes the information, and sends a response to your extremities (arms and legs).
- Your leg muscles must move your foot from the gas pedal to the brake pedal and begin to push the pedal.
- The vehicle continues to move as you respond until the car finally stops just before you hit the deer—or after you have demolished your car.

Here are a few more examples of emergency situations:

- Accidentally touching a hot stove
- Recognizing that your shirt sleeve is being caught on the drill press chuck
- Realizing that your shoe string is dangling over the PTO shaft that you should not be stepping across
- Pulling a tractor and load onto the highway and seeing a fast-moving vehicle coming your way
- Trying to unplug a corn picker and being pulled into the gathering chains

Emergencies occur anytime and anywhere. Remember, an emergency does not give you time to think about what you will do. You react to emergencies as they occur with no warning or time to plan or prepare for action.

Many factors affect your reaction time. Read further to find out why you cannot beat a machine in an emergency. Your life may depend upon this information.

Figure 2.3.a. **Never step across a PTO shaft** which is turning. A severe emergency can develop to which you must react. You cannot beat the machine.

We are not the "super heroes" of television fame. We are not faster than a speeding machine.

Learning Goals

- To recognize that personal reaction time is slower than the speed of a machine
- To work safely with attention to safe procedures and sound practices based upon knowledge of the limitations of human reaction time

Related Task Sheets:

Age-Appropriate Tasks	**2.4**
Mechanical Hazards	**3.1**
NAGCAT Tractor Operation Chart	**4.3**
Using PTO Implements	**5.4.1**

Cooperation provided by The Ohio State University and National Safety Council.

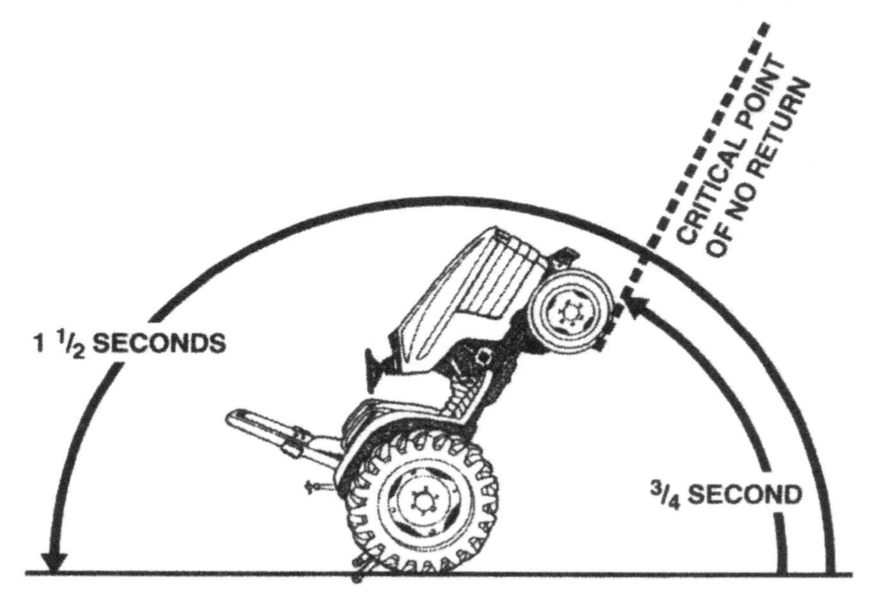

Figure 2.3.b. A tractor can upset in 3/4 of a second. Our reaction time will not prevent the injury and damage that can occur. Can you tell why the tractor would rear up as shown in this picture? *Safety Management for Landscapers, Grounds-Care Businesses, and Golf Courses, John Deere Publishing, 2001. Illustrations reproduced by permission. All rights reserved.*

If your reaction time is 3/4 of a second, a tractor tipping backward can reach the point of no return before you can react.

Figure 2.3.c. Know your own limitations which can affect your work reactions. *Farm and Ranch Safety Management, John Deere Publishing, 1994. Illustrations reproduced by permission. All rights reserved.*

Factors Affecting Reaction Time

Here are a few factors that affect your reaction time:

- Experience
- Age
- Fitness
- Fatigue
- Illness
- Pre-occupation
- Distraction
- Mood
- Weather
- Drugs/medication
- Alcohol and tobacco
- Machine vibrations
- Poor vision
- Poor hearing

Something To Think About:

- Experienced operators have gained knowledge of potential hazards. Beginning operators may not know when danger exists.
- Healthy, well-rested operators think through hazardous situations more clearly than fatigued workers.
- Distracted or daydreaming operators are less cautious than focused workers.
- Frustrated workers tend to make bad decisions.
- Medications, as well as drugs alcohol and tobacco, can slow your reaction time.
- Machine vibrations have been shown to fatigue operators and reduce reaction time.
- Poor vision and hearing can lead to poor reaction time.

Rotating Parts Are Everywhere

Working around or near shop equipment, machinery or tractors exposes the operator to more hazards than an office worker. Rotating parts are everywhere. Some examples are:

- Grinding wheels
- Drill presses
- Chain saws
- Lawn mowers
- Augers
- Belts and pulleys
- Chains and sprockets
- Gears
- Power take off shafts

All exposed rotating parts of farm tools and equipment spin faster than you can pull away should you become entangled.

PTOs and Reaction Time

Now is a good time to ask, "Are you faster than a speeding machine?" "Can you react faster than the machine and avoid injury or death?"

We have all been warned not to step over a turning PTO shaft, but PTO entanglements are still happening. A simple arithmetic problem can be used to explain what can happen should your pant leg be caught on an unguarded rotating shaft.

The unguarded PTO shaft is turning at 540 RPMs. You decide to step over it to save a few steps and seconds, rather than walk around the tractor or piece of equipment. You feel a tug on your pants leg and begin to pull away.

With a reaction time of 3/4 of a second (0.75), how many turns of the shaft will be tugging at your pants before you begin to pull away (if you can at all)?

First, convert 540 RPM to revolutions per second (RPS) by dividing 540 by 60 seconds.

$$540/60 = 9 \text{ RPS}$$

Second, multiply 9 RPS by your reaction time to get the revolutions of the PTO shaft before you begin to pull away.

$$9 \times 3/4 = 27/4 = 6.75$$

Or $9 \times 0.75 = 6.75$ revolutions before you react or begin to pull away.

Avoid Rotating Part Hazards

To avoid rotating part entanglements, try these practices.

1. Keep guards in place on rotating shafts and parts.
2. Stop the engine before dismounting the tractor.
3. Dress safely to avoid entanglements.
4. Think before you take a chance: "Is saving a few seconds or steps worth risking my life?"

Figure 2.3.d. **Do not use equipment with unguarded PTO shafts.**

Think, "What is the worst thing that can happen to me?" A few seconds of thought can prevent injury or death.

Figure 2.3.e. Be sure PTO shaft guards and stub shields are in place.

Safety Activities

1. If you are involved in an agricultural education mechanics program, ask the instructor if you can conduct a survey of electric motors on machines and small appliances (drills, portable saws, etc.) to chart the speed in RPM of those motors. The speed of the motor in RPM is found on the motor nameplate. Make a chart of the information as follows:

Motor /Machine	Speed of Motor in RPM
Table Saw	1740

2. Conduct activity 1 in the farm shop or in the home with any electrical appliance where you can view the electric motor nameplate information.

3. Using a stop watch, press the start button to start the timer, and as quickly as possible, press the stop button. See how fast you can do this simple task. Take several readings, record the results, and calculate the average time you needed to stop the timer. Although this is not a measure of reaction time to an emergency, you can use this measurement to make reaction time calculations in the following questions.

   Time it took you to start/stop the stop watch: __________seconds/fractions of a second

4. Solve this reaction time math problem.

   A drill press is rotating at 1800 rpm. If your reaction time is 1/2 second (0.5), how many revolutions of the drill press will occur before you react and pull your shirt sleeve away?
   ____________ Revolutions before reaction to pull away.
   Hint 1: Convert RPM to revolutions per second (RPS).
   Hint 2: There are 60 seconds in a minute.
   Hint 3: Multiply RPS (Hint 1) by your reaction time in Activity 1 or 2, or use 1/2 second reaction time.

5. A PTO shaft turns 540 RPM. Your reaction time is 1/2 second. If your shoelace is caught in the shaft, how many turns of the PTO shaft would occur before you react? Use the hints from Activity 4.
   ____________ Revolutions before reaction to pull away.

6. Make the same calculation from Activity 5 using a 1000 RPM PTO shaft as the speed of the machine.
   ____________ Revolutions before reaction to pull away.

References

1. Safety Management for Landscapers, Grounds-Care Businesses, and Golf Courses, John Deere Publishing, 2001. Illustrations reproduced by permission. All rights reserved.
2. Farm and Ranch Safety Management, John Deere Publishing, 2009.

Contact Information

National Safe Tractor and Machinery Operation Program
The Pennsylvania State University
Agricultural and Biological Engineering Department
246 Agricultural Engineering Building
University Park, PA 16802
Phone: 814-865-7685
Fax: 814-863-1031
Email: NSTMOP@psu.edu

Credits

Developed by WC Harshman, AM Yoder, JW Hilton and D J Murphy, The Pennsylvania State University. Reviewed by TL Bean and D Jepsen, The Ohio State University and S Steel, National Safety Council. Revised 3/2013

This material is based upon work supported by the National Institute of Food and Agriculture, U.S. Department of Agriculture, under Agreement Nos. 2001-41521-01263 and 2010-41521-20839. Any opinions, findings, conclusions, or recommendations expressed in this publication are those of the author(s) and do not necessarily reflect the view of the U.S. Department of Agriculture.

AGE-APPROPRIATE TASKS

HOSTA Task Sheet 2.4 **Core**

NATIONAL SAFE TRACTOR AND MACHINERY OPERATION PROGRAM

Introduction

Farming offers a unique opportunity for children and adolescents to learn the value of hard work, how to handle responsibilities, and how to set priorities.

Traditionally, farming has been a family affair with children working on their own farm. Larger farms may hire youth for work as well. While early labor has personal development benefits, there are also many risks involved for young workers.

This task sheet offers guidelines for matching youthful farm workers with farm tasks.

North American Guidelines for Children's Agricultural Tasks (NAGCAT)

These guidelines were developed under the direction of the Children's Center for Rural and Agricultural Health and Safety. The guidelines assist adults in assigning farm jobs to children age 7 to 16 years, living or working on farms. Employers should also be aware of the guidelines. Visit the NAGCAT website shown in the reference section for more information about the guidelines.

Here are some key points in your development toward adulthood.

Ages 12-13 (Early Teens)

Some of the traits shown by early teens include: being clumsy and rebellious, lacking focus, being easily distracted, and taking risks.

Typical death and injury risk scenarios for the age group 12-13 include:

- Machinery entanglements ("I cannot get caught; I am too fast.")
- Head and spine injuries from ATVs and motorcycles ("I want to be faster than my peers.")
- Falls from machines (extra riders) and structures
- Sprains due to working harder than growing muscles permit

Age-Appropriate Tasks for 12– and 13-Year-Olds:

- Hand raking and digging
- Limited power tool use with supervision
- Operating lawn mower or garden tractor
- Handling /assisting with animals
- Other low-risk tasks

Figure 2.4.a. Young people sometimes want to do everything an adult does.

All of us go through growth phases. Our physical and mental maturity may not match our work assignment.

Learning Goals

- To identify typical growth traits by age groups and how these traits may affect what jobs and tasks young workers should be assigned

Related Task Sheets:

Injuries Involving Youth	2.1
Reaction Time	2.3
NAGCAT Tractor Operation Chart	4.3

 Cooperation provided by The Ohio State University and National Safety Council.

Figure 2.4.b. Taking chances with motorized vehicles may be fun, but the odds of an injury or death increases greatly when risky behavior is substituted for correct operation of ATVs, dirt bikes, and tractors.

Physical growth can lead youth to believe that they can do more than they can mentally handle.

Figure 2.4.c. Young muscles and joints may be strained and sprained trying to move heavy loads. Animals pose their own special risks in being moved as well.

Ages 14-15 (Young Teens)

Some of the traits shown by young teens include: being moody and rebellious, taking risks, being mentally active, and having feelings of immortality.

Typical death and injury risk scenarios for the age group 14-15 include:

- Machinery entanglements with amputations from PTO, augers, turning parts, and power tools ("I cannot get caught; I am too fast.")
- Head and spine injuries from ATVs and motorcycles ("I want to be faster than my peers.")
- Falls from machines (extra riders) and structures
- Hearing loss from machinery
- Animal handling incidents
- Tractor overturns
- Roadway crashes or mishaps

Age-Appropriate Tasks for 14– and 15-Year-Olds:

- Equipment maintenance
- Manual feeding of livestock
- Operating non-articulated tractors for field work, including those with implements that are 3-point mounted, use remote hydraulics, and/or are PTO-powered
- Raking hay
- Operating a pressure washer

Ages 16-18 (Older Teens)

Some of the traits shown by older teens include: being aggressive and taking risks, feelings of being immortal and overconfident, and experimenting with adult independence and behaviors.

Typical death and injury risk scenarios for the age group 16-18 include:

- Machinery entanglements with amputations
- Falls from machines and structures
- Hearing loss from machinery
- Animal handling incidents
- Tractor overturns
- Roadway crashes or mishaps
- Added risks if experimenting with drugs and/or alcohol

Age-Appropriate Tasks for 16-18 Year Olds:

- Ordinary use of tractors, self-propelled machinery, augers, elevators, and other farm equipment
- Pulling oversized loads, simultaneous use of multiple vehicles, and application of chemicals with specific training and close supervision

NAGCAT Guidelines

There are a total of 62 age-appropriate guidelines in seven categories as follows:

Tractor Fundamentals Tasks

Tractor Operation Chart

Driving a Farm Tractor

Trailed Implements

3-Point Hitch Implements

Hydraulics

PTO -Connect/Disconnect

Haying Operations

6 guidelines dealing with hay harvest and transport

Implement Operations

10 guidelines dealing with fieldwork

General Activities

Using a front-end loader plus 9 guidelines for various farm equipment operations

Other Guideline Categories include:

Animal Care

Manual Labor

Specialty Production

A few guidelines have been translated into Spanish. Check the nagcat.org website for details.

Figure 2.4.d. Machinery, ATVs, and falls account for most farm injuries and fatalities. Youth who will work with machinery, ATVs and agricultural structures must be trained and have guidance to be safe workers.

Guidelines for age-appropriate tasks are useful tools and can lead to mature actions of the safety-conscious youthful worker.

Figure 2.4.e. Youthful workers assigned to supervised tasks can learn safe, productive agricultural work habits. The NAGCAT Guidelines help parents and employers determine what tasks are appropriate.

Cooperation provided by The Ohio State University and National Safety Council

Safety Activities

1. Use the NAGCAT website to locate the guidelines for operating 3-point hitch implements. Print the guideline, and answer all the questions for yourself. The page will include pictures like this:

Share this information with your parents and tractor safety instructor or leader.

2. Use the NAGCAT website to explore other guideline task sheets that may focus on jobs you will do.

3. Write a short story about a hazardous situation you have encountered and how you approached that hazard based upon your stage of development at that time. Did your youthful immaturity influence the outcome?

4. Ask your class or club members to relate stories of hazardous incidents they encountered and how they handled them.

Special Note: Youth who are age 12 or 13 may complete studies of safe tractor operation and complete the written exam, but cannot take the skills or driving exams nor receive a certificate under the Hazardous Occupations Order In Agriculture program.

References

1. www.nagcat.org/Click on guidelines/Select category, July 2013.
2. www.cas.psu.edu/Type in search box children and safety on the farm/Click on Children and Safety on the Farm, Murphy and Hackett, 1997.
3. www.extension.umn.edu/Click on Farm Safety and Health/Click on Is Your Child Protected from Injury on the Farm?, Shutske, April 2002.

Contact Information

National Safe Tractor and Machinery Operation Program
The Pennsylvania State University
Agricultural and Biological Engineering Department
246 Agricultural Engineering Building
University Park, PA 16802
Phone: 814-865-7685
Fax: 814-863-1031
Email: NSTMOP@psu.edu

Credits

Developed by WC Harshman, AM Yoder, JW Hilton and D J Murphy, The Pennsylvania State University. Reviewed by TL Bean and D Jepsen, The Ohio State University and S Steel, National Safety Council. Revised 3/2013

This material is based upon work supported by the National Institute of Food and Agriculture, U.S. Department of Agriculture, under Agreement Nos. 2001-41521-01263 and 2010-41521-20839. Any opinions, findings, conclusions, or recommendations expressed in this publication are those of the author(s) and do not necessarily reflect the view of the U.S. Department of Agriculture.

SEVERE WEATHER

HOSTA Task Sheet 2.5

NATIONAL SAFE TRACTOR AND MACHINERY OPERATION PROGRAM

Introduction

Agricultural work must be done during various weather conditions. Farm work does not stop for summer heat or winter cold. Crops must be harvested, livestock must be tended, and every daily routine completed. Hot, cold, rain or shine, the work continues. Safe work still must be observed under any weather-related conditions.

This task sheet will discuss how to recognize severe weather and the effects of such weather on the farm worker. Additional task sheets in Section 2.5 will present safety precautions for heat, cold, sun exposure, lightning and wind storms, and rain.

Summer Weather

Crop production activities begin with the arrival of the summer season. This is the time of year to expect higher temperatures, higher humidity, thunderstorms, lightning, and tornadoes. Attention to safe work practices may not permit attention to weather hazards. See Task Sheet 2.5.1.

High Temperatures—Exposure to high summer temperatures can cause illness. Heat cramps, heat exhaustion, and heat stroke are serious problems.

- Heat cramps—Symptoms are leg and stomach cramps.
- Heat exhaustion—Symptoms are cool, moist, pale or flushed skin, headache, nausea, dizziness, weakness, and exhaustion
- Heat stroke—Symptoms include red, hot, dry skin; changes in consciousness; rapid, weak pulse; and rapid, shallow breathing. Heat stroke can result in death if not treated immediately.

High Humidity—Excessive humidity means that moisture evaporation slows down. Perspiration helps to cool the body as it evaporates. In high humidity, the body continues to lose moisture, but the cooling effect is not felt.

Thunderstorms and Lightning—Cold-weather fronts bring cooler air into contact with warm air masses. Severe thunderstorms result; lightning can happen. On average 93 persons are killed each year by lightning.

Tornadoes—These small but violent storms can pack up to 250 mph wind gusts. They usually follow dark skies with clouds that look like a wall and wind that sounds like an approaching freight train. Tornadoes kill people and can cause millions in property damage.

Figure 2.5.a. Summertime forecasts of extreme weather must be heeded. Attention to machine safety is a top priority, but changing weather conditions must be observed as well.

Severe weather can occur anytime of the year.

Learning Goals

- To recognize the effect that severe weather plays in safe work practices

Related Task Sheets:

The Work Environment	**1.1**
Heat and Sun	**2.5.1**
Cold Weather	**2.5.2**
Lightning, Tornadoes, and Rain	**2.5.3**

Cooperation provided by The Ohio State University and National Safety Council.

Winter Weather

Winter chores on the farm must be done regardless of the weather. Winter cold brings different hazards. Frostbite, hypothermia, and loss of traction leads to hazardous work conditions. See Task Sheet 2.5.2.

Frostbite– This health hazard occurs when body tissue freezes. Medical attention is needed as soon as possible.

Hypothermia– This health issue involves a general cooling of the entire body. When the body cools down, normal processes cease to function properly. Gradual warming of the victim is necessary, as well as immediate medical treatment.

Loss of Traction– Winter weather affects footing—for both people and animals. Tractors that can pull heavy loads under normal circumstances now slip and slide. Observing extra care and taking extra time in moving machinery, livestock, and ourselves becomes more important on slippery surfaces.

Figure 2.5.b. Winter weather brings a different set of rules for work. Attention to farm chores may cause the worker to forget that the air is icy cold and that the skin can freeze.

Frostbite destroys body tissue.

Safety Activities

1. Call your nearest TV or Radio weatherperson and ask for an explanation of humidity in the atmosphere.
2. Use the Internet to define heat index (apparent temperature).
3. Use the Internet to define wind chill.
4. Contact you local emergency preparedness officials to learn what signals or warning sirens are used in your community to announce impending weather or other emergencies.
5. With your family, develop an emergency action plan for dealing with high wind or tornado conditions. Practice the plan at least once per year with the entire family.

References

1. Safety Management for Landscapers, Grounds-Care Businesses and Golf Courses, John Deere Publishing, 2001.

Contact Information

National Safe Tractor and Machinery Operation Program
The Pennsylvania State University
Agricultural and Biological Engineering Department
246 Agricultural Engineering Building
University Park, PA 16802
Phone: 814-865-7685
Fax: 814-863-1031
Email: NSTMOP@psu.edu

Credits

Developed by WC Harshman, AM Yoder, JW Hilton and D J Murphy, The Pennsylvania State University. Reviewed by TL Bean and D Jepsen, The Ohio State University and S Steel, National Safety Council. Revised 3/2013

This material is based upon work supported by the National Institute of Food and Agriculture, U.S. Department of Agriculture, under Agreement Nos. 2001-41521-01263 and 2010-41521-20839. Any opinions, findings, conclusions, or recommendations expressed in this publication are those of the author(s) and do not necessarily reflect the view of the U.S. Department of Agriculture.

HEAT AND SUN

HOSTA Task Sheet 2.5.1

NATIONAL SAFE TRACTOR AND MACHINERY OPERATION PROGRAM

Introduction

Agricultural work must be done during various weather conditions. Crops must be planted and harvested, livestock must be tended, and every daily farm routine completed. Hot or cold, rain or shine, the work continues. Safe work must be observed under all weather-related conditions.

This task sheet will discuss safe work in the heat and humidity of the summer season. Skin cancer, heat stroke, eye damage, and dehydration are health problems which farm workers must understand.

Health Risks From the Sun

Farmers are considered strong and hearty people. All farmers must pay attention to the problems posed by farm work during the summer season.

Health risks increase from overexposure to the sun and heat. These include:

- Sunburn/skin cancer
- UV light damage/eye damage
- Overheating/heat stoke
- Overheating/dehydration

Each area is discussed here.

Sunburn/Skin Cancer

Farm workers must spend a great deal of time working in the sun. Overexposure to the sun leads to sunburn, an actual burning of skin cells. Prolonged exposure to the sun over time is the most common cause of skin cancer. As the number of exposures to the sun increases, so does the chance of developing skin cancer.

Preventing Sunburn and Skin Cancer

Protect the skin from the harmful effects of the sun by dressing properly and using a sunscreen ointment. See Figure 2.5.1.a. Long sleeves, long pants, a neckerchief, and a broad-brimmed hat will protect the skin while working in the sun.

A SPF (Sun Protection Factor) sunscreen ointment with at least a 15 rating is recommended for areas which cannot be protected by clothing. The higher the SPF number, the more protection that is offered. Use sunscreen according to directions on the container.

The American Cancer Society provides information about skin cancer. Contact this organization through your local telephone directory or the Internet.

Please note: One sunburn experience will not cause cancer, but constant exposure or continuing exposure to the sun from working outdoors can increase the risk of skin cancer.

Figure 2.5.1.a. This drawing shows the recommended clothing and skin care precautions for summer sun exposure. *Safety Management for Landscapers, Grounds-Care Businesses, and Golf Courses, John Deere Publishing, 2001. Illustrations reproduced by permission. All rights reserved.*

A worker suffering from sunburn is not a productive worker.

Learning Goals

- To understand the health risks from summer heat and sun
- To prevent health risks from summer heat and sun

Related Task Sheets:

The Work Environment	**1.1**
Severe Weather	**2.5**
Personal Dress	**2.7**

Sunglasses are not just to make you look cool, but must protect the eyes.

Figure 2.5.1. b. Sunglasses are available in spectacle or clip-on versions. Look for the UV rating of 99-100 for "blockage" or "absorption." The word "protection" does not guarantee that the sunglasses will block or absorb UV rays.

Ultraviolet (UV) Rays and Eye Damage

The eye functions to control entering light and to focus an image on the optic nerve. Any damage limits the ability of the eye to function properly, and we lose some of our sight.

The sun produces different kinds of light:

- Ultraviolet (UV) invisible radiation
- Bright or intense light
- Blue light (a visible light)

The bright sun can damage the eye through the effects of UV radiation. This damage is called keratitis, an inflammation of the cornea of the eye. Sun-induced cataracts (a clouding of the lens of the eye) have been reported.

Blue light is visible light from the blue portion of the color spectrum. The intense glare from snow or water contains blue light. We cannot focus clearly in this intense light. Intense glare leads to eye strain and fatigue. Prolonged exposure to blue light is believed to age the retina of the eye. The result is an increased risk of blindness.

Protecting the Eyes

Protect the eyes from the harmful effects of the sun with the correct type of sunglasses. Sunglasses that provide blockage or absorption of the ultraviolet rays of the sun are best.

Sunglasses are rated according to their capability to block or absorb UV radiation. Look for terms such as "blockage" or "absorption," not only "protection" on the label. A UV rating of 100 is preferred.

Blue light blockers appear as tinted lenses in our glasses. These lenses alter the blue and green colors to reduce glare without making the world appear darker. To block the blue color, a yellow tint must be used. If you work often in bright, glaring conditions, these "sunglasses" can be helpful.

Types of Sunglasses

Several types of sunglasses are made to meet different needs.

- Regular lenses reduce brightness evenly
- Polarizing lenses reduce glare
- Photochromic lenses become darker in bright light
- Mirror lenses reflect light

Note: The price tag of sunglasses is not a measure of their blockage or absorptive value.

Cooperation provided by The Ohio State University and National Safety Council.

Dehydration

Sweating or perspiring is normal for a hot summer day. When the heat of the day is coupled with strenuous work, perspiration losses may equal or exceed water intake. *The body can lose as much as three gallons of water in a day.* Water serves as a coolant to our bodies.

When working on a hot day, a person can become fatigued or tired. Excessive sweating removes elements such as sodium, potassium, and chloride from our bodies. Water will not replenish minerals lost through perspiration. Sports drinks contain minerals which replenish our systems. Regular soft drinks do not replenish our nutrient needs. To replenish our mineral needs, we must eat properly before going to work and drink plenty of liquids while working.

Heat Stroke

Exposure to summer heat and humidity can cause serious illness. Health risks from heat occur when the body cannot cool down by sweating or make up the fluids and minerals lost through perspiration. Each year an average of 175 people die from the effects of summer heat.

Health problems from heat can include:

- *Heat Rash*– When sweat does not evaporate from the skin, the pores can become clogged. A rash develops. Cotton clothing can help to "wick" the moisture away from the skin. Use corn starch to treat the rash.
- *Heat Syncope*-Fainting from the heat can occur. Help the victim to lie down in a cool spot, and elevate their legs to improve circulation. Let them rest there.
- *Heat Cramps*- Leg and stomach cramps are caused by loss of body fluids due to sweating. Drink cool water often to cool the body. Massage the cramps.
- *Heat Exhaustion*– Loss of body fluids and salts from sweating and decreased blood flow to the brain can cause heat exhaustion. Symptoms include cool, moist, pale or flushed skin, headache, nausea, dizziness, weakness and exhaustion. Go to a cool place, lie down with feet elevated, and drink plenty of cool fluids. Medical help should be summoned.
- *Heat Stroke*-This is a medical emergency. The body's systems are failing. Symptoms include red, hot, and dry skin (perspiration has stopped); changes in consciousness; convulsions; delirium; rapid, weak pulse; and rapid, shallow breathing. The victim may become chilled. Some victims exhibit anger. **Heat stroke can be fatal if not treated immediately.** See page 4.

Effects of Humidity on Sweating

Evaporation rates are reduced with excessive humidity. Evaporation of water and sweat has a cooling effect. Without this cooling effect, high temperatures actually feel higher. Heat index charts show "apparent temperatures" comparing air temperature with humidity (Figure 2.5.1.d.).

Figure 2.5.1.c. Water is a nutrient we need. It cools the body, carries nutrients, and flushes waste from the body.

Heat stroke is an emergency calling for immediate medical care.

Heat Index—Apparent Temperature

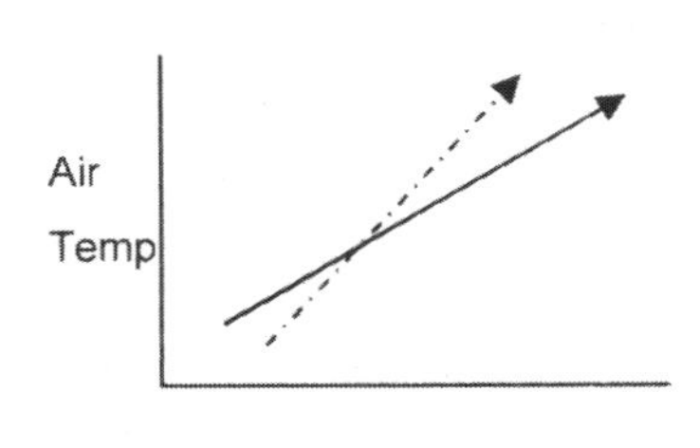

% Relative Humidity

Air temperature ⟶

Apparent Temperature - - - ->

Figure 2.5.1.d. The chart shows that as humidity levels increase with rising temperatures, the apparent temperature (heat index) may appear higher or lower than the reported temperature.

Treating Heat Stroke

Heat stroke is a medical emergency. Follow these treatment procedures immediately.

- Call for medical help at once.
- Remove the victim's outer clothing.
- Immerse the person in cold water. If no pool is available, sponge the person's body with water until help arrives.
- Do not give the person anything to drink.

Preventing Heat Illness

Follow these guidelines to prevent heat illness.

1. Drink water approximately every 15 minutes. Do not wait to be thirsty.
2. Avoid caffeinated and alcoholic drinks.
3. Wear appropriate summer clothing that fits loosely and reflects the sunlight.
4. Perform the most strenuous jobs during the coolest part of the day.
5. Take periodic breaks in the shade.
6. Adjust gradually to the heat.

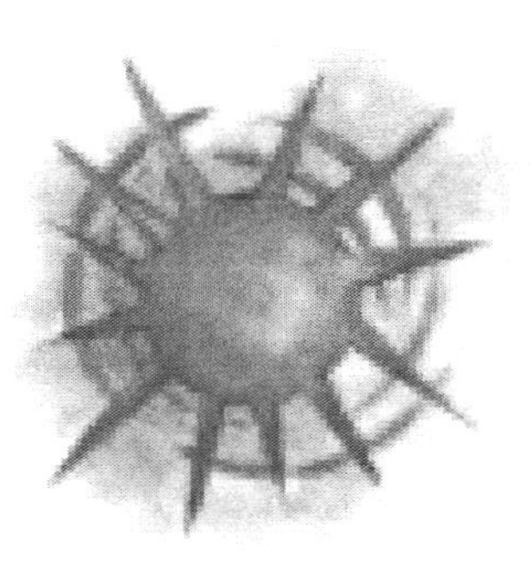

Figure 2.5.1. e. Take extra precautions while working in the summer sun and heat.

Safety Activities

1. Using the Internet, type "heat index chart" on any search engine, and locate information on a heat index (apparent temperature) chart. Then answer these questions.
   a. On a 90 degree day with a 70% relative humidity, the heat index is _______degrees.
   b. On a 95 degree day with a 50% relative humidity, the heat index is _______degrees.
   c. On a 85 degree day with a 85% relative humidity, the heat index is _______degrees.

2. Call your nearest TV or radio weatherperson and ask for an explanation of relative humidity.

References

1. Safety Management for Landscapers, Grounds-Care Businesses, and Golf Courses, John Deere Publishing, 2001. Illustrations reproduced by permission. All rights reserved.
2. www.marshfieldclinic.org/nfmc.
3. Gempler's Inc., (www.gemplers.com)

Contact Information

National Safe Tractor and Machinery Operation Program
The Pennsylvania State University
Agricultural and Biological Engineering Department
246 Agricultural Engineering Building
University Park, PA 16802
Phone: 814-865-7685
Fax: 814-863-1031
Email: NSTMOP@psu.edu

Credits

Developed by WC Harshman, AM Yoder, JW Hilton and D J Murphy, The Pennsylvania State University. Reviewed by TL Bean and D Jepsen, The Ohio State University and S Steel, National Safety Council. Revised 3/2013

This material is based upon work supported by the National Institute of Food and Agriculture, U.S. Department of Agriculture, under Agreement Nos. 2001-41521-01263 and 2010-41521-20839. Any opinions, findings, conclusions, or recommendations expressed in this publication are those of the author(s) and do not necessarily reflect the view of the U.S. Department of Agriculture.

COLD WEATHER

HOSTA Task Sheet 2.5.2

NATIONAL SAFE TRACTOR AND MACHINERY OPERATION PROGRAM

Introduction

Agricultural work must be done during various weather conditions. Farm work does not stop for winter cold or summer heat. Crops must be harvested, livestock must be tended, and every daily routine completed. Cold, hot, snow, ice, rain or shine, the work continues. Safe work habits must still be practiced under all weather-related conditions.

This task sheet will discuss how to recognize the effects of cold weather on the farm worker. Frostbite, hypothermia, and decreased traction pose hazards which farm workers must understand.

Winter Health Hazards

Winter chores on the farm must be done regardless of the weather. Winter weather offers different hazards with which to contend. Examples include frostbite and hypothermia. Our bodies may become accustomed to working in the cold, but exposure to low temperatures and wintry wind can be dangerous. For example, slippery conditions affect our ability to safely handle equipment and livestock.

Frostbite:

Frostbite occurs when body tissue becomes frozen. Skin that feels numb should send the message to the outdoor worker that the skin is too cold and in danger of further damage.

To prevent frostbite, pay attention to the low temperatures and how your skin is reacting. Covered skin is at risk for frostbite as well.

If a person develops frostbite, seek shelter and use warming towels or lukewarm water to warm the skin, Never use hot water. It can burn the skin. Severe cases of frostbite require immediate emergency medical treatment.

Hypothermia:

Hypothermia occurs as the body's temperature drops below 96 degrees. Exposure to severe cold causes this condition. Everyone is familiar with the hypothermia reported when someone falls through the ice. Extreme cold can produce weakness, drowsiness, or confusion, which can lead to further exposure and eventually death.

To prevent hypothermia, dress in layers to help trap air between the clothing. Air has an insulation value. Wear a head covering as well. Proper winter dress should keep you warm, not hot, and also fit well for safe work around equipment and livestock.

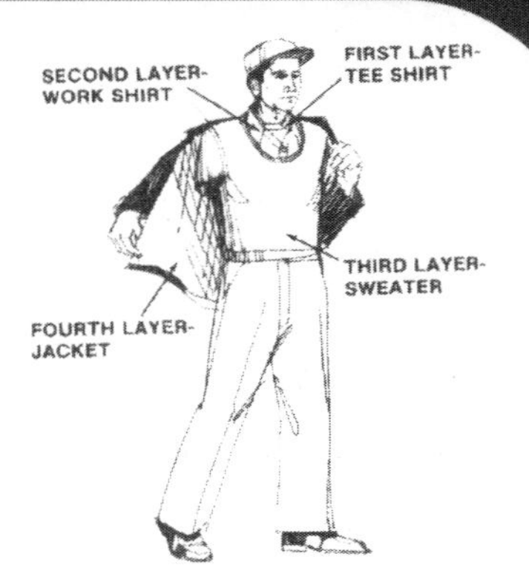

Figure 2.5.2.a. Layers of clothing offer the best cold weather protection. If the day's weather warms, outer layers can be removed. Synthetic fibers wick away the moisture of perspiration, while cotton materials absorb and hold moisture. Moisture next to the skin becomes chilled. *Safety Management for Landscapers, Grounds-Care Businesses, and Golf Courses, John Deere Publishing, 2001. Illustrations reproduced by permission. All rights reserved.*

High winter winds coupled with low temperatures may result in a wind chill advisory.

Learning Goals

- To understand the health risks from working in the winter cold
- To prevent health risks from working in the winter cold

Related Task Sheets:

The Work Environment	1.1
Severe Weather	2.5
Personal Dress	2.7
Working With Livestock	3.4

Cooperation provided by The Ohio State University and National Safety Council.

Loss of Traction:

Winter weather brings icy and muddy conditions. Footing is more difficult for people and livestock. Tractors that can pull heavy loads under normal circumstances may slip and slide. Livestock can slip and fall and be injured. Animals being moved on slippery surfaces can slip into the worker. Consider these extra precautions.

- Footwear must have treads that will provide traction.
- Use traction chains on tractor tires under extremely icy conditions.
- Operate the tractor carefully and more slowly than when weather conditions are dry.
- Recognize that vehicles traveling on public roadways may need greater distances to slow to a stop as they approach farm equipment sharing the road.
- Move livestock slowly to prevent the animal from falling or sliding into you.

Winter activities require slower, more deliberate movements to prevent injury.

Figure 2.5.2.b. Be extra careful while mounting the tractor steps in icy conditions. Drive carefully over slippery roads or fields.

Loss of traction means loss of control.

Safety Activities

1. Using the Internet, type "wind chill chart' on any search engine. Use this chart to answer these questions.
    a. On a 30-degree day with a 15 mph wind, the temperature will feel like ____degrees on your skin.
    b. On a 20-degree day with a 15 mph wind, the temperature will feel like ____degrees on your skin.
    c. On a 10-degree day with a 30 mph wind, the temperature will feel like ____degrees on your skin.
    d. If you are snowmobiling and the temperature is 10 degrees and you are driving at 40 mph, what is the wind chill factor in degrees? _________

References

1. Safety Management for Landscapers, Grounds-Care Businesses, and Golf Courses, John Deere Publishing, 2001. Illustrations reproduced by permission. All rights reserved.

Contact Information

National Safe Tractor and Machinery Operation Program
The Pennsylvania State University
Agricultural and Biological Engineering Department
246 Agricultural Engineering Building
University Park, PA 16802
Phone: 814-865-7685
Fax: 814-863-1031
Email: NSTMOP@psu.edu

Credits

Developed by WC Harshman, AM Yoder, JW Hilton and D J Murphy, The Pennsylvania State University. Reviewed by TL Bean and D Jepsen, The Ohio State University and S Steel, National Safety Council. Version 4/2004. Revised 7/2013

This material is based upon work supported by the National Institute of Food and Agriculture, U.S. Department of Agriculture, under Agreement Nos. 2001-41521-01263 and 2010-41521-20839. Any opinions, findings, conclusions, or recommendations expressed in this publication are those of the author(s) and do not necessarily reflect the view of the U.S. Department of Agriculture.

LIGHTNING, TORNADOES AND RAIN

HOSTA Task Sheet 2.5.3

NATIONAL SAFE TRACTOR AND MACHINERY OPERATION PROGRAM

Introduction

Agricultural work must be done during various weather conditions. Farm work does not stop because the weather forecast for the day calls for thunderstorms, rain, lightning strikes, or even a threat of tornadoes. Work may be interrupted by these events when they happen, but the work is not cancelled until the weather event occurs.

This task sheet discusses lightning, tornadoes, and rain and the risks they pose to safe farm work.

Lightning

Note: Field work puts stress on everyone, especially if the weather report predicts that stormy conditions will interfere with that effort. Your first priority is safe equipment operation. Knowledge of weather patterns and how they change can improve your safe work habits.

Sudden rainstorms are often preceded by violent lightning storms. Lightning is caused by a buildup of static electricity in the air. Positive charged molecules rise into the sky and negative charged molecules fall to the bottom of clouds. The negative charged particles are attracted to the positive charged particles in a flash of lightning.

Lightning fatalities rank second to floods in weather-related deaths. Lightning energy as high as 100 million volts and as much as 50,000 degrees F. is released within half a second. Lifelong disability and death can result from exposure to the extreme levels of electricity and temperature.

Myth 1: Lightning does not strike the same place more than once.

Truth: Lightning can strike in the same place many times.

Myth 2: Lightning occurs only under stormy skies.

Truth: Lightning can strike 10 miles from a storm.

Precautions to Take

Follow these precautions if severe thunderstorms are forecast.

- Check the weather forecast before starting to work.
- Observe threatening clouds and increasing winds that begin to develop.
- Use the "30-30 rule." If the time delay between seeing the flash of lightning and hearing the bang of thunder is less than 30 seconds, you should already be moving toward shelter. Lightning can strike 30 minutes before or after a visible storm.
- In an open field, seek low spots for shelter.
- Seek shelter at a location which is away from hilltops, trees or utility lines.
- Use closed buildings for shelter if possible. Do not use items connected to plumbing or house wiring.
- Tractors with cabs and vehicles can be used for shelter.

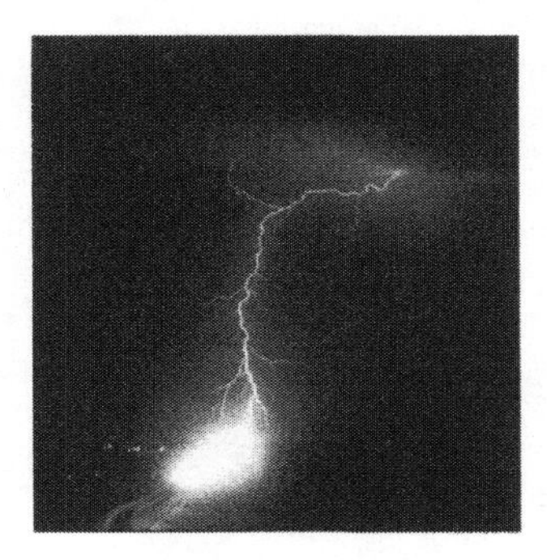

Figure 2.5.3.a. Lightning strikes can be fatal. Take shelter indoors if possible. Do not seek shelter beneath trees or near utility lines.

Hurrying to beat a storm during harvest increases the risk of injury.

Learning Goals

- To work safely in all types of weather conditions

Related Task Sheets:

The Work Environment	1.1
Severe Weather	2.5
Tractor Stability	4.12
Using the Tractor Safely	4.13

Cooperation provided by The Ohio State University and National Safety Council.

Tornadoes are possible from April to November in parts of the U.S.

Figure 2.5.3.b. While tornadoes do not occur in all areas of the United States, tornadoes and high winds lead to destruction, falling debris from trees, buildings, and utility wires. These storms can carry winds of up to 250 mph.

Wind and Tornadoes

A tornado is a violently rotating column of air extending from a thunderstorm to the ground. Eastward moving cold-weather fronts colliding with warm, moist weather form ideal conditions for high wind and tornadoes to develop. These conditions can occur rapidly. Some areas of the country are more prone to high wind and tornado conditions than others.

Tornadoes accompany thunderstorms. The following signs indicate a potential for a tornado.

- Dark, often greenish sky
- Large hail
- A cloud that looks like a wall
- A loud roaring sound

Be prepared to respond to these weather signals.

Remember these points in a tornado.

- Understand the radio and local siren warnings used to sound impending weather emergencies.
- If a tornado "watch" is issued, remain alert to storms. See page 3.
- If a tornado "warning" is issued, a tornado has been sighted or has appeared on weather radar. Move to safe shelter immediately. See page 3.
- Do not try to outrun a tornado. The speed and direction of a tornado can be deceiving.
- If caught outdoors in high winds or tornadoes, seek a ditch or low spot for protection. Lie face down with your hands over your head.
- If you find shelter in a building, go to the basement or to an inner room. Stay away from outside walls which may collapse, and stay away from windows which may shatter.

Tornado Myths and Truths

Myth 1: Tornadoes cause buildings to explode.

Truth: Violent winds and debris smashing into the building cause most of the structural damage.

Myth 2: Windows of the house should be opened to equalize pressure and minimize damage.

Truth: Opening the windows only opens the building to the damaging winds. Go to a safe place instead.

With early-warning systems in place throughout the U.S., tornado deaths have been reduced greatly. Know what the changing weather means to your safety.

National Oceanic and Atmospheric Administration (NOAA)

The NOAA agency of the federal government conducts weather and environmental observations around the world. NOAA information is used by National Weather Service forecasters to report weather patterns and events. NOAA satellite data benefits many groups. Aviation, maritime, and farm groups need up-to-the-minute weather information to assure safety and economic success.

Special NOAA weather radios can be purchased in many stores. These radios continuously broadcast updated weather warnings and forecasts. The radio's average range is 40 miles depending upon topography. Some NOAA radios have a feature that automatically sounds a tone when a watch or warning is issued in your area.

Rain and Rainstorms

Regular rainfall is necessary for crop growth. Periods of drought reduce yields and cause anxiety for farmers. Excessive rainfall delays planting and harvest and frustration again builds. Rain is necessary for success, but rain and rainstorms affect farm safety. Examine these points.

- Excessive rain causes reduced traction. Tractor steps may be mud covered. Fields may be slippery. Tractors can become stuck. See Task Sheet 4.13, Using the Tractor Safely.
- Excessive rain causes flooding. Crops can be damaged when soils become saturated.
- Saturated soils cannot hold more water. Flash flooding can occur. High water can sweep people and vehicles away.
- Rainy periods delay crop operations resulting in potential yield loss.
- Long periods of weather extremes frustrate farm growers. Unsafe acts can result as producers attempt to hurry to complete the work.

Think about these scenarios. Have you seen these effects of weather?

U.S. Weather Notification System

The National Weather Service issues daily forecasts and long-range weather outlooks. This service also decides when to issue severe weather watches. The notices include "watches" and "warnings."

Severe weather watch

This notice indicates conditions are favorable for the development of severe weather, such as tornadoes thunderstorms, blizzards, and potentially damaging wind or hail.

Severe weather warning

This notice indicates that a tornado, severe thunderstorm, or winter storm is in the immediate vicinity. People who are outdoors should find shelter as soon as possible.

Figure 2.5.3.c. NOAA and the National Weather Service tracks weather developments to give advance notice of weather problems to airlines, maritime operators, and to farmers.

Safety must never be a lower priority in farming due to weather factors.

Figure 2.5.3.d. Excessive rainfall delays crop operations, reduces traction, and builds frustration.

Cloud Types

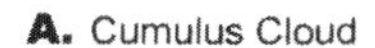

A. Cumulus Cloud

B. Stratus Cloud

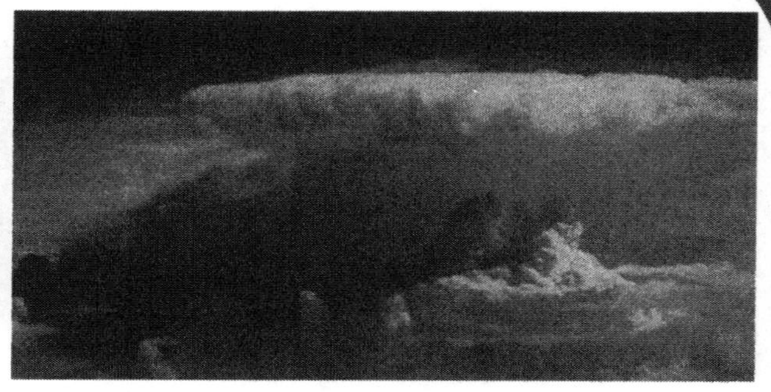

C. Cumulonimbus Cloud

Figure 2.5.3.e. Clouds can help predict weather. Cloud A is a cumulus cloud. These heaped or lumpy clouds indicate a period of fair weather. Cloud B is a stratus, layered cloud. They are full of ice crystals. These layered clouds can also form fog and mist. Cloud C is a cumulonimbus cloud. These towering clouds have an anvil shape at the top. They forecast rain, hail or storms. See if you can observe clouds and make a weather prediction.

Special Note: An individual is responsible for his or her own personal safety and has the right to take appropriate action when threatened by severe weather. No employer can force you to work in a dangerous situation.

Safety Activities

1. Use the Internet to learn more about lightning and tornadoes. Write a report for your teacher, leader, or for extra credit in science class.
2. Develop a family or farm emergency plan for severe weather if one does not exist.
3. If a weather emergency plan does exist, have the family or farm employees gather to review and practice the plan together.
4. After a rainstorm, clean the steps to each tractor and implement ladder to reduce slip and fall hazards.
5. Make a cloud project. You will need a large clear plastic jar, a small metal tray, ice cubes, and hot water.

   Step 1. Fill the jar 1/2 full of hot water (be careful).

   Step 2. Place some ice trays on a metal tray on top of the jar.

   Step 3. Observe the air space in the jar beneath the tray. Air and water vapor inside the jar next to the tray is cooled, condensing into water droplets (a cloud).

References

1. www.noaa.gov/(National Oceanic and Atmospheric Administration website)/Search the site for any weather information you desire.
2. www.nws.noaa.gov (National Weather Service site).
3. www.lightningsafety.com/(National Lightning Safety Institute website).
1. www.lightning.org/(Lightning Protection Institute website).

Contact Information

National Safe Tractor and Machinery Operation Program
The Pennsylvania State University
Agricultural and Biological Engineering Department
246 Agricultural Engineering Building
University Park, PA 16802
Phone: 814-865-7685
Fax: 814-863-1031
Email: NSTMOP@psu.edu

Credits

Developed by WC Harshman, AM Yoder, JW Hilton and D J Murphy, The Pennsylvania State University. Reviewed by TL Bean and D Jepsen, The Ohio State University and S Steel, National Safety Council. Revised 3/2013

This material is based upon work supported by the National Institute of Food and Agriculture, U.S. Department of Agriculture, under Agreement Nos. 2001-41521-01263 and 2010-41521-20839. Any opinions, findings, conclusions, or recommendations expressed in this publication are those of the author(s) and do not necessarily reflect the view of the U.S. Department of Agriculture.

HOUSEKEEPING

HOSTA Task Sheet 2.6

NATIONAL SAFE TRACTOR AND MACHINERY OPERATION PROGRAM

Introduction

Tractors and machinery are not the only sources of occupational hazards on a farm. Work areas must also be considered for creating danger to the worker. Recognizing housekeeping needs, including storage, use, and cleanup practices, are a must for the safety of every worker.

This task sheet discusses the relationship between good housekeeping and safety.

Importance of Housekeeping

Lack of housekeeping creates hazards. Picking up, wiping up, sweeping up, and removing scraps and waste all help to control hazards. Storing objects properly makes the work area safer. Unorganized and unplanned methods of work often indicate an unsafe place to work and increases the opportunity for injuries.

Several topics are important when discussing good housekeeping on the farm. They are:

- Worksite adequacy
- Environmental hazards
- Storage needs
- Cleanup practices

Worksite Adequacy

The worksite must be safe from the beginning of the workday. Observe these points:

- Are aisles and passages wide enough and high enough for safe movement?
- Is there adequate lighting?
- Is there adequate ventilation?
- Are there slip-resistant floors and ramps?
- Are pits and floor openings covered?
- Are sharp edges eliminated?
- Are exits defined and clear of obstruction?
- Are hoists sized to the needs of the business?
- Are sink and toilet facilities clean and sanitary?

Young workers cannot change the physical layout of the farm shop or storage areas. But young workers can develop the skills and safe attitudes necessary to maintain the facilities. Shop cleanup is a valuable job skill. Some things you can do to make facilities safe and healthy are:

- Report unsafe work areas.
- Report burned-out lights.
- Put tools, materials and unused supplies in their correct places.
- Sweep floors.
- Clean oil and grease spills from floors.
- Clean sinks and toilet facilities.

As one farmer says to young workers, "If you have time to lean, you have time to clean." If you are not assigned to a specific job, make yourself a valuable employee by doing some housekeeping chores.

Figure 2.6.a. Even on a temporary basis, this is not proper pesticide storage or disposal. While the operator is in the field, children or animals could come in contact with deadly pesticides.

A clean farmstead is a safe farmstead.

Learning Goals

- To recognize how good housekeeping helps prevent health and safety hazards

Related Task Sheets:

The Work Environment	**1.1**
Personal Protective Equipment	**2.10**
Respiratory Hazards	**3.3**
Respiratory Protection	**3.3.1**
Agricultural Pesticides	**3.5**
Electrical Hazards	**3.6**
Fire Safety	**3.7**
Fire Prevention and Control	**3.7.1**
Hay Storage Fires	**3.7.2**
Farmstead Chemicals	**3.13**

Cooperation provided by The Ohio State University and National Safety Council.

Figure 2.6.b. Welding can cause sparks to fly into nearby combustible material. The spark can ignite fine particles quickly with the resulting fire causing thousands of dollars of damage. How would you prevent welding sparks from igniting nearby materials?

Environmental Hazards

The farm environment has many health risks that can be reduced by good housekeeping. Chemicals, dusts, molds, welding rays, noise, heat, cold, and excessive moisture are common. Each poses a special problem. Chemicals, molds, heat, cold, and noise will be discussed in other task sheets.

Dust

If you have started a fire in a fireplace, you know that you must use kindling materials to get the fire started before adding the larger firewood. Dust, which is dry, has a low kindling temperature. Dust can burn explosively much like the fumes from gasoline.

You cannot remove all dust from the farmstead. No one would ask you to do that task, but unnecessary dust buildup near sources of fire increases the risk of fire. Some cleaning near these sources could prevent a fire.

Dust explosions have occurred in feed mills and at grain storage elevators. The explosion usually occurs due to electrical sparking igniting the dust particles. Sparks from welding can also ignite dust and chaff. You may notice that special dust and moisture-proof motors and controls are used to prevent fires and explosions in many agricultural applications.

Welding Rays

Defective welding helmets, cracked welding lenses, and torn welding curtains can create eyesight damage risks. You can repair welding helmet lenses and welding curtains as part of your beginning level housekeeping work routine. This would increase your value as an employee.

Excessive Moisture/Slippery Floors

Water, oils, or other substances cause floors to become slippery. Take a few minutes to clean the walkway. Use a floor-drying compound or sand to reduce slippage and to clean the area. You should place a warning sign or barricade at the location until the floor is dry and safe.

Definitions:
Kindling temperature is the lowest temperature at which a solid fuel will ignite and begin to burn when brought near a source of heat.

Electric sparking is the spark of electricity occurring when two conductors (leads, wires, contacts) come close together, and an electrical current jumps across the gap.

Cooperation provided by The Ohio State University and National Safety Council.

Storage

Proper storage of materials creates an organized and safe work space. No one wants to waste time looking for tools or materials. Safe storage prevents lost work time from injuries. Improper storage can lead to a risk of fire.

Heavy and Long Objects- Heavy and long objects must be stored correctly to prevent trip, fall, or falling object hazards. Long stock, such as wood or pipe, should be stored on racks designed to hold long pieces. Long stock stored under benches and sticking out may cause a leg injury to persons passing by. Heavy objects should be stored as close to the floor or ground as possible to prevent the chance of being hit by falling objects.

Fuels and Lubricants- Fuel storage is an important housekeeping chore. Liquid fuels have a *flash point*. A flash point is that point at which temperatures are high enough to ignite a gaseous fuel source. The fuel may also be *volatile*. Volatility is a property of fuels in which they produce vapors that easily ignite.

To keep the fuel area as safe as possible, there are several good housekeeping rules to follow. These include:

1. Keep caps on all fuel containers.
2. Use only approved diesel and gasoline storage containers. Green or yellow-colored containers are used to store diesel fuel. Red-colored containers are used for gasoline storage.
3. Keep areas around refueling stations free of fuel spills.
4. Use an approved absorbent compound to clean up fuel spills.

Cleanup

Work areas cannot be perfectly clean at all times, but they can be made safer to work in at all times. Cleaning as your work progresses will eliminate major cleaning chores later and will make for a safer work space.

Use these ideas for cleanup.

1. Clean all spilled material immediately. Avoid cleaning procedures that would make those materials become air-borne inhalation hazards.
2. Place oil, grease, paint, and solvent–soaked rags in metal containers to reduce fire risks.
3. Use hand cleaners and disinfectants before eating or drinking.
4. Dispose of animal health equipment tools and supplies as directed. See Figure 2.6.c.
5. Manure and mud are slippery. Both can be brought into the shop area on machinery. Clean manure and mud from alleyways and high-traffic areas to decrease the risk of falls.

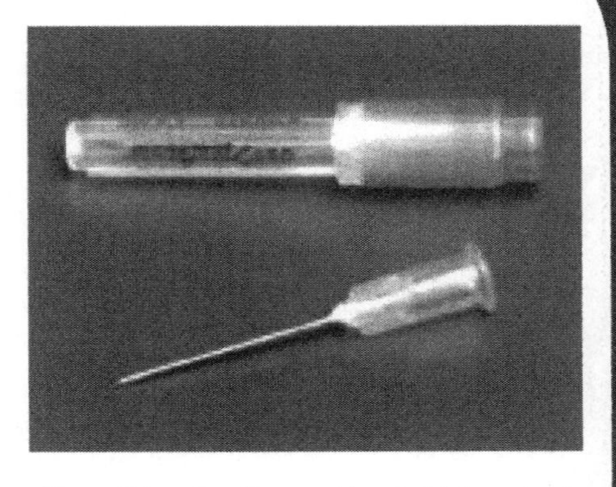

Figure 2.6.c. Needles used to administer animal vaccines are sharp, and the user can be accidentally stuck. If left uncapped, or if not disposed of properly, they create a hazard to people and livestock as well. Follow the manufacturer's directions for safe storage and disposal of needles and syringes.

Oily rags thrown in a pile can be a fire hazard. Store them in a closed metal container.

Safety Activities

1. Define these terms:
    a) kindling point
    b) flash point
    c) volatility

2. Survey a farm shop, and make a list of those housekeeping items that you judge to be potentially hazardous.

3. Survey the school's agricultural shop or industrial technology shop (with the instructor's permission), and list those housekeeping items that are potentially hazardous. If your school does not have such an area, ask the chemistry teacher to show you the storage facilities for that subject area.

4. Bring some very dry, fine dust from the barn or farm shop to a safe place where air currents are minimal. Sprinkle small amounts of the dust over a lighted candle. What happens?

5. Bring some very dry, fine metal filings from the farm shop to a safe place where air currents are minimal. Sprinkle small amounts of the metal filings over a lighted candle. What happens?

References

1. Safety Management for Landscapers, Grounds-care Businesses and Golf Courses, John Deere Publishing, 2001.
2. Farm and Ranch Safety Management, John Deere Publishing, 2009.

Contact Information

National Safe Tractor and Machinery Operation Program
The Pennsylvania State University
Agricultural and Biological Engineering Department
246 Agricultural Engineering Building
University Park, PA 16802
Phone: 814-865-7685
Fax: 814-863-1031
Email: NSTMOP@psu.edu

Credits

Developed by WC Harshman, AM Yoder, JW Hilton and D J Murphy, The Pennsylvania State University. Reviewed by TL Bean and D Jepsen, The Ohio State University and S Steel, National Safety Council. Revised 3/2013

This material is based upon work supported by the National Institute of Food and Agriculture, U.S. Department of Agriculture, under Agreement Nos. 2001-41521-01263 and 2010-41521-20839. Any opinions, findings, conclusions, or recommendations expressed in this publication are those of the author(s) and do not necessarily reflect the view of the U.S. Department of Agriculture.

Cooperation provided by The Ohio State University and National Safety Council.

PERSONAL DRESS

HOSTA Task Sheet 2.7 **Core**

NATIONAL SAFE TRACTOR AND MACHINERY OPERATION PROGRAM

Introduction

When a person goes to work, they should dress for the work they will do. You would not look like a good candidate for work if you showed up at a farm in your sandals. Some workplaces have dress codes. Think about your safety as you dress for work.

This task sheet discusses personal dress choices for safe work. Ask your employer if specific work dress is expected.

What Should I Wear?

Some young workers might rebel about the idea that someone is going to tell them what to wear to work. The latest fashions or stylish clothes will not make you a better worker. Dressing safely will make you a smarter worker because it increases your chances of avoiding injury or death on the job.

Know what each job you perform requires and dress accordingly. During the summer, mowing fields or baling hay may mean several hours in the sun. Over exposure to the sun is a serious hazard for young workers. A long-sleeved shirt, a hat that protects the ears and neck, and sun block are all part of safe dressing.

Here are some other approved safety practices for how you should dress for work.

1. Wear snug-fitting clothes which are in good repair. Loose clothes with dangling threads, ripped sleeves and cuffs, and drawstrings can be caught in machinery or snag on tractor parts.
2. Leave jewelry at home. Jewelry can be caught in machine parts or snagged on the tractor as you mount or dismount.
3. Wear hard shoes with slip-resistant treads. Sandals or sneakers offer little protection from livestock trampling, briars, nails, welding sparks, falling lumber or other objects. Check to see if steel-toed work boots are necessary.
4. Tie shoes snugly. Loose shoe strings can be caught in rotating parts.
5. Tie long hair out of the way. Tying or covering long hair will prevent the hair from being pulled into turning parts of machinery and save you from being scalped.
6. Wear long pants that are the correct length. Long pants, which fit properly and are in good repair, will protect your legs from sunburn, splinters, briars and thistles. Sloppy fitting clothes can easily become entangled in machinery or snagged on tractor parts.

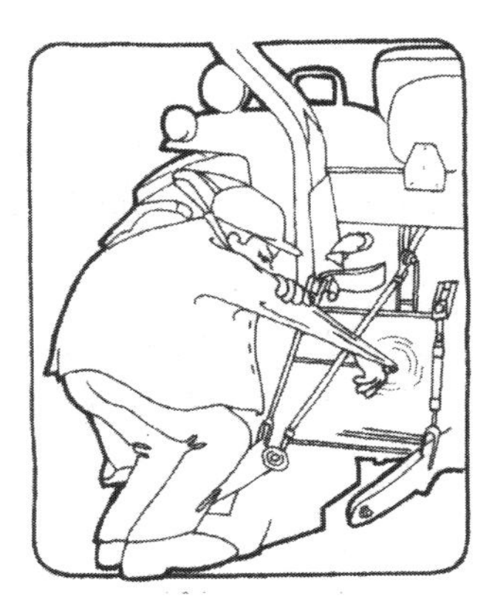

Figure 2.7.a. Not dressing safely for the job means that you expose yourself to the risk of injury or death.

Dressing properly for the work to be done is the first step in preparing to work safely.

Learning Goals

- To dress safely for work

Related Task Sheets:

Cooperation provided by The Ohio State University and National Safety Council.

A Well-Dressed Worker

DO WEAR RECOMMENDED HEAD, EYE, AND HEARING PROTECTION.

DO WEAR SUN-SAFE HATS THAT PROTECT YOUR EARS AND NECK AS WELL AS YOUR FACE.

DO WEAR CLOTHING THAT FITS RIGHT AND IS IN GOOD CONDITION. FRAYED AND RAGGED EDGES ARE DANGEROUS.

DO REMOVE JEWELRY BEFORE GOING TO WORK.

DO WEAR JACKETS AND SWEATSHIRTS THAT ARE PROPERLY SIZED AND DO NOT HAVE DRAWSTRINGS.

DO WEAR CLOSE-FITTING PULLOVERS

DO WEAR GLOVES THAT FIT WELL, PROVIDE GOOD GRIP, AND ARE IN GOOD CONDITION.

DO WEAR JEANS THAT ARE THE RIGHT LENGTH. CUFFS CAN BE DANGEROUS.

DO CHOOSE CLOTHING WITH BANDED CUFFS. AVOID CUFFS THAT NEED BUTTONING.

DO WEAR STURDY BOOTS WITH GOOD SLIP-RESITANT TREAD.

DO KEEP SHOELACES SHORT OR TUCK THEM INTO YOUR SHOES.

If you do not know what clothing to wear for a job, ask your employer.

Figure 2.7.b. Safely dressed workers wear the clothing and equipment needed to do the job without risking danger to themselves. *Safety Management for Landscapers, Grounds-Care Businesses, and Golf Courses, John Deere Publishing, 2001. Illustrations reproduced by permission. All rights reserved.*

Safety Activities

1. Find the following words in the word search.

Dangling sleeve	Long hair tied
Shoe strings	Drawstrings
Loose cuffs	Snug Clothes
Hard shoes	No jewelry

S	E	H	T	O	L	C	G	U	N	S	E
V	G	Y	A	K	M	Z	W	G	H	P	V
F	W	N	A	R	N	W	Q	O	U	O	E
D	N	U	I	H	D	D	E	C	P	F	E
E	D	F	K	R	X	S	Q	V	D	X	L
T	G	Z	K	X	T	A	H	Y	Z	Y	S
D	H	O	P	R	I	S	L	O	R	L	G
A	G	R	I	W	H	Z	W	L	E	F	N
F	Z	N	Q	P	J	L	E	A	G	S	I
J	G	J	V	P	C	W	Y	X	R	M	L
S	F	F	U	C	E	S	O	O	L	D	G
K	Y	P	Q	J	Q	O	N	A	Q	H	N
D	G	U	O	T	V	H	R	G	V	E	A
L	O	N	G	H	A	I	R	T	I	E	D

References

1. Safety Management for Landscapers, Grounds-Care Businesses, and Golf Courses, John Deere Publishing, 2001. Illustrations reproduced by permission. All rights reserved.
2. Farm and Ranch Safety Management, John Deere Publishing, 2009.

Contact Information

National Safe Tractor and Machinery Operation Program
The Pennsylvania State University
Agricultural and Biological Engineering Department
246 Agricultural Engineering Building
University Park, PA 16802
Phone: 814-865-7685
Fax: 814-863-1031
Email: NSTMOP@psu.edu

Credits

Developed by WC Harshman, AM Yoder, JW Hilton and D J Murphy, The Pennsylvania State University. Reviewed by TL Bean and D Jepsen, The Ohio State University and S Steel, National Safety Council. Revised 3/2013

This material is based upon work supported by the National Institute of Food and Agriculture, U.S. Department of Agriculture, under Agreement Nos. 2001-41521-01263 and 2010-41521-20839. Any opinions, findings, conclusions, or recommendations expressed in this publication are those of the author(s) and do not necessarily reflect the view of the U.S. Department of Agriculture.

HAZARD WARNING SIGNS

HOSTA Task Sheet 2.8 **Core**

NATIONAL SAFE TRACTOR AND MACHINERY OPERATION PROGRAM

Introduction

Uniform safety signs are designed to promote and improve personal safety in agricultural workplaces. Safety signs have been developed to warn of farm machinery hazards, but there are also safety signs that apply to non-machinery hazards. Signal words, sign format, and color combinations all play a role in safety signs.

This task sheet discusses uniform hazard warning signs that farm workers should observe and understand. Use specific owners' manuals to learn more about them.

Safety Alert Symbol

This symbol was created to draw attention to the need for safety. The symbol means:

Attention!

Become Alert!

Your safety is involved!

The safety alert symbol is used with agricultural, construction, and industrial equipment. The primary uses of the symbol are in an owner's manual and on hazard warning signs.

Good Hazard Warning Signs:

- Include the "safety alert" symbol
- Warn a person of the nature and degree of hazard or potential hazard
- Provide recommended safety precautions or evasive actions to take
- Provide other directions to eliminate or reduce the hazard

DANGER–The most serious potential hazard. These are RED.

WARNING– Show a lesser degree of potential hazard. These are ORANGE.

CAUTION– Indicates a need to follow safety instructions. Usually are YELLOW.

Figure 2.8.a. What message does this safety alert sign have for the operator? Try writing the message in as few sentences as possible. Which method—pictorial or written—conveys the message more quickly?

Pictorial hazard warning signs provide safety alerts to readers and non-readers of any language.

Learning Goals

- To quickly gather hazard potential and safe operation information by understanding hazard warning signs

Related Task Sheets:

Tractor Instrument Panel	4.4
Tractor Controls	4.5
Tractor Operation Symbols	4.5.6
Preventative Maintenance and Pre-Operation Checks	4.6

Cooperation provided by The Ohio State University and National Safety Council.

A pictorial quickly presents a potential hazard situation and a possible result of ignoring this potential danger. When these "picture" messages are seen, ask the question, "What is the worst thing that can happen to me?"

Figure 2.8.b. Potential for crushing hazard from shifting overhead load exists.

Figure 2.8.c. Potential for electric shock hazard exists.

Figure 2.8.d. Potential for crushing the feet from an overhead hazard exists.

Pictorials pose the potential hazard to us, as well as the consequences of ignoring the hazard warning.

Figure 2.8.g. This warning sign placed near the radiator warns the operator of what hazard?

Pictorials

A pictorial is a graphical representation intended to convey a message without the use of words. It may represent:

- Hazards
- Hazardous situations
- Precautions to avoid a hazard
- Results of not avoiding a hazard
- A combination of these messages

Pictorials may be used in addition to or in place of a word message. The pictorial should quickly help a person to recognize a hazard. Many pictorials have been developed and are shown and explained here. Learn what each pictorial is trying to communicate. This could help you respond to or avoid a serious injury. Use the reference section to find a complete exhibit of pictorials for farm work.

Figure 2.8.e. This sign warns of a potential PTO entanglement.

Figure 2.8.f. This safety sign warns of a potential fall hazard. Always use handholds. Falls account for a large number of agricultural injuries and fatalities each year.

Pictorials (Continued)

Figure 2.8.h. A potential high pressure hydraulic hose leak is a hazard which could force oil beneath your skin. Check hydraulic leaks with a mirror or piece of metal instead of your hand or fingers.

Figure 2.8.i. This safety sign warns of the potential to be run over by a tractor. Use the seat belt while operating the tractor equipped with a ROPS. Do not stand to drive. Passengers should not be allowed to ride the tractor. Extra riders are at a great risk for injury or death.

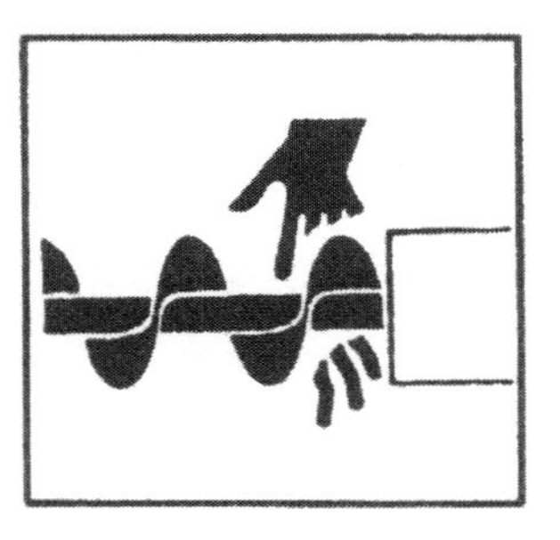

Figure 2.8.j. Potential entanglement in a rotating auger with cutting is shown. Moving parts need guards. If unguarded areas are encountered, the agricultural worker must use extreme caution.

Figure 2.8.k. Electrical contact with overhead power lines and the high lift bucket shows the potential for electrocution to the operator.

Figure 2.8.l. Possible slippery area with potential fall hazard is shown in this warning sign.

Figure 2.8.m. This safety pictorial shows the potential for thrown objects and the need for safety goggles. High noise levels indicate the need for ear protection.

Hazard warning signs placed on tractors and machinery serve as quick, easy sources of information. They do not replace an owner's manual. The warning signs make the information readily available.

Figure 2.8.n. Use the handholds and face the steps when mounting the tractor. What hazard warning sign would you expect to see in this situation?

Hazard signs and symbols provide the most direct information nearest the potential hazard site. Use them!

In the space above, draw a safety sign that warns someone of the potential to be entangled in a belt drive. Check the asae.org website to compare results.

Cooperation provided by The Ohio State University and National Safety Council.

Safety Activities

1. Use the Internet websites shown in the reference section, locate the safety signs standard S441.3, and print out the .pdf file. Use the information for a class or group discussion.

2. Safety signs are constantly being developed. ASAE Standards from question Number 1 also give the rules for developing safety signs. Choose a potential hazard, and design a safety sign for that situation. Perhaps someday your sign will be used as an industry standard.

3. Tell your leader, teacher, or employer what the safety alert signal words mean:
   Caution
   Warning
   Danger

4. Draw a picture of the safety sign or symbol for each of these:

   A. Hand entanglement in a chain and sprocket drive

   B. Hot engine coolant temperature

   C. Falling into machinery, such as an auger

5. Draw the Safety Alert symbol here.

6. Develop a hazard warning sign for a potential dog bite on a farm. Draw your sign here.

References

1. Safety Management for Landscapers, Grounds-Care Businesses and Golf Courses, John Deere Publishing 2001.
2. Farm and Ranch Safety Management, John Deere Publishing, 2009. Illustrations reproduced by permission. All rights reserved.
3. American Society of Agricultural and Biological Engineers, ANSI/ASAE, S441, Safety Signs, St. Joseph, MI.

Contact Information

National Safe Tractor and Machinery Operation Program
The Pennsylvania State University
Agricultural and Biological Engineering Department
246 Agricultural Engineering Building
University Park, PA 16802
Phone: 814-865-7685
Fax: 814-863-1031
Email: NSTMOP@psu.edu

Credits

Developed by WC Harshman, AM Yoder, JW Hilton and D J Murphy, The Pennsylvania State University. Reviewed by TL Bean and D Jepsen, The Ohio State University and S Steel, National Safety Council. Revised 3/2013

This material is based upon work supported by the National Institute of Food and Agriculture, U.S. Department of Agriculture, under Agreement Nos. 2001-41521-01263 and 2010-41521-20839. Any opinions, findings, conclusions, or recommendations expressed in this publication are those of the author(s) and do not necessarily reflect the view of the U.S. Department of Agriculture.

HAND SIGNALS

HOSTA Task Sheet 2.9 **Core**

NATIONAL SAFE TRACTOR AND MACHINERY OPERATION PROGRAM

Introduction

Perhaps you have experienced the shouting and hand-waving that seems to fit many farm chores. Noise from machinery and/or distance between workers often leads to a communication breakdown. An increased risk for hazardous situations can occur.

This task sheet presents 11 standard hand signals adopted by the American Society of Agricultural Engineers (ASAE) and three signals for public road use.

Memorize and use these hand signals. Teach them to others. You will save time and establish safe communications.

Hand Signals

ASAE Figure 1: **This Far To Go**

Place palms at ear level facing head and move inward to show remaining distance to go.

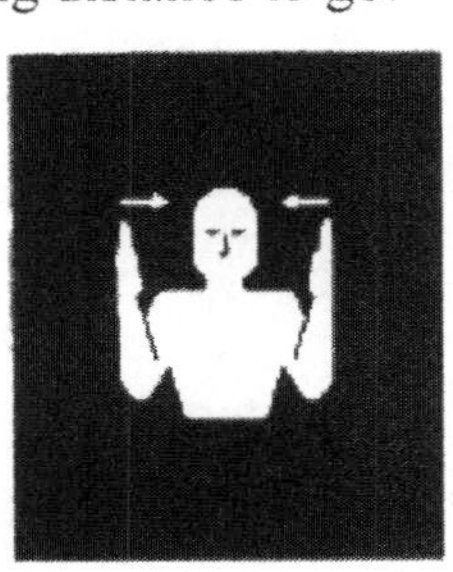

Example: Use this signal to assist a tractor operator in backing a loaded wagon or hitching to a wagon.

ASAE Figure 2: **Come To Me**

Raise the arm vertically overhead, palm to the front, and rotate in large horizontal circles.

Example: Someone has opened the gate for the cows to be brought forward: You will signal in this manner.

ASAE Figure 3: **Move Toward Me—Follow Me**

Point toward person(s), vehicle(s), or unit(s). Signal by holding arm horizontally to the front, palm up, and motioning toward the body.

Example: Use this signal to motion an equipment operator to move toward you to position or move equipment in a crowded area where side visibility is poor.

Figure 2.9.a. You are assigned to bale hay today. Someone approaching you on the ground to deliver a message would not be able to shout loud enough for you to hear. What signals could they use?

Hand signals provide standard communication to all workers.

Learning Goals

- To use the 11 standard hand signals to communicate actions to be taken with the tractor and equipment
- To use standard hand signals for highway use

Related Task Sheets:

Tractor Hazards	**4.2**
Tractor Controls	**4.5**
Using the Tractor Safely	**4.13**
Operating the Tractor on Public Roads	**4.14**

Cooperation provided by The Ohio State University and National Safety Council.

Figure 2.9.b. Public Road Hand Signals. Other hand signals provide means of communicating in traffic situations. Use these signals for public road travel or anywhere others may be following you. These signals are standard highway signals to the general public as well. *Safety Management for Landscapers, Grounds-Care Businesses, and Golf Courses, John Deere Publishing, 2001. Illustrations reproduced by permission.*

Noisy equipment and distance between workers makes hand signals a necessity. How many of these hand signals do you use?

ASAE Figure 4: **Move Out—Take Off**

Face the desired direction of movement; hold the arm extended to the rear; then swing the arm overhead and forward in the direction of desired movement until the arm is horizontal with palm down.

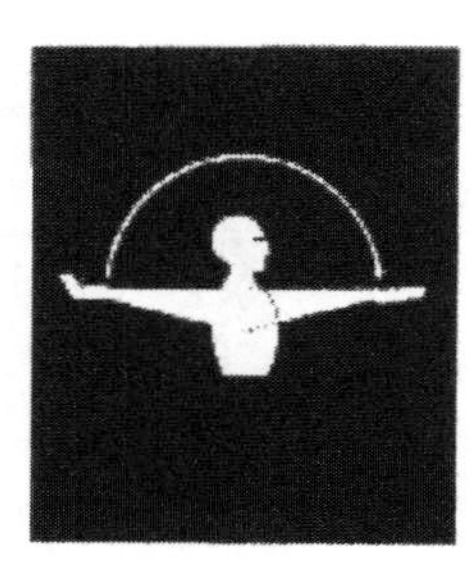

Example: You have hitched the machine for the operator and connected the PTO; signal the person to move out for field work.

ASAE Figure 5: **Stop**

Raise the hand upward to the full extent of the arm, palm to the front. Hold that position until the signal is understood.

Example: The tractor and forage wagon are now positioned for unloading into the silage blower. You signal the operator to stop.

Cooperation provided by The Ohio State University and National Safety Council.

ASAE Figure 6: **Speed It Up—Increase Speed**

Raise the hand to the shoulder, fist closed; thrust the fist upward to the full extent of the arm and back to the shoulder rapidly several times.

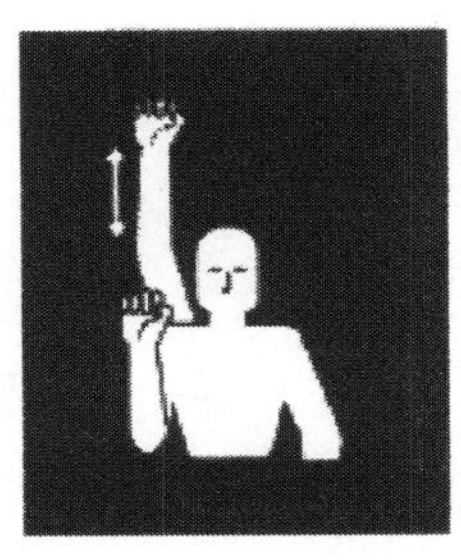

Example: Move the unit out now; the way is clear. We need to move on.

ASAE Figure 7: **Slow Down—Decrease Speed**

Extend arm horizontally sideward with palm down; wave arm downward at 45 degrees minimum several times. Do not move arm above horizontal.

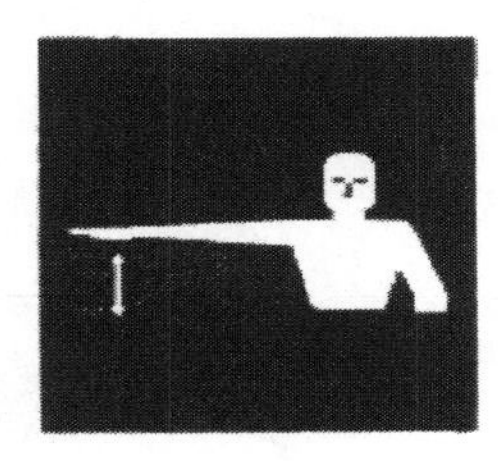

Example: You are going too fast; slow down.

ASAE Figure 8: **Start the Engine**

Move arm in circular motion at waist level to simulate cranking engine.

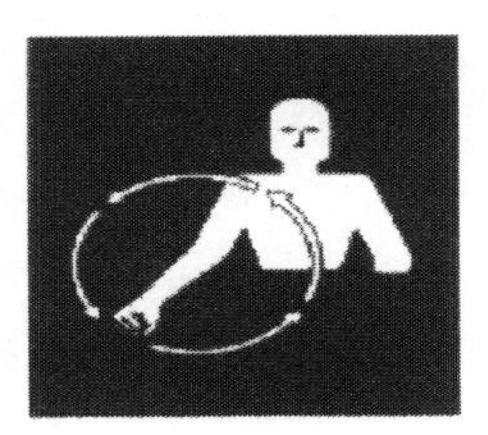

Example: You need to signal the operator to start the engine after some adjustment has been made.

ASAE Figure 9: **Stop the Engine**

Draw right hand, palm down, across the neck in a "throat-cutting" motion left to right.

Example: You need to have the operator stop the engine for some adjustments to the machinery.

ASAE Figure 10: **Lower Equipment**

Use circular motion with either hand pointing to the ground.

Example: Use this signal to have operator lower high lift or machine header.

ASAE Figure 11: **Raise the Equipment**

Make circular motion with either hand at head level.

Example: Use this signal to have operator raise high lift or machine header.

Learn the 11 standard hand signals and use them. Then teach them to all your fellow workers. Perhaps even your employer will not know them.

 Cooperation provided by The Ohio State University and National Safety Council.

Safety Activities

1. Identify each hand signal and give examples of when to use each signal.

Identifies:________________________________

An example is:________________________________

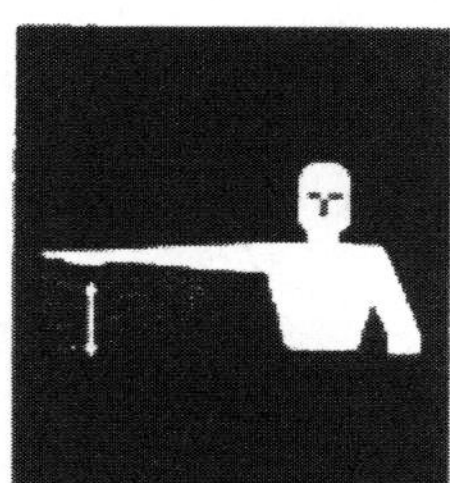

Identifies:________________________________

An example is: ________________________________

Identifies________________________________

An example is:________________________________

2. Demonstrate all 11 hand signals to your leader, teacher, parents, or employer.

3. Demonstrate the hand signals to be used when you are traveling with the transport disk in highway traffic.
   - Right Turn
   - Left Turn
   - Stop

References

1. American Society of Agricultural and Biological Engineers, ANSI/ASAE S351, Hand Signals for Use in Agriculture, St. Joseph, MI.
2. Safety Management for Landscapers, Grounds-Care Businesses, and Golf Courses, John Deere Publishing, 2001. Illustrations reproduced by permission. All rights reserved.

Contact Information

National Safe Tractor and Machinery Operation Program
The Pennsylvania State University
Agricultural and Biological Engineering Department
246 Agricultural Engineering Building
University Park, PA 16802
Phone: 814-865-7685
Fax: 814-863-1031
Email: NSTMOP@psu.edu

Credits

Developed by WC Harshman, AM Yoder, JW Hilton and D J Murphy, The Pennsylvania State University. Reviewed by TL Bean and D Jepsen, The Ohio State University and S Steel, National Safety Council. Revised 3/2013

This material is based upon work supported by the National Institute of Food and Agriculture, U.S. Department of Agriculture, under Agreement Nos. 2001-41521-01263 and 2010-41521-20839. Any opinions, findings, conclusions, or recommendations expressed in this publication are those of the author(s) and do not necessarily reflect the view of the U.S. Department of Agriculture.

PERSONAL PROTECTIVE EQUIPMENT

HOSTA Task Sheet 2.10 **Core**

NATIONAL SAFE TRACTOR AND MACHINERY OPERATION PROGRAM

Introduction

Items of personal protective equipment (PPE) are designed to protect you from injury and illness. Use PPE to prevent injury or damage to your head, eyes, ears, body and feet.

PPE is the last line of defense against workplace injuries—ranging from bruised toes, to the loss of an eye, to death from a falling object hitting you on the head.

This task sheet discusses personal protective equipment, including the symbols that show the need for this equipment.

Eye Protection

Flying objects, chemicals, dust, and crop debris can all be eye hazards in agricultural work. Always use eye wear approved by the American National Standards Institute (ANSI). Certified safe eyewear is marked ANSI Z87.1.

Eye protection may involve safety glasses, goggles, chemical goggles, or face shields. Protection from the front and side must be considered. High-impact hazards require different protection than splash hazards.

Industrial safety glasses are recommended when you see this symbol. Industrial safety glasses protect against flying and pointed projectiles and may come with brow and side-protection panels.

Goggles with impact-resistant lenses are recommended when you see this symbol. Goggles protect against splashes from all types of hazardous liquids.

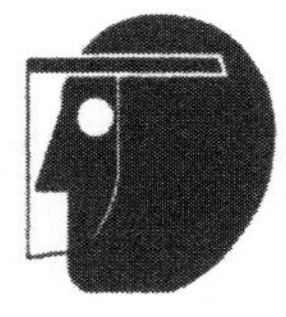

Face shields are recommended when you see this symbol. Face shields protect against splashing and crop debris, but are not designed for high-impact hazards (projectiles). Use industrial safety glasses under the face shield for complete protection.

Figure 2.10.a. What PPE should be used to complete the work with lead acid storage batteries?

Prepare to work safely by using the recommended personal protective equipment (PPE) for that job.

Learning Goals

- To learn when to use specific types of personal protective equipment
- To recognize the symbols that indicate specific types of personal protective equipment

Related Task Sheets:

Injuries Involving Youth	2.1
Personal Dress	2.7
Hazard Warning Signs	2.8
Mechanical Hazards	3.1
Noise Hazards and Hearing Protection	3.2
Respiratory Hazards	3.3
Respiratory Protection	3.3.1

Cooperation provided by The Ohio State University and National Safety Council.

Figure 2.10.b. What types of agricultural jobs is this person dressed to do? Discuss the possibilities with your club, class, leader or mentor.

Dust masks are different than cartridge masks. Match the filter mask to the job. If assigned to a job requiring a respirator, ask for guidance.

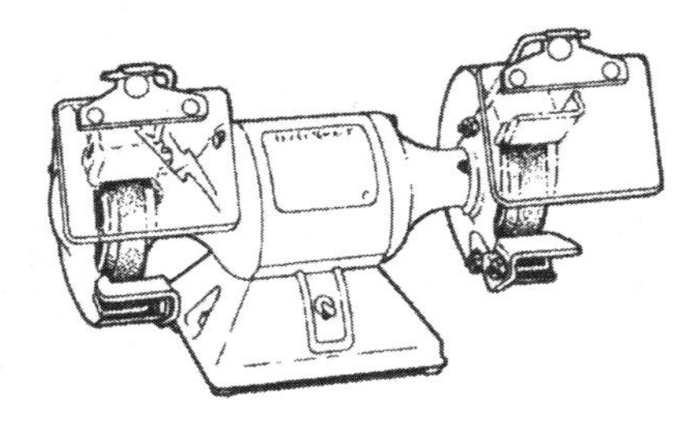

Figure 2.10.c. What PPE would you recommend using with this shop tool while grinding a part for the tractor? *Safety Management for Landscapers, Grounds-Care Businesses, and Golf Courses, John Deere Publishing, 2001. Illustrations reproduced by permission. All rights reserved.*

Respiratory Protection

Protection of the lungs is vital to our health. Agricultural work exposes the worker to vapors, fumes, and dust. Using a National Institute for Occupational Safety and Health (NIOSH) certified respirator is important. Older devices will be identified with a "TC" number written on the respirator (Example TC-23). Newer respiratory protection devices will be identified with a N95, N99, or N99.97 representing the percentage of particles which the filter can trap. See Task Sheet 3.3.1 for further information on respiratory protection devices.

Respirators are either:

- Air purifying, or
- Air supplying

Air purifying respirators filter dust, vapors and fumes out of the air you breathe. A single strap dust mask is not an approved respirator and offers little breathing protection.

Air supplying respirators are the type firefighters wear when fighting fires. ***Never attempt to work with an air supplying respirator without extensive training.***

A NIOSH-approved dust mask is recommended when you see this symbol. An approved dust mask will always have two straps. Make sure that the mask fits snugly around your mouth and nose.

A cartridge type mask is recommended when you see this symbol. Air purification from chemical fumes or vapors is necessary. Specific cartridges must be used, and the mask must fit snugly. Eye protection may be needed as well.

Cooperation provided by The Ohio State University and National Safety Council.

Head Protection

Work spaces where you could bump your head while working are bump cap areas. Workplaces where someone is working above you are hard hat areas. ANSI certified bump caps or hard hats will be marked with the ANSI Z89.1 code.

When you see this symbol, bump caps will be needed.

When you see this symbol, hard hats are required for head protection.

Hearing Loss Protection

Exposure to noise levels varies with jobs and activities. Sound level is measured in decibels. Normal conversation measures 60 decibels (dB), while a jet airplane at take-off measures over 120 dBs. Prolonged exposure to loud noises leads to hearing loss. ***Hearing loss is permanent unless you wear a hearing aid***. Protect your hearing with ANSI-approved ear protection devices.

Ear plugs or acoustic muff style protective devices are two types of hearing protection. Ear plugs fit into the ear, while acoustic ear muffs fit over the ear itself. The preferred ear protection device covers the ear and ear canal.

Hearing protection is recommended when you see this symbol. If you cannot hear a person who is standing 3 feet away and who is talking in a normal voice, hearing protection is needed.

Protective Clothing

Foot Protection

Steel-toed shoes or boots with steel shanks are recommended when you see this symbol. Working with a chain saw and logs, cattle and horses, lumber and concrete block, barrels, or 55-gallon drums are a few farm tasks that require foot protection.

Hand protection is recommended when you see this symbol. Leather gloves are for handling rough or abrasive materials. Neoprene, nitrile, rubber or barrier-laminate gloves should be used for handling pesticides and solvents (leather does not resist chemicals).

Snug-fitting long sleeves and long pants are recommended when you see this symbol. General rules for clothing include shirttails tucked in, jackets zipped or buttoned, and draw strings removed from clothing.

Never stuff cotton into the ears to reduce noise levels.

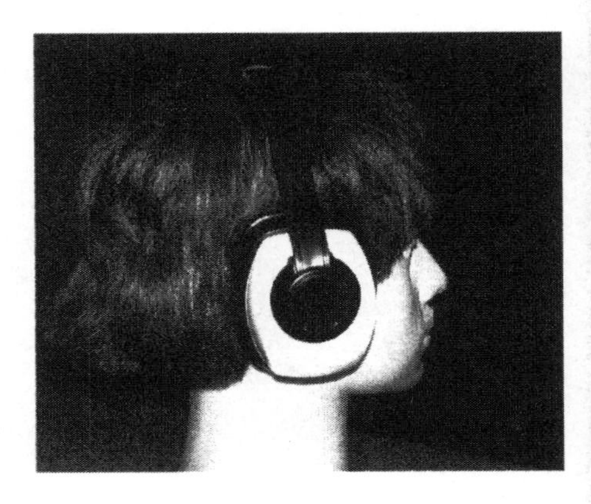

Figure 2.10.d. Working with machinery may require hearing loss protection as worn by this worker.

Safety Activities

1. Match the hazard with the PPE needed (You may select more than one answer).

   A. Operating a tractor with a faulty muffler 1.___

   B. Checking battery fluid level 2.___

   C. Grinding a broken bolt 3.___

2. Where have you seen PPE symbols on your farm or the farm on which you are employed?
3. Invite a sales or product representative from a safety equipment supply company to demonstrate the correct use of a variety of personal protective equipment.
4. Collect a sample of personal protective equipment and give a presentation on the proper use and care of the equipment.

References

1. Safety Management for Landscapers, Grounds-Care Businesses, and Golf Courses, John Deere Publishing, 2001. Illustrations reproduced by permission. All rights reserved.
2. Farm and Ranch Safety Management, John Deere Publishing, 2009.

Contact Information

National Safe Tractor and Machinery Operation Program
The Pennsylvania State University
Agricultural and Biological Engineering Department
246 Agricultural Engineering Building
University Park, PA 16802
Phone: 814-865-7685
Fax: 814-863-1031
Email: NSTMOP@psu.edu

Credits

Developed by WC Harshman, AM Yoder, JW Hilton and D J Murphy, The Pennsylvania State University. Reviewed by TL Bean and D Jepsen, The Ohio State University and S Steel, National Safety Council. Revised 3/2013

This material is based upon work supported by the National Institute of Food and Agriculture, U.S. Department of Agriculture, under Agreement Nos. 2001-41521-01263 and 2010-41521-20839. Any opinions, findings, conclusions, or recommendations expressed in this publication are those of the author(s) and do not necessarily reflect the view of the U.S. Department of Agriculture.

FIRST AID AND RESCUE

HOSTA Task Sheet 2.11 **Core**

NATIONAL SAFE TRACTOR AND MACHINERY OPERATION PROGRAM

Introduction

Knowledge of first aid and rescue should be part of everyone's safety experience. Hazards and risks can be reduced by careful planning and safe work habits, but injuries can still occur. What can you do if an injury or fatality occurs where you work?

This task sheet discusses first aid and rescue basics, however, it will not make you a professional emergency rescue worker.

Important: Enroll in a CPR and first-aid course and keep your skills current.

Preparations

In addition to safe equipment, a safe work site should include:

- A person trained in CPR and first-aid procedures
- A first-aid kit and supplies
- An emergency plan, including telephone numbers for services such as 911
- A location or site map available for emergency responders

Let us examine these points in more detail.

CPR Training

Cardiopulmonary Resuscitation (CPR) is used to provide manual ventilation (air intake) and chest compressions to stimulate the patient's heart and lung operation until medical help arrives or the victim begins to breathe on his or her own. Injured victims or those persons suffering from a heart attack or stroke can be assisted by CPR techniques.

CPR classes are offered by the American Heart Association or the American Red Cross in most communities. CPR is best learned in the classroom and with practice under the supervision of a qualified instructor. CPR guidelines change periodically. Once trained stay updated.

First-Aid Kit

See page 3 for details.

Emergency Contacts

In the event of an emergency, a call to 911 or to emergency medical service (EMS) personnel must be made quickly. Telephone numbers should be posted near the phone or stored in your cell phone. Include these numbers:

- Fire department
- Police department
- Ambulance service
- Poison control center
- Chemtrec 1-800-424-9300 (for chemical spills)
- Electric and gas companies

Be prepared to give directions to the site of the accident. Many times people panic and cannot remember their address, phone number, or directions to the farm. Have this detailed information posted by the phone with the emergency phone numbers. Farm maps should be provided to emergency responders for their files.

Figure 2.11.a. Often during an emergency, panic-stricken people will forget their own address. Written directions posted next to emergency phone numbers can be of help.

Write down the directions to the farm to read to emergency responders.

Learning Goals

- To learn how to prepare for emergency situations
- To learn how to respond to farm injury emergencies

Related Task Sheets:

National Ag Safety and Health Resources **1.5**

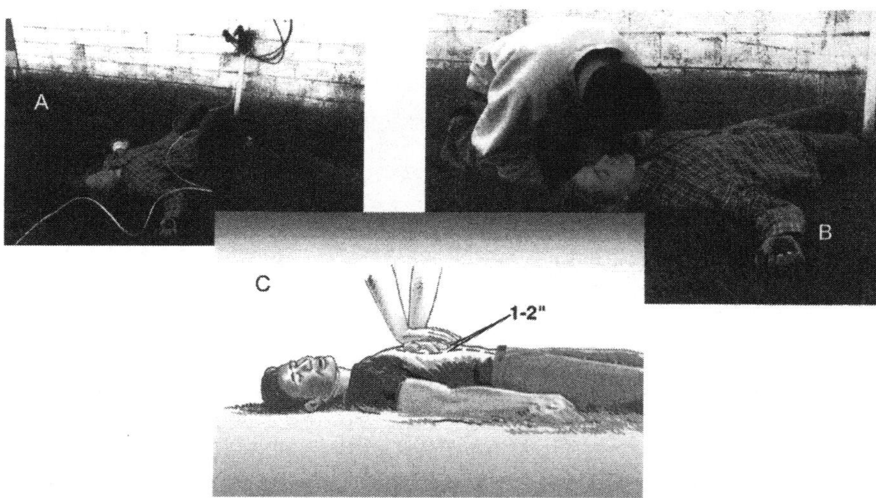

Figure 2.11.b. A farm emergency has occurred. You approach the victim to render aid. What must you do? In this case (A) you must be sure that you are not going to be electrocuted. Turn off the power at the main power switch. If you don't know how to turn off the power, do not touch the victim. If you can disconnect the current, then clear the victim's airway and be sure that the victim is breathing (B). If the victim is not breathing (C), CPR should be administered. Chest compressions and mouth-to-mouth resuscitation must be done properly to assist the person to breathe and maintain a heart beat. You may wish to enroll in CPR and first aid training to respond to these types of emergencies.

Have you been trained in first aid and CPR?

First-Aid Basics

Non Life-Threatening Injury

First-aid practices for minor cuts, abrasions, splinters, insect stings, snake bites, and burns are easily completed. First-aid kits will consist of disinfectants, bandages, and light wraps useful until medical help is secured. Exercise care to keep dirt out of open wounds and do not apply any ointment or creme to burns.

Life-Threatening Injury

Trauma, electrocution, severe bleeding, severe burns, and pesticide exposure may be a matter of life and death. The victim needs immediate medical care and may be unconscious. You may become the initial caregiver. See page 3, Farm Family Emergency Response. What can you do? *Without training, your emergency response may be inappropriate and may create a liability issue.*

Follow the CABs of first aid after assessing the overall situation. Do not put yourself or the victim in more danger. Here are the CABs of first aid.

- C (Circulation)

  Blood must flow throughout the body to carry oxygen to the cells. Without oxygen, brain damage can occur in minutes. Cardiopulmonary Resuscitation (CPR) will be needed if the victim cannot breathe on his or her own. CPR involves regular chest compressions and breathing assistance. *You must be CPR trained to provide this service.*

- A (Airway)

  The victim must be able to breathe. Lay the victim flat on his/her back after checking that there are no broken bones or spinal injuries which could cause further harm. Be sure that the airways (nose, mouth, and throat) are clear. Remove any material from the mouth. Tilt the head and lift the chin to open the airway. Loosening the shirt collar and belt may improve breathing.

- B (Breathing)

  Determine if the person is responsive. Shout, "Are you okay?" If there is no response, mouth-to-mouth resuscitation may be needed.

Did you learn how to conduct mouth-to-mouth resuscitation in Junior High health class?

First-Aid Equipment Needs

General purpose first-aid kits are readily available. A small, well-maintained first-aid kit should be placed on every tractor, farm truck, and major piece of equipment. Larger kits should be located in the farm shop or at home. The small kits should contain at a minimum:

- Sterile first-aid dressings and compresses of various sizes
- Roller bandages
- Adhesive tape
- Disinfectant soap or wound cleanser
- Tweezers
- Scissors
- Latex gloves
- Directions for requesting emergency assistance

Farm Family Emergency Response

A farm family member is often the first person on the accident scene. Fear, panic, crying, and shock can occur. These emotional responses may delay getting help for the victim. Discuss farming hazards, and practice emergency procedures to better handle emergencies.

Discovery of a victim of an agricultural accident requires immediate action. Three actions are needed.

- Activating emergency medical services (EMS)
- Stabilizing the scene
- Providing patient care

Activating EMS

You must quickly and calmly determine whether to remain at the site or to seek help. Discovering a farm accident means a call for assistance is needed. It is recommended that each farm have a site map located at the farm entrance. Rescue teams can then assess the location and identify potential hazards.

Stabilizing the Scene

Controlling hazards at the scene that could harm you or cause further harm to the victim is called "stabilizing the scene." Tractors and machinery can roll further. Fire and explosions can occur. Hazardous materials could spill, or toxic fumes can exist. Be cautious. You may rush to help the victim and become a victim as well.

If the scene cannot be stabilized, but you can still safely approach the victim, try to remove them from the danger. If you suspect spinal injury to the victim, there is a risk of paralysis or death if you move them. Take time to think about the risk to the victim.

Your decisions are important. Think about them, read about these situations, and enroll in CPR and first-aid classes to increase your decision-making skills in emergency matters.

Providing Patient Care

If you are not trained in CPR, your actions may be limited to assuring that the victim is breathing and that bleeding is controlled. Review the airway information on page 2.

Arteries carry blood away from the heart in pulses. Severed arteries spurt blood. You must apply pressure to that point to stop the bleeding.

Talk with the victim to help keep the victim calm. Do not attempt to move the victim. Further injury can result.

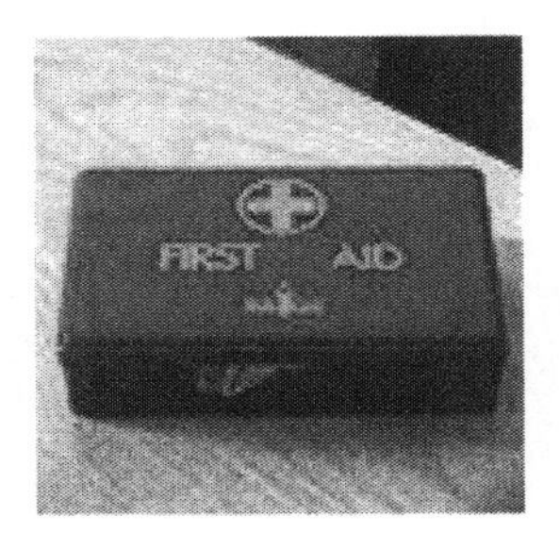

Figure 2.11.c. A first-aid kit should be kept within easy access of tractors and equipment. Replace those items which have been used so that the supplies are always available.

Sometimes the best help for an injured victim is for you to seek more help.

Figure 2.11.d. Severed arteries spurt blood with pressure. In this case, the artery of the upper arm must be pressed firmly against the upper arm bone to stop bleeding. Ask your physician or teacher to show you how this is done.

Safety Activities

1. Conduct a farm survey to identify the locations of first-aid kits. Are they complete? Have supplies been replaced?
2. Conduct a survey of all persons on a local farm to find out how many have been trained in first aid and CPR.
3. Complete a CPR course sponsored by a local agency, such as the American Heart Association or the American Red Cross.
4. If you have CPR certification, remember to enroll in a refresher course.
5. Complete a lifeguard certification program.
6. Join the local Junior Volunteer Fire Program of your local VFD to learn skills in fire safety and rescue.
7. Produce a poster showing the steps needed to perform mouth-to-mouth resuscitation.
8. Many schools and shopping centers now have automated external defibrillators (AEDs) to use if someone has a heart attack. Learn more about these devices and how they work.
9. Conduct a training session on responding to an emergency, such as a tractor turnover, machinery entanglement, or grain bin entrapment. Make sure that all family members and employees understand what to do in an emergency.
10. Offer to set up a farm accident rescue program for the local VFD and EMS groups. Seek adult sponsorship to help you do this.
11. Learn about pressure points used to stop arterial bleeding. Post a drawing of the body's pressure points in the farm shop.
12. Post detailed directions to your farm next to your telephone or in the directory of your cell phone. The directions should begin at your local emergency medical service.
13. Organize a day on the farm where everyone can learn and practice how to shut off every engine/motor in the event of an emergency.

References

1. Farm and Ranch Safety Management, John Deere Publishing, 2009.
2. First on the Scene, 1989, Northeast Regional Engineering Services (NRAES), Ithaca, NY.
3. www.osha.gov/Type"first aid" in search box.
4. Farm Family Emergency Response Program, College of Agricultural Sciences, Department of Agricultural and Biological Engineering, Penn State University, University Park, PA.

Contact Information

National Safe Tractor and Machinery Operation Program
The Pennsylvania State University
Agricultural and Biological Engineering Department
246 Agricultural Engineering Building
University Park, PA 16802
Phone: 814-865-7685
Fax: 814-863-1031
Email: NSTMOP@psu.edu

Credits

Developed by WC Harshman, AM Yoder, JW Hilton and D J Murphy, The Pennsylvania State University. Reviewed by TL Bean and D Jepsen, The Ohio State University and S Steel, National Safety Council. Revised 3/2013

This material is based upon work supported by the National Institute of Food and Agriculture, U.S. Department of Agriculture, under Agreement Nos. 2001-41521-01263 and 2010-41521-20839. Any opinions, findings, conclusions, or recommendations expressed in this publication are those of the author(s) and do not necessarily reflect the view of the U.S. Department of Agriculture.

MECHANICAL HAZARDS

HOSTA Task Sheet 3.1 **Core**

NATIONAL SAFE TRACTOR AND MACHINERY OPERATON PROGRAM

Introduction

There are many hazards in agriculture associated with mechanical equipment. Knowing every hazard of every machine is very difficult. For this reason, agricultural safety and health professionals group them in ways that help the operator recognize the different types of hazards regardless of the machine.

Your ability to recognize these hazardous components is the first step in being safe.

This task sheet identifies groups of hazards, what the danger is, where the hazards may be found, and gives instruction for avoiding them.

Pinch, Wrap and Shear Points

A ***pinch point*** hazard is formed when two machine parts move together and at least one of the parts moves in a circle (Figure 3.1.a). These types of hazards are often found in power transmission systems such as belt drives, chain drives and gear drives. *Avoid pinch points by keeping machine guards in place.*

Any type of rotating machine component can be considered a ***wrap point***. The rotating components are often shafts such as the PTO. Individuals can be caught in a wrap point by their loose clothing or long hair. *Guards can protect the operator from wrap points. Attention to dress and care of long hair is important as well.*

A ***shear point*** occurs when the edges of two machine parts move across or close enough to each other to cut a relatively soft material. One of the two objects can be stationary or moving while the second is moving. Hedge trimmers are a good example of a shear point.

Figure 3.1.a. Pinch points can be found on most machines.

Shielding the worker from the shear point is difficult on many agricultural machines. *The best precaution to take for preventing injury is to shut off the machine before making repairs or adjustments.*

Learning Goals

- To identify the mechanical hazards associated with agricultural machinery
- To avoid mechanical hazards

Related Task Sheets:

Reaction Time	**2.3**
Hazard Warning Signs	**2.8**
Making PTO Connections	**5.4**
Using Power Take-Off (PTO) Implements	**5.4.1**

Cooperation provided by The Ohio State University and National Safety Council.

PTO Stub

- Transfers power from the tractor to the machine
- Rotates at 540 rpm (9 times/sec.) or at 1,000 rpm (16.6 times/sec.)
- Some tractors have a stub shaft guard that screws onto the PTO stub.

Master Shield

- Protects the operator from the PTO stub
- Is often damaged or removed and never replaced.

Figure 3.1.b. A PTO stub and a master shield on a tractor. A PTO is a wrap point hazard that causes countless injuries and deaths each year.

Awareness is the best protection from hazards that cannot be eliminated or shielded.

Crush, Pull-in and Burn Points

Crush points are formed when two objects are moving toward each other, or when one object is moving toward a stationary object, and the gap between the two is decreasing. The most common example of a crush point is formed when an implement is attached to a tractor's drawbar. Most often the tractor is moving toward a stationary implement, and the gap between the tractor's drawbar and the implements hitch is decreasing. *Do not permit another person to stand between the tractor and implement while hitching.*

Pull-in points occur most often where crops are fed into harvesting machinery. Rotating parts that come in close contact with each other, such as feed rolls, often form pull-in points. Pull-in points can also be formed by moving components, such as feed chambers on square balers. *To avoid being pulled into a machine, shut down the engine and the PTO before making repairs or adjustments.*

Hot mufflers, engine blocks, pipes, and fluids (fuel, oils, chemicals) are all examples of possible ***burn points*** on tractors, self-propelled machinery, and pulled machinery. Machine inspection, servicing, and maintenance are the most common types of activities that may result in exposure to a burn point hazard. *To avoid being burned, do not touch the engine or machine parts you are inspecting. Place your hand near the surface of the part to determine if heating has occurred.*

Figure 3.1.c. Pull-in points are found on harvesting machinery.

Freewheeling Parts

When parts of a machine continue to move after the power to the machine has been turned off, they are called *freewheeling parts.* These hazards exist because many machines require a large amount of rotational force to keep them running smoothly under irregular loading. Bringing this rotational force to a sudden stop is almost impossible. A baler is an example of the freewheeling hazard.

To avoid injury from freewheeling parts, stop the tractor engine, disengage the PTO, and wait for the machine to stop completely before making repairs or adjustments.

Figure 3.1.d. The flywheel on a small square baler is an example of a freewheeling part. The flywheel keeps the baler running smoothly if a large amount of hay is suddenly taken into the bale chamber. Notice that part of the PTO driveline is unguarded.

Figure 3.1.e. Mowers are a frequent source of thrown objects.

Thrown Objects

Thrown object hazards occur as normal machine operations discharge materials into the surrounding environment. These hazards are formed by rotating fan or knife blades that are used to cut, grind or chop materials. The blades can throw small or large objects, such as glass, metal, rocks, sticks or other vegetation. A common example of a thrown object hazard is the material that it discharged from a rotary mower.

To avoid injury from thrown objects, be sure the machine is at a complete stop before nearing the discharge area. Keep the work area clear of bystanders. Wear eye protection when working with this type of hazard.

The ability to identify hazards is the first step in avoiding them.

Stored Energy

Stored energy hazards occur when energy that is confined is released unexpectedly. This hazard is present in pressurized systems and their components. Example include springs, hydraulic, pneumatic, and electrical systems.

Avoid the hazard of stored energy by knowing which parts which may be spring loaded. Relieve hydraulic system pressure when the job is completed. Ask for a demonstration of where you might encounter this potential hazard.

Figure 3.1.f. Hydraulic systems often have stored energy.

Safety Activities

1. Draw a line from the Mechanical Hazard to the correct definition.

Pinch Point ·	• Hot mufflers, engine blocks, pipes, and fluids (fuel, oils, chemicals) are all examples of this type of hazard on tractors, self-propelled machinery, and pulled machinery.
Freewheeling Part ·	• A hazard formed when two machine parts move together and at least one of the parts moves in a circle.
Pull-in Point ·	• This type of hazard occurs when machine parts continue to move after the power to the machine is turned off.
Shear Point ·	• Any type of rotating machine component can be considered this type of hazard.
Crush Point ·	• These types of hazards occur when a machine discharges materials into its surrounding environment.
Stored Energy ·	• A hazard formed when the edges of two objects move across or close enough to each other to cut a relatively soft material.
Burn Point ·	• These hazards are caused by energy that is confined and then released.
Wrap Point ·	• A hazard formed when two objects are moving toward each other or when one object is moving toward a stationary object, and the gap between the two is decreasing.
Thrown Objects ·	• Rotating parts that come in close contact with each other, such as feed rolls, often form these points. They can also be formed by moving components, such as feed chambers on square balers.

2. Find an old and a new machine on your farm or at a local dealership, and identify as many mechanical hazards as you can. Compare the two machines.

References

1. Farm and Ranch Safety Management, John Deere Publishing, 2009.
2. Murphy, D.J. 1992. *Safety and Health for Production Agriculture.* St. Joseph, MI: ASAE.
3. American Society of Agricultural and Biological Engineers, ANSI/ASABE S318 Safety for Agricultural Equipment, St. Joseph, MI.

Contact Information

National Safe Tractor and Machinery Operation Program
The Pennsylvania State University
Agricultural and Biological Engineering Department
246 Agricultural Engineering Building
University Park, PA 16802
Phone: 814-865-7685
Fax: 814-863-1031
Email: NSTMOP@psu.edu

Credits

Developed by WC Harshman, AM Yoder, JW Hilton and D J Murphy, The Pennsylvania State University. Reviewed by TL Bean and D Jepsen, The Ohio State University and S Steel, National Safety Council. . Revised 3/2013

This material is based upon work supported by the National Institute of Food and Agriculture, U.S. Department of Agriculture, under Agreement Nos. 2001-41521-01263 and 2010-41521-20839. Any opinions, findings, conclusions, or recommendations expressed in this publication are those of the author(s) and do not necessarily reflect the view of the U.S. Department of Agriculture.

NOISE HAZARDS AND HEARING PROTECTION

HOSTA Task Sheet 3.2

NATIONAL SAFE TRACTOR AND MACHINERY OPERATION PROGRAM

Introduction

Farm equipment can generate high noise levels. High sound levels pose serious health risks to the people who work long hours around this equipment. Hearing damage seldom occurs with one loud noise. Hearing damage results from an exposure to loud noises over an extended period of time.

This task sheet will examine the problem of noise hazards and how to protect your hearing.

What Is Noise?

Sound is created by anything that causes pressure waves in the air. Different wave sizes, or frequencies, are formed by different levels of shock to the air. Unwanted sound is called "noise."

All sound, including noise, is measured in *decibels.* The unit of measurement is shown by the designation dB(A). A decibel meter is a tool that measures the dB level. The "A" represents the sound scale used for the measurement.

Not all sound levels are a hazard. Knowing typical sound levels of various sources of sounds helps us understand if the sound level is unsafe. Consider the following decibel level information.

Decibel Level Chart

dB(A) Level	Sound Source
15	A whisper
50	Gentle breeze or babbling brook
60	Normal talk level
85	Tractor at idle engine speed
90	Chopping silage (no cab) or lawnmower at full throttle
100	Tractor at work or table saw in use
110	Stereo with headphones set at mid-volume
120	Bad muffler or rock concert
140	Shotgun blast or jet engine

Sound levels that cause hearing loss begin at about 85 dB(A). Hearing loss occurs more quickly with louder noise. See Table 3.2 for time exposure to various sound levels which can lead to hearing loss.

OSHA standards consider sound measured at 85 decibels or higher as damaging to the eardrum and therefore a risk to hearing.

Figure 3.2.a. A straight pipe used for the exhaust or a worn-out muffler will increase noise levels coming from the engine. Muffler condition should be part of a safety audit.

You don't adapt to loud noise; you lose your ability to hear loud noise.

Learning Goals

- To recognize when sound levels can become a threat to hearing
- To use correct hearing protection devices

Related Task Sheets:

The Work Environment	1.1
Personal Dress	2.7

Cooperation provided by The Ohio State University and National Safety Council.

Permissible Noise Exposures:

Duration Per Day (hours)	Sound Level, dB(A)
8	90
6	92
4	95
2	100
1	105
1/2	110
1/4	115

Table 3.2. Exposure time limits to sound levels decrease as the db(A) level increases. Use the chart on page 1 to answer the following questions. What is the sound level at your high school dance or at a rock concert? How long should you be exposed to that intensity of sound pressure level?

Is loud music or farm equipment causing you to lose your hearing?

How Does Hearing Loss Occur?

Sound waves have pressure. High frequency sound waves have greater pressure than lower frequency sound waves. This pressure pushes on the ear drum.

Hearing loss occurs over a period of time. Deafness and loss of hearing usually occur with the high frequency sounds and not the lower frequency sounds.

Hearing is lost as auditory nerve endings are exposed to the same frequency of sound for extended time periods. The nerves lose their ability to recover from that hostile frequency. The ability to hear that sound frequency is then decreased forever.

Sound levels may be nearing the danger point for hearing loss if you notice any of these:

- Ears ringing
- Noises in your head
- Your own speech sounds muffled
- You have to shout to be heard by someone working next to you

By the time you recognize any of these events, some hearing loss has occurred.

Hearing loss accumulates over time and cannot be reversed. Hearing aid assistance may be necessary. Many older farmers have developed hearing problems over time. Hearing loss in the young also occurs. With the knowledge gained from this task sheet, the younger farm worker should avoid unnecessary hearing loss.

Cooperation provided by The Ohio State University and National Safety Council.

Protection of Hearing

Reduction of excessive noise is the first step to hearing protection. Hearing protection starts in the farm shop by keeping the exhaust and muffler system of the tractor in good repair. Machine parts that are not well-lubricated or adjusted also cause loud noises.

What farm tasks have you encountered that require hearing protection?

Reduction of excess noise levels may require a sound proofing barrier between the ear and the source of the noise. Sound-proof tractor cabs are designed to reduce sound levels. Compressor rooms may need to be sound-proofed as well. Sound-insulating building materials can reduce noise levels.

Where on your farm is the highest noise level likely to be found?

Types of Ear Protection

Commercially available hearing protection devices are recommended. There are two devices to use. They are:

- Acoustical Muffs
- Ear Plugs

Acoustical Muffs

Acoustical muffs, or ear muffs, are effective in reducing sound level at the ear. They cover the ear and ear canal to provide a barrier to sound. They do not block out all sounds, therefore, conversation for information and safety purposes is readily heard.

Ear Plugs

Ear plugs are made to fit into the ear opening. A snug, tight fit is necessary for effective sound reduction. Ear plugs can be a source of ear infection; so they must be kept clean and sanitized. Do not share ear plugs with others as ear infection can be spread in this way.

There are two types of ear plugs:

- Formable Plugs

  These plugs are compressed before inserting into the ear. They expand to fill the ear canal. One size fits all.

- Preformed Plugs

  These plugs come in many sizes and must be fitted to the individual's ear. They usually have a cord attached between each plug making them more difficult to lose.

Ear-protection devices are ranked by their Noise Reduction Rating (NRR). An NRR31 rating signifies that noise will be reduced by as much as 31 decibels under ideal conditions. For example, in a 100 dB(A) work area, a device with a NRR of 31dB would reduce the effective sound level to 69dB.

Be sure that the hearing-protection device reduces sound to a safe level. Typical ratings are shown.

Device	dB NRR
Ear Muffs	21-31
Ear Plugs	26-33
Combined	Add 3-5 db

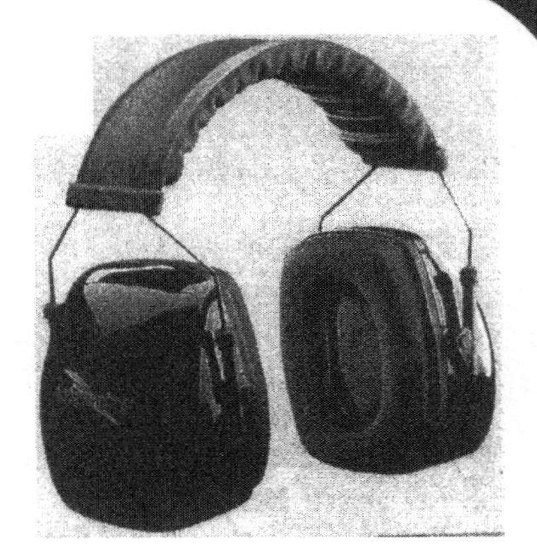

Figure 3.2.b. Acoustical ear muffs offer the greatest level of hearing protection because they cover the entire ear and ear canal.

Cotton stuffed into the ears does not offer hearing protection!

Figure 3.2.c. Ear plugs offer hearing protection, but not as much as full-ear coverage protection devices. *Safety Management for Landscapers, Grounds-Care Businesses, and Golf Courses, John Deere Publishing, 2001. Illustrations reproduced by permission. All rights reserved.*

Safety Activities

1. Obtain a decibel meter (available at electronics stores if your school or club does not have one), measure and record the decibel levels of the following farming operations:

   A. Tractor being used to agitate liquid manure
   B. Tractor being used to operate ensilage blower
   C. Chain saw in use
   D. Milk-cooling equipment compressor

2. Using a supply catalog, such as Gempler's or NASCO, make a list of the various ear-protection devices, their NRR, and their costs.

3. Call a hearing-protection salesperson and a hearing-aid dealer and request hearing-protection literature, or invite them to make a presentation to your group, family, or coworkers.
4. Have a hearing test done as a baseline test to compare your hearing results on an annual basis.
5. Make arrangements with the school nurse or a volunteer nurse to conduct hearing tests for local farmers.

References

1. Safety Management for Landscapers, Grounds-Care Businesses, and Golf Courses, John Deere Publishing, 2001. Illustrations reproduced by permission. All rights reserved.
2. www.gemplers.com/ Type in search box key word(s), hearing protection/Choose a site.
3. www.howstuffworks.com/Type in search box key word decibel/Choose a site.
4. www.osha.gov.

Contact Information

National Safe Tractor and Machinery Operation Program
The Pennsylvania State University
Agricultural and Biological Engineering Department
246 Agricultural Engineering Building
University Park, PA 16802
Phone: 814-865-7685
Fax: 814-863-1031
Email: NSTMOP@psu.edu

Credits

Developed by WC Harshman, AM Yoder, JW Hilton and D J Murphy, The Pennsylvania State University. Reviewed by TL Bean and D Jepsen, The Ohio State University and S Steel, National Safety Council. Revised 3/2013

This material is based upon work supported by the National Institute of Food and Agriculture, U.S. Department of Agriculture, under Agreement Nos. 2001-41521-01263 and 2010-41521-20839. Any opinions, findings, conclusions, or recommendations expressed in this publication are those of the author(s) and do not necessarily reflect the view of the U.S. Department of Agriculture.

RESPIRATORY HAZARDS

HOSTA Task Sheet 3.3

NATIONAL SAFE TRACTOR AND MACHINERY OPERATION PROGRAM

Introduction

The daily activities of farming generate dust and dirt. Working with crops, livestock, and equipment creates more dust and dirt. The worker is placed in conditions perfect for the growth of microorganisms, such as fungi and molds. The worker is often exposed to hazardous gases and vapors. Farm shop work can create respiratory hazards as well. Oxygen-deficient areas present the risk of death.

Continual exposure to breathing hazards creates long-term health problems. Farm workers can suffer from breathing difficulties, such as asthma, "farmers lung," and organic dust toxicity syndrome (ODTS).

This task sheet discusses the problem of respiratory hazards. Respiratory-protection equipment and practices will be discussed in Task Sheet 3.3.1.

Dusts, Mists, and Fumes

Particulates are airborne particles of material that can be measured. Dusts, mists, and fumes make up a group of various-sized particles. They are measured in microns. A micron is 1/25,400th of an inch (50 micron-size particles are visible). Particle sizes over 5 microns are heavy enough to settle quickly without posing a respiration hazard. Finer materials are the major concern to lung health.

Dusts—Dusts include the solid particles (0.1– 25 microns in size) created by handling, crushing, grinding, and moving materials such as rock, metal, wood, and crops.

Crop production exposes the worker to dust particles from the crop, spores from microorganisms growing on the crop, and the fine, airborne particles of soil stirred by field work. Many particle sizes are produced. Fine chopped crop particles can be inhaled into the lungs (respirable dust). As plant materials break down, molds and fungus are also inhaled.

Livestock production exposes the worker to dirt, dust, mites, fungus, and the dry scaly skin found on or around the animal or bird or in its housing area. Antibiotics added to livestock feeds can also pose a respiration hazard.

Mists—Liquid droplets suspended in the air represent mists as a respiration hazard. Paint sprays and cutting oil become airborne breathing hazards.

Fumes —Material that becomes airborne during welding (metal, welding rod, and flux) are examples of fumes. See page 2 for a discussion of toxic gases and vapors.

Figure 3.3.a. Dust from agricultural work can lead to eye and lung irritation. Respiratory protection, such as filter masks, are discussed in Task Sheet 3.3.1.

Coal miners can get "black lung" from breathing coal dust. Farmers can get "green lung."

Learning Goals

- To recognize respiratory hazards associated with agriculture

Related Task Sheets:

The Work Environment	**1.1**
State Agricultural Safety and Health Resources	**1.4**
Personal Protective Equipment	**2.10**
First Aid and Rescue	**2.11**
Silos	**3.9**
Grain Bins	**3.10**
Manure Storage	**3.11**
Anhydrous Ammonia	**3.12**

Cooperation provided by The Ohio State University and National Safety Council.

Carbon monoxide vapors from engines can kill. This gas is colorless and odorless.

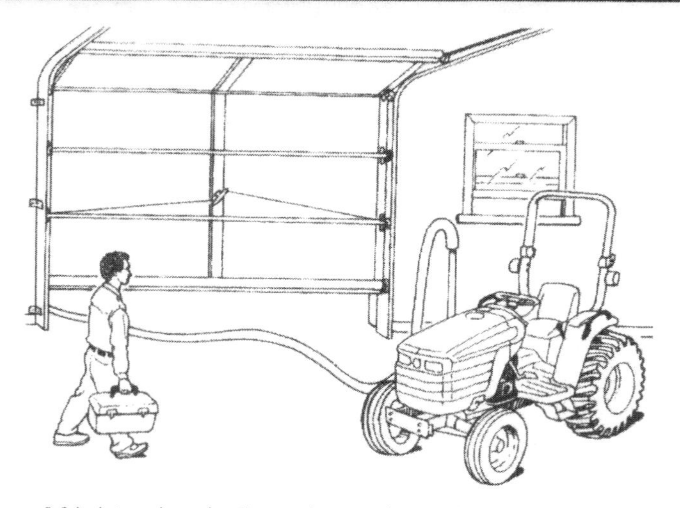

Figure 3.3.b. Internal-combustion engines produce carbon monoxide gas in the exhaust. This colorless, odorless gas can asphyxiate or suffocate a person working on the engine in an enclosed area. Be sure to ventilate the exhaust gases to the outside of the building, or work with plenty of air flow into the building. *Safety Management for Landscapers, Grounds-Care Businesses, and Golf Courses, John Deere Publishing, 2001. Illustrations reproduced by permission. All rights reserved.*

Gases and Vapors

Manure Gases

Manure breaks down chemically when held in storage pits. Hydrogen sulfide, carbon dioxide, ammonia, and methane gases are produced in the manure. These gases intensify in their concentration and are trapped in the manure. The oxygen level of the storage pit or tank becomes too low to support life.

To move the manure from storage to field application, the manure must be agitated and pumped to a spreader unit. The gases are then released into the air.

With equipment breakdowns, unsuspecting farm workers, co-workers, and family members have entered the unventilated, low-oxygen level, confined areas and have been killed by suffocation. Oftentimes multiple fatalities have occurred attempting a rescue.

Stay out of manure storage facilities!

Manure gases can cause asphyxiation, eye and nose irritation, or can be explosive (methane). See Task Sheet 3.11 for more details.

Silo Gases

The silage fermentation process produces deadly nitrogen dioxide gas. This yellow brown gas is heavier than air and settles to a low point in the silo or feed room. Workers entering unventilated silos are often overcome with this gas. A few survive the exposure with lung damage, but many victims perish. See Task Sheet 3.9 for further discussion on silo safety.

Farm Shop Gases

The farm shop exposes workers to respiratory hazards during jobs such as welding, painting, and engine repair. Ventilation is needed for each of these tasks.

Check with the owner of the shop as to what safety procedures to follow to activate ventilation fans.

Welding

Ventilation is necessary during all welding processes. Galvanized metal emits zinc smoke fumes during welding. These fumes can be fatal to inhale. Weld gases such as acetylene can be explosive in high concentrations. The arcing of a light switch can cause acetylene vapors to explode.

Engines

Engines produce deadly carbon monoxide gas. This colorless, odorless gas can asphyxiate the worker who operates an engine in an enclosed area. Do not operate an internal combustion engine inside a closed building!

Solvents and Paint Thinners

Vapors from paint thinners or solvents are released into the air and can be explosive. Paint thinners also produce symptoms of nausea when inhaled. Skin damage is possible. Read the labels on solvents and thinners to learn about ventilation requirements.

Lung Disease

Inhalation of dusts, mists, fumes, vapors, gases, and smoke causes irritation to the respiratory system. Repeated, prolonged exposure can cause more severe problems. Two of the problems are described here.

Farmer's Lung– Farmer's Lung is an allergic reaction caused by inhaling moldy hay, straw, and grain. When the lungs cannot remove the material, an allergy can develop. Repeated exposure further increases lung tissue damage and allergic reaction. Symptoms are similar to those of pneumonia.

Organic Dust Toxicity Syndrome (ODTS)- ODTS is caused by a reaction to inhaling molds from spoiling grain and forage. ODTS usually does not cause permanent lung damage. Symptoms include cough, fever, chills, body aches, and fatigue. Symptoms can last 1-7 days.

Asthma

Do you know someone who has asthma? They probably use an inhalant (medicine in an aerosol tube) to provide breathing relief. National statistics show an increase in the number of persons suffering from asthma

What is asthma? Asthma is a disease of the respiratory system. It is not known how people develop asthma. The small air tubes of the lungs tend to make more mucous than normal. The air tubes tend to swell, and the muscles around the air tubes tighten when an asthma attack occurs.

Asthma can be triggered by several causes. Some of them are:

- Allergies
- Infection (colds and bronchitis)
- Weather changes
- Smoke
- Physical exercise

Allergies such as exposure to dusts, mists, fumes, vapors, and gases irritate the lungs and can bring on an asthma attack. All of these irritants can be found in agriculture. Weather changes can lead to colds and bronchitis. Hot, humid weather as well as winter cold is a factor in asthma. Cigarette smoking or standing in the smoke of a burning fire is an irritant to the lungs also. Sports activities and physical work can also trigger an asthma attack.

If you are an asthma sufferer, there are two recommendations.

1. Avoid those factors that trigger an asthma attack.
2. Follow your doctor's advice and prescription program.

Since repeated exposure to lung irritants reduces respiratory health, asthma can develop. Take the necessary precautions to protect your lungs from developing asthma and other respiratory problems.

Respiratory-protection devices will be discussed in Task Sheet 3.3.1. Be sure to use the knowledge from this task sheet to select the proper respiratory protection for the materials with which you are working.

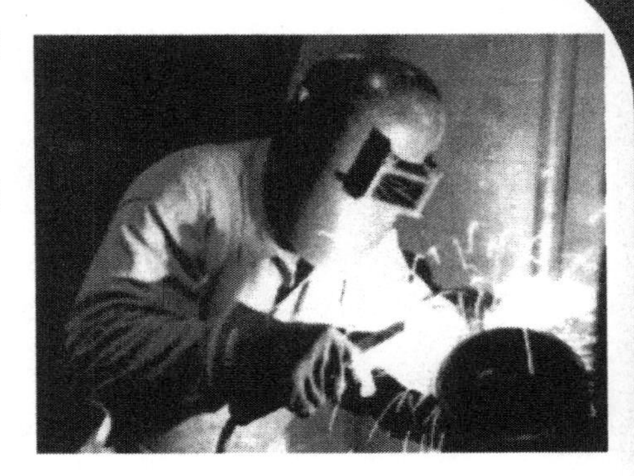

Figure 3.3.c. Welding produces fumes. As the metal melts, and the welding rod and flux covering is burned, fumes are produced. These fumes can cause irritation to the nose and lungs.

When you can't breathe, nothing else matters.®
American Lung Association

Figure 3.3.d. Silo gas can leave a person unconscious or dead. It is difficult to rescue a victim from inside of a farm silo.

Safety Activities

1. Visit the American Lung Association website (www.lungusa.org) to learn more about lung disease.

2. Invite a respiratory therapist to speak to you, your 4-H club, or FFA chapter about lung disease and its prevention.

3. Visit the website www.gemplers.com. Locate the respiratory-protective devices for the following situations, and then make a chart of the device, use, and price:

Device	Used For:	Price Range	NIOSH Rating
____________	Welding Respirator	____________	____________
____________	Dust/Mist Respirator	____________	____________
____________	Nuisance Odor Respirator (livestock odors)	____________	____________
____________	Full-Face Respirator	____________	____________

4. Interview older farmers in the community about their experiences with "farmers lung" and ODTS, then write a news article to submit to an agricultural publication or newspaper in your state.

5. Interview people in your community who are welders. Ask them what they do to protect their lungs.

References

1. Safety Management for Landscapers, Grounds-Care Businesses, and Golf Courses, John Deere Publishing, 2001. Illustrations reproduced by permission. All rights reserved.
2. Any Internet search engine. Type in asthma. Scroll to various sites to learn about asthma.
3. www.gemplers.com.
4. www.lungusa.org

Contact Information

National Safe Tractor and Machinery Operation Program
The Pennsylvania State University
Agricultural and Biological Engineering Department
246 Agricultural Engineering Building
University Park, PA 16802
Phone: 814-865-7685
Fax: 814-863-1031
Email: NSTMOP@psu.edu

Credits

Developed by WC Harshman, AM Yoder, JW Hilton and D J Murphy, The Pennsylvania State University. Reviewed by TL Bean and D Jepsen, The Ohio State University and S Steel, National Safety Council. Revised 3/2013

This material is based upon work supported by the National Institute of Food and Agriculture, U.S. Department of Agriculture, under Agreement Nos. 2001-41521-01263 and 2010-41521-20839. Any opinions, findings, conclusions, or recommendations expressed in this publication are those of the author(s) and do not necessarily reflect the view of the U.S. Department of Agriculture.

RESPIRATORY PROTECTION

HOSTA Task Sheet 3.3.1

NATIONAL SAFE TRACTOR AND MACHINERY OPERATION PROGRAM

Introduction

Many people think that farming means working in the clean, fresh air. Farming, however, has many respiratory (breathing) hazards. Some air will be dirty. Some air can be lethal (deadly) to breathe.

This task sheet discusses respiratory-protection devices to be used in agricultural work. Specific devices must be used with the correct work hazard to reduce lung damage. Failure to use the correct device can be the same as having no protection at all.

Breathing Hazards

The first step in selecting a respirator is to determine what the hazard is. Three categories of respiratory hazards can be found on the farm. They are:

- Particulates (dusts, mists, fumes)
- Gases and vapors
- Oxygen-deficient atmospheres

Particulates

Particulates are airborne particles of sizes that can be measured. Dusts, mists, and fumes are the types of these various-sized particles. Dusts are the largest-size particles. Dust may be dirt, but also can be spores from moldy hay, silage, or grain. Mists are suspended liquid droplets held in the air from mixing, cleaning, and spraying operations. Fumes are particles of airborne solid evaporated metals such as from welding tasks.

Gases and Vapors

Chemical reactions of materials with the air produce gases and vapors. Gases are released from chemical reactions, such as manure decomposition, silage fermentation, and the exhausts of internal combustion engines. The gaseous products of these reactions exist during normal temperatures of the reaction.

Vapors are gases from substances that are normally solid or liquid. Evaporation from liquids, such as pesticides, paints, adhesives, and solvents become vapors. These become airborne breathing hazards.

Oxygen-Deficient Atmospheres

The air we breathe normally contains about 21% oxygen.

Some agricultural storage areas are oxygen-free by design or by the chemical reaction going on inside of them.

- Sealed silos are kept free of oxygen to keep certain bacteria from spoiling the silage.
- Controlled Atmosphere (CA) storages of fruit and vegetables lower the oxygen levels to maintain food quality and storage times.
- Manure storage, especially covered pits, become oxygen-deficient due to manure decomposition depleting the oxygen supply.

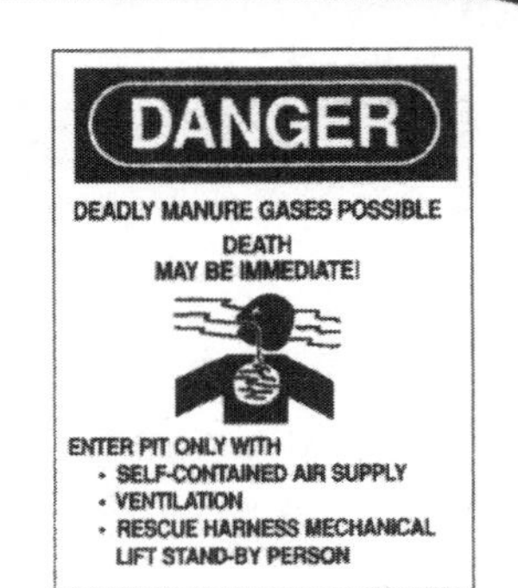

Figure 3.3.1.a. Safety signs warn us of immediate danger. This sign tells us that respiratory protection is required. What other safety practices does this warning sign recommend?

Try a different work practice to reduce breathing hazards. If you are still at risk, use a respirator.

Learning Goals

- To be able to select the correct respiratory protection for use in specific agricultural work

Related Task Sheets:

The Work Environment	1.1
Worker Protection Standards	1.2.4
Personal Protective Equipment	2.10
Common Respiratory Hazards	3.3
Agricultural Pesticides	3.5
Confined Spaces	3.8
Silos	3.9
Grain Bins	3.10
Manure Storage	3.11
Anhydrous Ammonia	3.12

Cooperation provided by The Ohio State University and National Safety Council.

Poor air environments can be immediately dangerous to life and health.

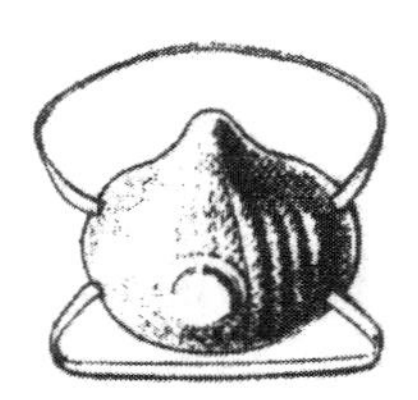

A. Disposable toxic dust mask

B. Chemical cartridge mask

C. Powered air purifying respirator (PAPR)

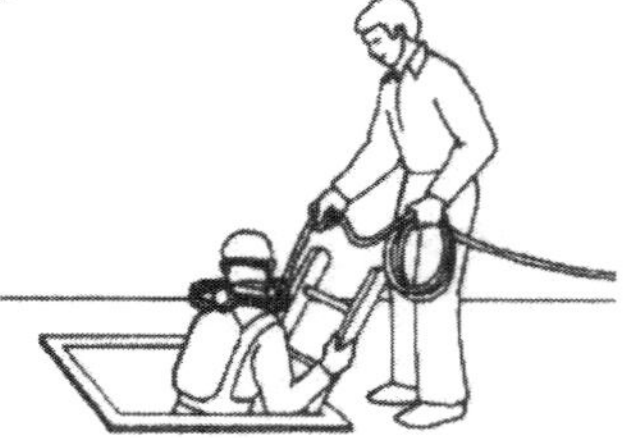

D. Self-contained breathing apparatus (SCBA)

Figure 3.3.1.b. Respirators are available in two main categories. Air-purifying respirators such as the disposable toxic dust mask (A), the chemical cartridge mask (B), and the powered air-purifying respirator (PAPR), (C) shown above. Air-supplying respirators or self-contained breathing apparatus (SCBA), (D) provide clean fresh air from an outside source.

Types of Respirators

There is no such thing as an all-purpose respirator. Specific respirators are used for specific contaminants. A disposable dust mask will not filter chemicals. A self-contained breathing apparatus (SCBA) is not needed to load hay on a wagon.

Respirators can be placed in two categories:

- Air-purifying respirators
- Supplied-air respirators

See Figure 3.3.1.b.

Air-purifying respirators are equipped with filters. The user breathes through these filters. The respirator filters may be disposable or may be replaced according to the material to be filtered (Figures 3.3.1.c and 3.3.1.d.).

Replacement-filter respirators should have filters replaced when your breathing becomes labored, the mask loses its shape or no longer fits your face, or you taste or smell the substance. A mechanical filter for particulates is not a replacement for a chemical-replacement filter.

Gas masks filter chemicals through a cartridge canister filter system. They have a full-face piece. Do not use the gas mask-type respirator in an oxygen limited area as they do not supply oxygen to the user.

Powered Air Purifying Respirators (PAPR) have a motorized blower to force air through a filter to the wearer. A constant stream of air is placed over the user's head and face. They have the appearance of a hard hat with a face shield.

Air-supplying respirators bring an outside source of air to the wearer. These respirators are used in those areas where the oxygen levels are so low that they are considered immediately dangerous to life or health (IDLH).

Air-supplying respirators are of two types:

- Air-line respirator
- Self-contained breathing apparatus (SCBA)

Air-line respirators supply air to a respirator face piece through a hose connected to an air pump or tank. *Self-Contained Breathing Apparatus (SCBA)* devices have a portable air tank that must be carried on the back like those worn by scuba divers and firefighters. Air-supplying respirators are expensive, and the user must learn and practice how to use them.

Use and Care of a Respirator

Respirators must be properly cared for if they are to protect your lungs. The device must snugly fit your face to provide lung protection. The respirator must not expose you to harmful residues either. The respirator must be cleaned. Filters must be changed often.

A properly fitted respirator will make an air-tight seal around your mouth and nose but still allow you to breathe. Poorly fitted respirators provide little or no protection. Dirty filters will prevent you from breathing normally.

Respirators must be clean before use. Clean the respirator body with warm soapy water and rinse thoroughly. Change the filters also. Clean the straps as well.

Use disposable filter masks just one time; then dispose of them.

Selecting a Respirator

Approved respiratory protection equipment should have NIOSH (National Institute for Occupational Safety and Health) shown on the device. Letter and number designations can be found. Look for the designation to be sure that the respirator is approved. Older labels will show the MSHA/NIOSH TC# or approval number. For example, a TC-23C respirator is used for pesticides. There may be older respiratory-protection devices to be found where you are employed.

Newer labels on respirators will show the NIOSH approval number and describe the new NIOSH-approved respirator. An example would be the NIOSH TC-23C dual-cartridge half mask with disposable filter used for pesticides and ammonia.

Under current standards, air-filtering masks or respirators are rated according to the filter's efficiency in reducing solid particles of dust, mists, and fumes. Respirators are rated as being 95%, 99%, and 99.97 percent effective at filtering dust particles.

Filters are also rated according to time-use limitations in using the filter for protection against oil-based chemicals or pesticides in the atmosphere. The following designations are found:

N= Not resistant to airborne oils. Becomes plugged quickly.

R= Resistant to airborne oils for up to 8 hours

P= Oil proof– Possibly resistant to airborne oils for more than 8 hours. Change filters after 40 hours of use or every 30 days, whichever is first.

The air-purifying disposable filter mask in Figure 3.3.1.c. could have a N95 rating. The filter respirator in Figure 3.3.d. may have a N99.97 NIOSH rating. This assures you that the filter offers 99.97% protection from exposure to particulates. There are no 100% filters in theory.

The work situation dictates the respirator to be used; not what happens to be hanging on the shop wall.

Use a respirator for its intended use only, and take proper care of the respirator as well.

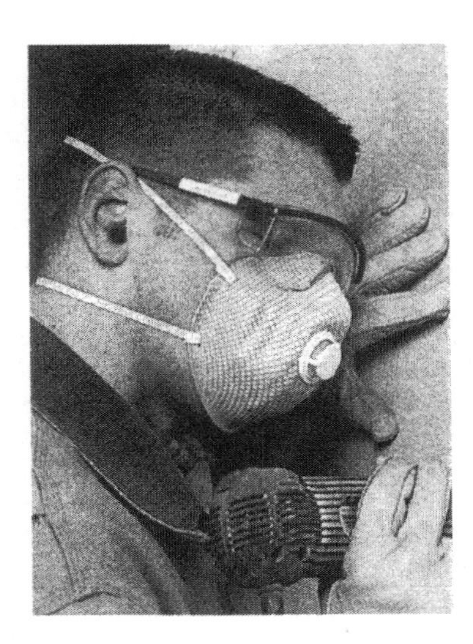

Figure 3.3.1.c. A double-strap respirator provides for a snug fit over the mouth and nose. If a respiratory protective device does not fit snugly, it cannot offer effective respiratory protection from small particles that can damage your lungs. A beard may cause the respirator to fit improperly.

Figure 3.3.1.d. The chemical cartridge respirator mask has a replaceable filter to trap dust, chaff, and larger particles. These respirators do not supply oxygen. These respirators do not filter toxic dust and vapor materials.

Figure 3.3.1.e. The nuisance dust mask is the simplest form of protection. These devices do not filter out small particles of dust that cause respiratory disease. You can identify a nuisance mask by its single strap. *Safety Management for Landscapers, Grounds-Care Businesses, and Golf Courses, John Deere Publishing, 2001. Illustrations reproduced by permission. All rights reserved.*

A handkerchief over the nose will not filter gases, fumes, or small particles!

Safety Activities

1. During a farm visit, list as many places as you can that are oxygen-limited structures or locations.
2. Are all silos oxygen-limiting? Why or why not?
3. Visit a local orchard to find out more about controlled atmosphere (CA) storage of apples. Write a report on CA storage.
4. Using a vendor's catalog such as Gemplers, Inc, locate the respiratory-protective devices, and make a chart including the efficiency rating (95, 99, 99.97) and the respirator's rating for exposure to oils in the atmosphere (N, R, P) for each of the devices.
5. Match the recommended respirator type with the situation where that respirator would be used.

A._____ Air-purifying filter mask with double straps	1. Oxygen-limited area, such as a manure pit.
B._____ Chemical cartridge face shield and respirator	2. Nuisance dust areas, such as sweeping a shop.
C______SCBA	3. Pesticide mixing and filling area.

References

1. www.cdc.gov/niosh (Search the site for respirator use information)
2. www.gemplers.com.
3. Farm Respiratory Protection, Fact Sheet E-36, College of Agricultural Sciences, Department of Agricultural and Biological Engineering, Dennis J, Murphy and Cathleen M. LaCross.
4. Farm and Ranch Safety Management, John Deere Publishing, 2009

Contact Information

National Safe Tractor and Machinery Operation Program
The Pennsylvania State University
Agricultural and Biological Engineering Department
246 Agricultural Engineering Building
University Park, PA 16802
Phone: 814-865-7685
Fax: 814-863-1031
Email: NSTMOP@psu.edu

Credits

Developed by WC Harshman, AM Yoder, JW Hilton and D J Murphy, The Pennsylvania State University. Reviewed by TL Bean and D Jepsen, The Ohio State University and S Steel, National Safety Council. Revised 3/2013

This material is based upon work supported by the National Institute of Food and Agriculture, U.S. Department of Agriculture, under Agreement Nos. 2001-41521-01263 and 2010-41521-20839. Any opinions, findings, conclusions, or recommendations expressed in this publication are those of the author(s) and do not necessarily reflect the view of the U.S. Department of Agriculture.

WORKING WITH LIVESTOCK

HOSTA Task Sheet 3.4 **Core**

NATIONAL SAFE TRACTOR AND MACHINERY OPERATION PROGRAM

Introduction

Working with livestock can be pleasurable and rewarding. To observe a litter of piglets being born, to assist with the birth of a dairy calf, or to train a young horse to lead by halter can be very satisfying. Working with animals is a major task in farming.

Working with livestock can also be dangerous. Animals have their own patterns of behavior. How well you understand animal behavior will be important to working safely with livestock.

This task sheet discusses what you will need to know to safely work with livestock.

Working With Livestock

Farm youth learn to work at an early age. Small children are routinely assigned to feed calves, heifers, pigs, and poultry. Junior livestock programs in rural counties help youth learn how to feed, care for, and market their animal project. Responsibility, confidence, and animal handling skills are gained by doing this work.

Statistics show us that working with livestock is also hazardous. Study these injury facts.

- In 2006 over 23,00 youth injuries occurred on farms with 39% being work-related.
- Falls, animals, and off-road vehicle use were three major sources of injury.
- Livestock and dairy farms led the injury list followed by crops farms.

Working with livestock does expose the youthful farm worker to an increased risk of injury.

Livestock hazards are also recognized as part of the Hazardous Occupations Order in Agriculture (HOOA). In these regulations, youth under age 16 are prohibited from working in a yard, pen, or stall with:

- Cows with newborn calves
- Bulls, boars, or stud horses kept for breeding purposes
- Sows with nursing pigs

Not all livestock jobs are hazardous for young people. Caring for poultry, milking cows, cleaning barns and equipment storage buildings, and riding, driving, or exercising horses are considered acceptable tasks, depending on the age and experience of the youth. Adult supervision of small children doing these tasks is recommended under North American Guidelines for Children's Agricultural Tasks (NAGCAT).

If you are employed by a local farmer to work with livestock, the expectation is that you will be trained and supervised by that person to safely do that work.

Figure 3.4.a. A cow with a newborn calf is usually protective of her offspring.

Livestock are linked to one in every five injuries on the farm.

Learning Goals

- To recognize hazards associated with caring for livestock
- To learn how to work with livestock

Related Task Sheets:

Injuries Involving Youth	2.1
Age-Appropriate Tasks	2.4
Personal Dress	2.7

Cooperation provided by The Ohio State University and National Safety Council.

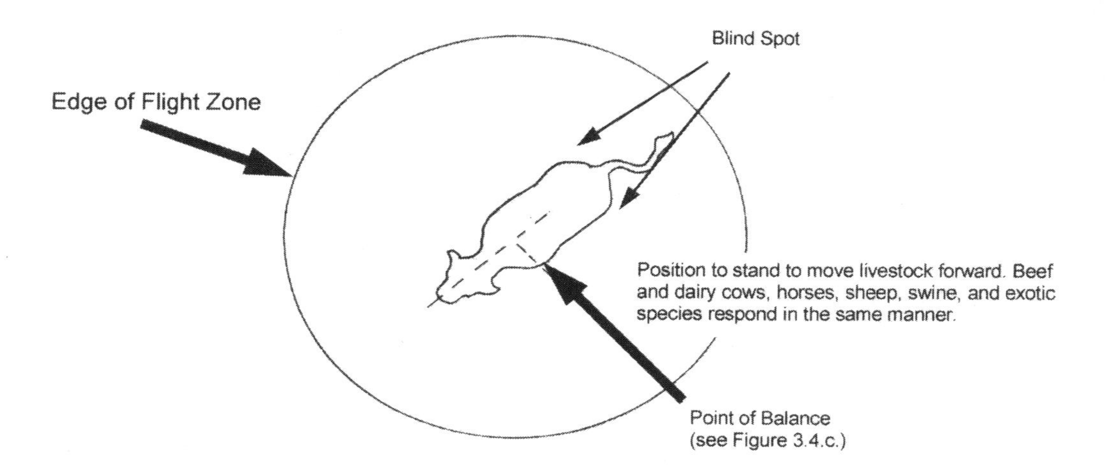

Figure 3.4.b. Flight Zone. Animals have a "personal" space. That space varies with how tame or wild the animal is. An excited animal has a larger flight zone. When you enter the flight zone, livestock turn to move away. If you surprise an animal by entering the blind spot of the flight zone from the rear, you may be kicked.

Roy's mother thought it was cute to teach the feeder calf to butt her hand with its head. When the FFA steer project weighed 1200 lbs., the animal still liked to butt his head. But that wasn't fun any longer.

Animal Behavior Facts

Animals have certain patterns of behavior which are instinctive and other behaviors that develop from habit. Cattle are "creatures of habit." Milking time finds cows lining up at the holding pen. The sound of feeding equipment being started is enough to bring animals to the feeder.

Understanding animal behavior is the first step in working safely with animals. Here are some animal behavior facts.

- Female species are maternal. They will try to protect their young from danger.
- Older male animals are more aggressive and unpredictable. Male hormones cause this.
- Animals tend to group together for safety. Single animals are more dangerous and difficult to handle.
- Animals are territorial. They may challenge an intruder that comes into their space.
- Animals tend to follow a leader when being moved. If no animal makes a move, the group tends not to move.
- Animals become acclimated to particular locations, sights, smells, and sounds. When moved to new and strange surroundings, livestock will react tentatively.
- Animals have a zone of comfort within which they will behave normally. Intrusion into that space will cause the animals to move to re-establish the comfort zone.
- Animals have poor depth perception and cannot see behind them. They will turn to keep you within their sight.

 Cooperation provided by The Ohio State University and National Safety Council.

Moving Animals

Getting livestock to move is a matter of understanding the animals "flight zone" and "point of balance." Animals will move easily if these two ideas are used with calm movement and the least amount of confusing noise.

Animals have a personal space just like people. The size of that space depends upon the animal's tameness, the excitement level, and the angle that you approach the animal. If you move into the animal's *flight zone* (Figure 3.4.b.), the animal will turn to move away from you. If you move outside the flight zone, the animal will turn to look at you. If the animal feels trapped in a corner and has limited vision, the animal will kick to warn you to stay away.

The animal will move according to your position at its *point of balance* (Figure 3.4.c.). The point of balance is the animal's shoulder. All species of animals will move forward if the person is behind the point of balance. All animals will back up or turn away if the person is in front of the point of balance.

Using the point of balance works for moving larger groups of animals as well. Use this knowledge to move animals without prods, "hot-shots," or shouting and screaming. People are smarter than animals and should use their thinking skills in working with livestock. *Hint:* Watch a livestock show. The leader will move in front and to the back of the point of balance to move his or her animal easily.

Precautions to Take

Livestock chores are not hazardous if the animal's behavior is understood. There are precautions to follow to assure that the work is a pleasant experience free of injury.

Plan to use these safety measures when working with livestock.

- Plan for an escape route when working with livestock. Pens and corrals should have people pass-through openings for escape purposes.
- Wear steel-toed, nonskid shoes—not sandals or sneak-ers—when working with live-stock.
- Avoid the hind legs of animals.
- Use squeeze chutes to hold animals securely for veterinarian procedures.
- Approach livestock so that they can see you coming.
- Move cattle in well-lighted areas, not shadowy places.
- Avoid quick movements and loud noises.
- Be patient.
- Keep animal-handling facilities in good repair with no sharp projections.
- Ask for help to move or work with an animal if the animal is excited or nervous.
- If the animal becomes nervous and agitated, wait 30 minutes before attempting to work with the animal again.

Figure 3.4.c. The animal's shoulder is the point of balance for movement. Stand behind the shoulder, the animal moves forward. Stand in front of the shoulder, the animal stops moving forward.

When working with animals, give yourself a route of escape. Do not corner the animal.

Figure 3.4.d. In days gone by, milking a cow was a different chore than it is today. What might be unsafe about this scene?

Safety Activities

1. Use a basketball and a tennis ball to represent an animal and a person respectively. Roll the tennis ball against the basketball to determine if the larger ball can be moved easily. Then roll the basketball against the tennis ball to determine if the tennis ball can stop the basketball. What did you observe?

2. Most animals are territorial. What does this mean? Make a list of incidences you have observed where an animal exhibited territorial habits and how they acted/reacted.

3. Use the Internet to locate your state's Land Grant University, College of Agriculture website. Search this site for any information you can find on how to construct animal-handling facilities for moving animals (chutes) and holding animals (squeezes). Make a sketch of the plans with dimensions.

4. Inspect a farm's facilities for handling livestock. How many pass-through gates are available?

5. Ask a friend who has a halter-broke animal to exhibit at the county fair. Ask your friend to show you how easily an animal will move backward or forward based on a person's slight movement front or back of the point of balance.

6. Practice moving a group of animals slowly and quietly by using knowledge of flight zone and point of balance.

7. Make a poster of the flight zone of a beef animal, a dairy cow, a hog, or a horse to show others how to safely move around animals.

8. Inspect all animal pens and alleyways where you will work for sharp obstructions (nails, sheet metal, etc), broken boards, and damaged gates. Report your findings to the owner. Suggest to the owner that they be repaired. Perhaps this is something that you can do as an employee of that farm.

References

1. Injuries to Youth on Farms and Safety Recommendations, Farm Youth Injuries 2006, cdc.gov/niosh/docs/2009-117, August 31, 2010
2. www.nagcat.org/Click on guidelines/Select category, December 2010.
3. Cooperative Extension Service publications of your State Land Grant University.
4. www.necasag.org/library/facts/Search for Livestock Handling Fact Sheet, April 2009.

Contact Information

National Safe Tractor and Machinery Operation Program
The Pennsylvania State University
Agricultural and Biological Engineering Department
246 Agricultural Engineering Building
University Park, PA 16802
Phone: 814-865-7685
Fax: 814-863-1031
Email: NSTMOP@psu.edu

Credits

Developed by WC Harshman, AM Yoder, JW Hilton and D J Murphy, The Pennsylvania State University. Reviewed by TL Bean and D Jepsen, The Ohio State University and S Steel, National Safety Council. Revised 3/2013

This material is based upon work supported by the National Institute of Food and Agriculture, U.S. Department of Agriculture, under Agreement Nos. 2001-41521-01263 and 2010-41521-20839. Any opinions, findings, conclusions, or recommendations expressed in this publication are those of the author(s) and do not necessarily reflect the view of the U.S. Department of Agriculture.

AGRICULTURAL PESTICIDES

HOSTA Task Sheet 3.5 **Core**

NATIONAL SAFE TRACTOR AND MACHINERY OPERATION PROGRAM

Introduction

Modern farming relies on many chemicals to produce and preserve an abundance of high-quality food. Fertilizers, pesticides, cleaners and sanitizers, crop preservatives, fuels and solvents are chemicals. Each of these chemicals poses a hazard. Youth younger than age 16 are prohibited from using many agricultural pesticides.

This task sheet discusses agricultural chemicals from a youth information standpoint. Older workers can be called upon to handle and apply most chemicals. *If asked to work with restricted use (Category I and II) agricultural chemicals, tell your employer that you are under age 16 and are prohibited by law from doing so. See Task Sheet 1.2.2.*

Pesticide Use Restrictions

At age 15, you have been hired to work at the neighboring farm. You have passed the safe tractor and machinery certification program. On your first day of work, the farmer has assigned you to rinse pesticide containers for return to the dealer and to burn pesticide bags. This may sound like a safe job for you to do, but is the job actually safe?

Hazardous Occupations Order in Agriculture regulations cover more than just tractor and machinery operation activities.

The agricultural chemical portion of the regulation clearly states, "Youth workers under the age of 16 are prohibited from handling or applying (including cleaning or decontaminating equipment, disposal or return of empty containers, or serving as a flagman for aircraft) agricultural chemicals classified as Category I of toxicity (identified by the word "poison" and the "skull and crossbones" on the label) or Category II of toxicity (identified by the word "warning" on the label). Categories of chemical toxicity and their signal words will be explained on page 2 of this task sheet.

Effects of Pesticides on People

Agricultural pesticides may come in dust form, granular particles, liquid concentrates, or solutions. They appear innocent and safe, but they are complex chemical compounds with very serious effects on humans.

Exposure to pesticides produces a variety of symptoms. Symptoms may include headache, nausea, stomach cramps, diarrhea, chills, fever, fainting, and possibly paralysis and/or death. Some persons mistake pesticide poisoning for what they call the "summer flu."

Ag chemical exposure can lead to a variety of symptoms–including paralysis and/or death.

Learning Goals

- To understand that 14- and 15-year-old workers cannot use some agricultural chemicals
- To understand the warning signs and symbols used on agricultural pesticides

Related Task Sheets:

Hazardous Occupations Order in Agriculture	1.2.1
Worker Protection Standards	1.2.4
Personal Dress	2.7
Personal Protective Equipment	2.10
Lead Acid Batteries	4.6.2

Cooperation provided by The Ohio State University and National Safety Council.

KEEP OUT OF REACH OF CHILDREN

DANGER POISON

PELIGRO

Figure 3.5.a. The most toxic of chemicals will display the signal words "danger-poison," along with the skull and crossbones. Peligro is Spanish for danger. Young farm workers are prohibited from working with or being exposed to these chemicals.

Learn the signal words used on pesticide containers.

TOXIC

Figure 3.5.b. The word toxic means deadly. Toxic materials can produce illness-like effects, or they may be deadly.

Signal Words and Categories

Every chemical label must display signal words. These industry standard words tell the user the toxicity of the product. Toxicity means how deadly the product is to people.

Signal words found on agricultural chemicals include:

- Danger-Poison (skull and crossbones included)
- Danger
- Warning
- Caution

These words and symbols indicate the product's potential risk to the user.

Danger-Poison

Category I chemicals show the "Danger-Poison" signal word. A skull and crossbones is included on the label. These chemicals may be corrosive (can burn) to the eyes and skin and lungs. Less than a teaspoon of the chemical can kill a 150-pound person. Most of these chemicals are "restricted use" materials due to increased risk to human health and/or the environment. They require certification to purchase and use.

Danger

These Category I chemicals can cause severe skin irritation and eye damage.

Warning

Category II chemicals use the signal word "Warning." Skin and eye irritations that could last longer than one week can result from exposure to these products. A tablespoon of some Category II chemicals can be fatal. These pesticides are considered as restricted-use pesticides.

Caution

Chemical labels using the signal word "Caution" are much less toxic products to use. Mild skin and eye irritation results from exposure to these chemicals. Nearly one pint of the material would have to be swallowed to be fatal to a 150-pound person. Pesticides sold over-the-counter to consumers use the signal word Caution.

 Cooperation provided by The Ohio State University and National Safety Council.

Ag Pesticide Exposure

Exposure to agricultural chemicals is not necessarily a harmful event, but exposure over time can be harmful. Exposure can be minimized by wearing personal protective equipment (PPE).

The handling and application of pesticides is prohibited for youth younger than age 16.

Chemical exposure can occur in four ways:

- Oral (mouth)
- Dermal (skin)
- Inhalation (lungs)
- Ocular (eye)

Let us examine these more closely.

Oral Ingestion (by the mouth)

Pesticides can contaminate the hands through the handling of the container. Small amounts of the chemical may end up on cigarettes, chewing tobacco, food, or drinks touched by contaminated hands. Ingestion of pesticides through food is a common means of ingestion. Hands could also be an oral source of exposure.

Dermal (Skin) Exposure

Pesticides may be taken in through the skin. Even the act of urinating with pesticide-covered hands causes pesticide exposure. Some persons mistakenly think that tough, calloused hands reduce the entry of the pesticides through the skin. Even by wiping the sweaty forehead or the back of the neck, dermal exposure occurs to those more sensitive tissues.

Touching treated surfaces or handling empty containers may cause dermal exposure. Walking through a recently treated field can lead to dermal exposure to pesticides.

Inhalation (Breathing) Exposure

Breathing pesticide or agricultural chemical mists, vapors, or dusts exposes the lungs to the product. Exposure can occur while mixing granular and powder forms of pesticides and during the burning of empty containers. Inhalation exposure provides the fastest route of exposure into the bloodstream.

Ocular (Eye) Exposure

Splashing of liquid chemicals and dust from granular pesticides during handling, mixing or rinsing of containers is a source of risk to the eyes.

Pesticide labels provide specific requirements for the personal protective equipment (PPE) which will give maximum protection and reduce pesticide exposure. *PPE use does not make it legal for youth younger than age 16 to handle or apply pesticides.*

Improper handling of agricultural pesticides can result in the production of toxic fumes and vapors.

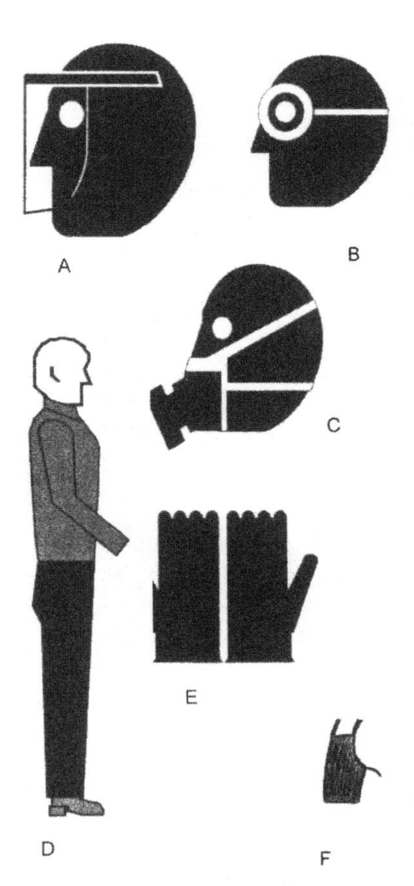

Figure 3.5.c. Face shields (A) and/or goggles (B), respirators (C), long sleeves and pants (D), chemical-resistant gloves (E) and aprons (F), should be used when handling pesticides, strong detergents, sanitizing chemicals, degreasers, and battery acid. Read the chemical label for the personal protective equipment (PPE) to use.

Safety Activities

1. Make an agricultural chemical inspection of a farm with the owner's permission. Make a list of all the chemicals that you find and the signal words that are included on the label. DO NOT HANDLE CONTAINERS WITH MATERIALS SPILLED OVER THE OUTSIDE OF THEM.

2. Solve this crossword puzzle.

Ag Chemicals

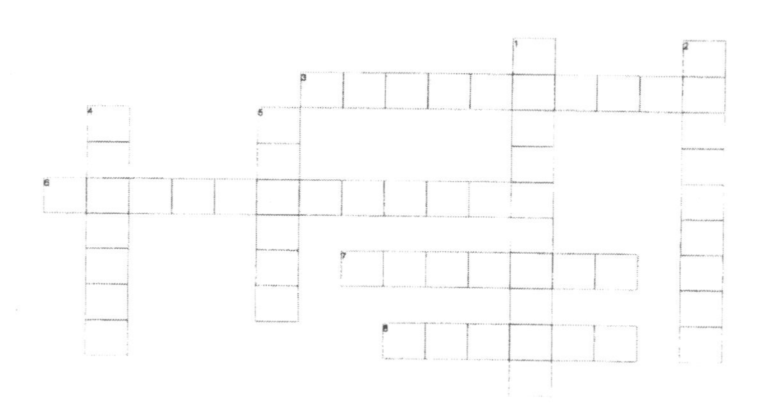

ACROSS

3. Breathing exposure hazard
6. protects the hands from chemicals
7. a signal word
8. exposure through the eyes

DOWN

1. protects face and eyes from chemicals
2. Oral intake of pesticides
4. Signal word for moderate toxicity
5. Pesticide exposure through skin

Use these words: Inhalation, face shield, ingestion, caution, warning, rubber gloves, dermal, ocular

References

1. Safety Management for Landscapers, Grounds-Care Businesses and Golf Courses, John Deere Publishing, 2001.
2. The Pennsylvania Pesticide Applicators Certification Core Manual, 2008 The College of Agricultural Sciences, Penn State University, University Park, PA.

Contact Information

National Safe Tractor and Machinery Operation Program
The Pennsylvania State University
Agricultural and Biological Engineering Department
246 Agricultural Engineering Building
University Park, PA 16802
Phone: 814-865-7685
Fax: 814-863-1031
Email: NSTMOP@psu.edu

Credits

Developed by WC Harshman, AM Yoder, JW Hilton and D J Murphy, The Pennsylvania State University. Reviewed by TL Bean and D Jepsen, The Ohio State University and S Steel, National Safety Council. Revised 3/2013

This material is based upon work supported by the National Institute of Food and Agriculture, U.S. Department of Agriculture, under Agreement Nos. 2001-41521-01263 and 2010-41521-20839. Any opinions, findings, conclusions, or recommendations expressed in this publication are those of the author(s) and do not necessarily reflect the view of the U.S. Department of Agriculture.

ELECTRICAL HAZARDS

HOSTA Task Sheet 3.6 **Core**

NATIONAL SAFE TRACTOR AND MACHINERY OPERATION PROGRAM

Introduction

Agriculture uses electricity as a tool. Jobs that were once labor intensive are now done with the help of electrical devices. The dairy industry uses compressors, vacuum pumps, refrigeration units, motors, and controls for all kinds of tasks. Grain producers use crop driers with fans and augers. Swine and poultry producers rely heavily on controlled ventilation and automatic feeding systems. There are many other examples.

This task sheet discusses the hazards posed by electricity. Beginning level farm workers will use many of the systems mentioned. Each year 30 to 40 persons are electrocuted on farms. Being safe with electricity is a work skill that must be mastered.

Electrical Hazards

Using electrical current and electrical equipment can lead to several hazards including electric shock, heat, and fire.

Electric Shock Hazard

When a person becomes part of an electric circuit, they are a conductor of the electrical current. Since electricity cannot be seen, the hazard is often overlooked until too late. Bodily injury and death can occur.

Current flowing through the body will affect the body in some manner. A slight tingling sensation may be felt. A shock may be felt which can result in muscular contractions that can "knock" the victim away from the circuit. Electric shock may "lock" the muscles to where release from the circuit is impossible. In severe cases, heart muscle rhythms are disrupted and death results.

Electrical current also produces heat which can burn body tissues both externally and internally.

Heat and Fire

Electricity can be the source of heat to ignite flammable materials. Current flow in a conductor produces heat because of the conductor's resistance to the flow of electricity. Increased heat in electrical conductors can be expected when:

- The wire size is too small to carry the current (trying to run an electric motor on a lamp cord)
- The electrical load is too great (operating a hair dryer, curling iron, and toaster on the same circuit)
- The electrical load is too far away from the electrical source (a 1/2 electric drill motor operated at the end of a 100- foot extension cord)
- The electrical connections are loose, and increased resistance develops

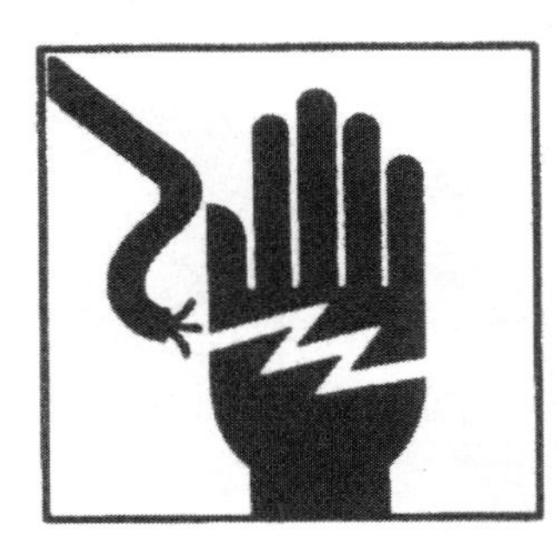

Figure 3.6.a. Contact with an electrical current can kill a person. The heart muscles will lock up if the electrical current passes through the heart when all the heart valves are closed (rest phase).

Electricity can kill. Be careful!

Learning Goals

- To understand electrical hazards
- To safely work with electrical equipment used in agriculture

Related Task Sheets:

Housekeeping	**2.6**
Fire Hazards	**3.6.2**

 Cooperation provided by The Ohio State University and National Safety Council.

Ground Fault Circuit Interrupters (GFCIs) are the best source of worker protection in damp areas.

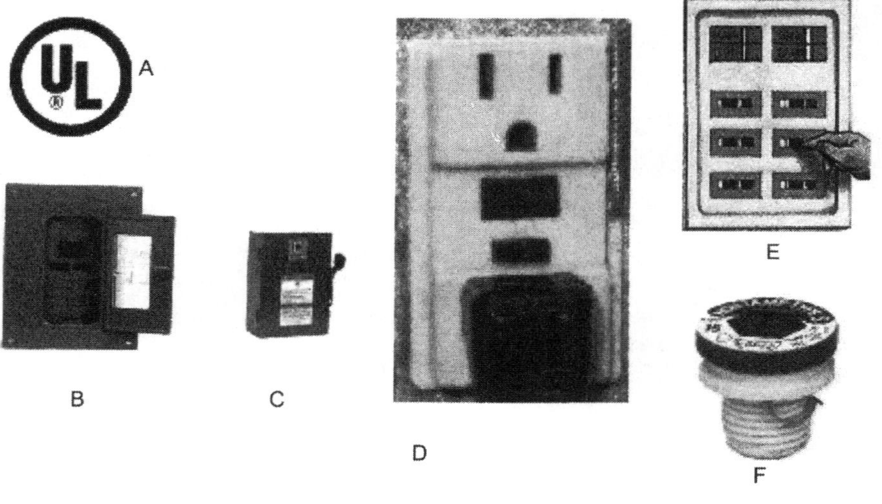

Figure 3.6.b. Electrical components are built to strict safety standards and tested for reliability by the Underwriters Laboratories. Look for the UL label (A) above. You may have to operate circuit breaker boxes (B,C) as switches for electrical tools. If the appliance suddenly stops working, then circuit breakers (E), fuses (F), or GFCIs, (D) may have broken the circuit to protect you and the wiring.

Electrical Devices You May Use

Work assignments on the farm may require use of electrical appliances and tools. The following describes the electrical equipment you may be called upon to use. *Note: A qualified electrician will be necessary to work with the electrical system beyond what is described here.*

Distribution Panel– The circuit breaker or fuse box contains many circuits. This is the location of circuit breaker devices to stop current flow to an electrical circuit. You may be assigned to go to the distribution panel (sometimes called circuit breaker panel or fuse box) to turn a circuit on or off.

Circuit Breakers and Fuses– These devices found in the distribution panel protect the wires of the circuit from overheating. Overloads cause fuses to “blow” and circuit breakers to “trip” to electrical flow. Three common protective devices are:

- Fuses
- Circuit breakers
- Ground fault circuit interrupters (GFCIs)

Fuses are either a screw-in or cartridge type. A metal strip melts when the circuit is overloaded and interrupts the circuit. The fuse must be replaced. Shut off the “main” power switch before changing fuses. See page 3 for more details.

Circuit breakers look like switches. When a bi-metal strip (two different metals) is heated from electrical overload, the metal becomes distorted in shape and causes the circuit breaker to cut out. The overload problem must be corrected and the switch returned to the on position. See page 3 for more details.

Ground Fault Circuit Interrupters can look like an electrical outlet or a circuit breaker. These GFCI devices break the circuit in microseconds when a difference in current is sensed. These devices are used where moisture is found. Milking parlors and milk rooms, swimming pools, kitchens, laundries, and outdoor receptacles should have GFCI protection. A red reset button and test light area make GFCI devices different than a regular outlet.

If fuses, circuit breakers, and GFCI devices are constantly “blowing,” ask your employer to check the situation before you continue.

Switches and Receptacles– Switches energize circuits. Receptacles connect the appliance to the circuit. Careless use can damage the receptacle and appliance. If you are assigned to a job where the electrical switch and/or receptacle is damaged, ask the employer to make the repairs.

Underwriters Laboratories

Electrical components must meet the Underwriters Laboratories (UL) standards. Look for the UL symbol to be sure that the device has approved safety construction. See Figure 3.6.b. above.

 Cooperation provided by The Ohio State University and National Safety Council.

Overhead Power Lines

Many overhead power lines do not have insulating covers. They normally carry high or higher current than building circuits. The person, or machine the person is moving or operating, becomes part of the electrical distribution grid. Contact with these wires can lead to a fatality.

Many deaths on farms are due to contact with overhead wires. Elevators, augers, metal ladders, and irrigation pipes must be moved. These objects are good conductors of electricity, and the operator is usually in direct contact with them through the tractor and implement. See Figure 3.6.c.

To prevent this hazard situation:

- Lower augers and elevators for transport.
- Take notice of low-hanging wires.
- Use a "spotter" while moving equipment under utility wires.

Recognizing Electrical Hazards

You do not have to be an electrician to be safe around electrical circuits. Use these ideas to be a valuable and safe employee.

Circuit Breakers and Fuses– If circuits are constantly breaking (shutting off), the circuit is overloaded. Tell your employer. Do not put foil or a copper penny in the fuse socket to eliminate the fuse. Even larger capacity fuses add to the dangers. A jumper wire to bypass the circuit breaker is not a good idea.

Grounding– Three-prong appliance plugs assure that the circuit is grounded. *Do not cut off the third prong (round prong) to make the plug fit.* A two-prong adapter with ground strap should be used.

Lock-outs– Distribution panels or fuse boxes, (Figure 3.6.b) can be fitted with a lock. Lock these boxes to prevent children and visitors from contacting the wiring inside of them. When working with an electrical circuit that is out of sight of the fuse box, lock the fuse box or controller so that another person does not accidentally energize the circuit while you are working.

*Hostile Farm Conditions–*Dust, moisture, corrosive materials, gases (manure), and physical damage is hard on electrical equipment. Report broken or damaged electrical equipment to your employer (Figure 3.6.d.).

Extension Cords– Many times extension cords are used to operate equipment. Use heavy-duty cords when using heavy-duty tools. Extension cords should not be used as permanent wiring. Do not jerk the extension cord from the wall receptacle by pulling on the cord. Be careful not to cut through the extension cord insulation. Report damaged extension cords immediately.

Underground Utilities– Phone, electrical, gas, satellite TV, and dog training wires may be buried. For public utility locations, call before digging. Check with www.digsafe.com, a national directory for the phone number in your state. The service is free.

Figure 3.6.c. Lower the hay elevator or grain auger to avoid contact with overhead power lines. *Farm and Ranch Safety Management, John Deere Publishing, 1994. Illustrations reproduced by permission. All rights reserved.*

Machinery contact with overhead wires causes many farm fatalities due to electrocution.

Figure 3.6.d. Improper electrical wiring methods and materials placed into a hostile farm environment can lead to fires and electrocution of people and livestock.

Safety Activities

1. With the permission of the farmer/owner, conduct a electrical safety survey of a farm in your area. Use this chart to complete the survey.

	Area to Inspect	How Many Found	Where Found
A.	Lock-out devices with locks attached	____	____
B.	Electric boxes or controls damaged by hostile farm conditions	____	____
C.	Low-Hanging Power Lines	____	____

2. Research the topic "stray voltage" to learn how a dairy cow can experience being electrically shocked in a barn setting.

3. Find out why a toaster wire heats up to toast our bread and an electric iron heats up to iron the wrinkles from our clothes. For help, access the website www.howstuffworks.com.

4. Research the topic "Ground Fault Circuit Interrupters" (GFCIs). How does this device work and where should the device be used?

References

1. Farm and Ranch Safety Management, John Deere Publishing, 2009. Illustrations reproduced by permission. All rights reserved.
2. www.howstuffworks.com/Type in search box,"how power distribution grids work."
3. OSHA Publication 3075, Controlling Electrical Hazards, 2002. (Available free via Internet order through OSHA.gov)

Contact Information

National Safe Tractor and Machinery Operation Program
The Pennsylvania State University
Agricultural and Biological Engineering Department
246 Agricultural Engineering Building
University Park, PA 16802
Phone: 814-865-7685
Fax: 814-863-1031
Email: NSTMOP@psu.edu

Credits

Developed by WC Harshman, AM Yoder, JW Hilton and D J Murphy, The Pennsylvania State University. Reviewed by TL Bean and D Jepsen, The Ohio State University and S Steel, National Safety Council. Revised 3/2013

This material is based upon work supported by the National Institute of Food and Agriculture, U.S. Department of Agriculture, under Agreement Nos. 2001-41521-01263 and 2010-41521-20839. Any opinions, findings, conclusions, or recommendations expressed in this publication are those of the author(s) and do not necessarily reflect the view of the U.S. Department of Agriculture.

FIRE SAFETY

HOSTA Task Sheet 3.7

NATIONAL SAFE TRACTOR AND MACHINERY OPERATION PROGRAM

Introduction

Fires are common in the home, in the shop, barn, or silo, and around farm machinery and automotive vehicles. Grease can catch fire in the kitchen or shop. Flammable materials can be ignited when welding or metal cutting is done nearby. Dust and crop debris can be ignited in or on machinery. Spontaneous combustion can occur in stored damp hay, with improperly stored silage, or in piles of oily rags. Electric circuits can overheat and cause fires.

Many people panic when a fire occurs. Panic is not necessary if you understand what causes fires, fire prevention, and fire extinguishing methods.

This task sheet provides information on fires in agricultural buildings and structures. Task Sheet 3.7.1 will discuss fire prevention and control.

Definitions

Auto-ignition: The situation where flammable materials stored near an open flame or where heat can build up results in a fire risk.

Combustible: The capacity to be burned makes a material combustible.

Flammable/Nonflammable: These terms are used interchangeably with the term "combustible."

Flash Point: A point at room temperature where a solvent will produce vapors in enough concentration to ignite when brought near a source of heat.

Kindling Point/Ignition Point: The lowest temperature at which a solid material will ignite and begin to burn when brought near a source of heat.

Spontaneous Combustion: The phenomenon in which a material unexpectedly bursts into flames without apparent cause. See Task Sheet 3.7.2.

Vapors: Vapors are the gas form of substances that are normally in the solid or liquid form.

Volatility: The tendency of a liquid to vaporize or evaporate into the air. Gasoline is volatile.

The Fire Triangle

The Fire Triangle: Three things are necessary for a fire to start and to continue to burn. They are: fuel, heat, and air. Fuels can be a variety of materials. See pages 2-4. Heat sources can be electrical, open flame, sparks, and chemical reactions. Oxygen is part of the chemistry that supports a fire. Without any one of these factors, a fire cannot exist.

Fires are classified according to the fuel that burns. A letter designation system is used. Categories of fire common to agriculture and rural residences are Class A, B, and C. See pages 2-4.

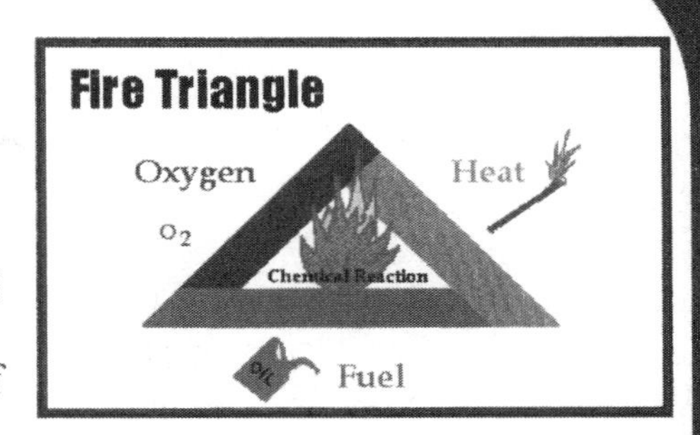

Figure 3.7.a To support a fire, three items must come together with a chemical reaction. Heat, fuel, and oxygen (air) must be available to support a fire.

Three things are needed for a fire to exist: fuel, heat, and air.

Learning Goals

- To understand the three factors which support a fire
- To understand the three classes of fire

Related Task Sheets:

Housekeeping	**2.6**
Agricultural Pesticides	**3.5**
Electrical Hazards	**3.6**
Fire Prevention and Control	**3.7.1**
Hay Storage Fires	**3.7.2**

Cooperation provided by The Ohio State University and National Safety Council.

Fires are classified by letters representing the fuel involved.

Class A

Class B

Class C

Figure 3.7.b. Fires are classified by letters representing the fuels which support them.
Class A fires involve wood, paper, rubbish, and plastic.
Class B fires involve burnable liquids like grease, oil, and fuels.
Class C fires involve electrical sources such as motors, wiring, switches, and connections.

Class A Fires

Class A fires involve wood, paper, rubbish, plastic, and crop materials. These fuels have a "kindling point" or "ignition point." Kindling point is the lowest temperature at which the substance will ignite and begin to burn. Small pieces of wood burn more quickly than a large fire log for example. A fireplace in a home must have a fire started with small pieces of kindling wood.

The kindling point of Class A materials varies with the material, its thickness, and moisture content. You cannot start a campfire with the largest fire log because it has a high kindling point and would need much more heat than a match could provide.

Dust from Class A materials can also burn quickly and violently. Dust has a low kindling point. At high levels of concentration, dust can even explode. Sparks from electric motors can cause the fire.

Dust explosion provides proof that smaller particles burn more quickly than larger particles. *Do you think that very fine, metal filings can burn also?* How could you prove whether or not this is true?

Figure 3.7.c. Dusty, dirty conditions in agriculture contribute to increased fire hazards. What materials in this picture are considered Class A fuel sources?

Class B Fires

Class B fires involve liquid materials which have the ability to produce vapors. These vapors can burn. When liquids give off enough vapors to burn, the fuel has a "flash point."

Three fuels can serve as examples of vapor-producing liquids. Gasoline is the most volatile liquid fuel and produces vapors which burn quickly and violently (low flash point). Diesel fuel and paint thinners produce less vapors (high flash point). Diesel fuel and paint thinners burn slowly when an open flame is placed directly near the fuel surface. Acetylene gas for welding and cutting is the product of a chemical reaction involving liquid elements producing gas. These vapors burn explosively.

Heavier Than Air or Lighter Than Air?

Some fuel vapors are heavier than air and settle to the lowest point nearby. Gasoline, propane, and diesel fuel serve as examples. Gasoline (the most volatile fuel) vapors are heavier than air and settle to a low point in a shop or enclosed space. Propane vapors are lighter than air and rise into the atmosphere. Diesel fuel is less volatile with the vapors being held near the surface of the fuel itself.

Precaution: When working on a vehicle inside of a shop or garage, do not permit gasoline to be spilled. Vapors may travel across the floor and be ignited by hot water tank pilot lights, welding sparks, and the sparks from a dropped, broken portable shop light.

Heavier Than Water or Lighter Than Water?

Class B liquid material fuels have weight or density. Some fuels may float on water, while others may sink beneath the surface. Gasoline and diesel fuel float on the surface of water, while grease sinks beneath the water. Fuel spilled on a body of water could be ignited and burned on top of the water.

Precaution: A major fuel spill on a farm pond or slow-moving stream should be reported to local fire officials immediately.

Vapors Concentrated in the Air:

As vapors of gaseous products gather in an enclosed space, they may be ignited by simply turning on a light switch. There is a momentary arcing of electrical current behind the light switch unless the switch is a snap action device. Acetylene gas leaking from a cylinder into a closed storage room can explode when the light switch is turned on. Acetylene tanks should be drained properly. Ask your employer about this hazard.

Hint: Smell the air in the acetylene storage area before "flipping" the light switch to turn on the lights. If it is safe, you should not be able to smell acetylene vapors. You can prevent an explosion by smelling the air first!

Because of the volatility of Class B fuels, auto-ignition may occur near open flames or in storage areas where heat can build up.

Figure 3.7.d. Vapors are produced by cleaning solvents, gas, and paints. When these vapors pass off into the atmosphere, they are called volatile.

Vapors and dust can be explosive.

Class C Fires

Class C fires involve electricity. These fires have electricity as the source of both fuel and heat. Motors, wiring, switches, and controls can overheat. The overheating is usually caused by an electrical overload. Electrical parts can catch fire. Nearby flammable objects can be ignited.

Electricity generates heat. Increased heat in electrical wiring can be expected when:

- The wire size is too small (trying to run an electric motor on a lamp cord)
- The electrical load is too great (operating a hair dryer, curling iron, and toaster on the same circuit)
- The electrical load is too far away from the electrical source (a 1/2 horsepower electric drill motor operated at the end of a 100-foot extension cord)
- The electrical connections are loose
- The electrical equipment is malfunctioning

Electrical equipment also can create sparks during its operation. Class A and B fires can be ignited by electrical overloads and sparking.

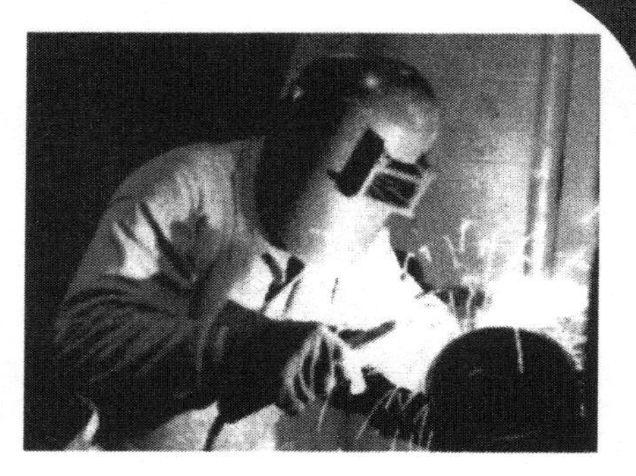

Figure 3.7.e. Welding sparks from electric arc welders and oxyacetylene gas welding equipment can create the spark that ignites nearby flammable materials or vapors.

Safety Activities

1. Review the fire safety lessons you learned in elementary school. What does Stop, Drop, and Roll mean?
2. Learn about the correct method of using a fire blanket. If you had to help someone who had caught on fire, would you know what to do?
3. Conduct a survey of a local farm to locate the placement, condition, and number of fire extinguishers on the tractors and other machinery and in the buildings. Make a report of your findings by making a chart or map.
4. Join a Junior Volunteer Fire Department.
5. Use the Internet to learn more about fire hazards and fire safety on farms. Type the phrase "fire safety" into any search engine.
6. In a safe location, secure a lighted candle so that it does not fall over. Lightly sprinkle fine metal shavings over the flame. Do the metal filings burn? Hint: The metal filings can be secured from a science teacher or by sweeping the area around a shop grinder. Do not use oily filings.

References

1. www.ask.com/Type fire safety in the search box.
2. Any Internet search engine/Type fire safety in the search box.
3. Farm and Ranch Safety Management, John Deere Publishing, 2009.

Contact Information

National Safe Tractor and Machinery Operation Program
The Pennsylvania State University
Agricultural and Biological Engineering Department
246 Agricultural Engineering Building
University Park, PA 16802
Phone: 814-865-7685
Fax: 814-863-1031
Email: NSTMOP@psu.edu

Credits

Developed by WC Harshman, AM Yoder, JW Hilton and D J Murphy, The Pennsylvania State University. Reviewed by TL Bean and D Jepsen, The Ohio State University and S Steel, National Safety Council. Revised 3/2013

This material is based upon work supported by the National Institute of Food and Agriculture, U.S. Department of Agriculture, under Agreement Nos. 2001-41521-01263 and 2010-41521-20839. Any opinions, findings, conclusions, or recommendations expressed in this publication are those of the author(s) and do not necessarily reflect the view of the U.S. Department of Agriculture.

FIRE PREVENTION AND CONTROL

HOSTA Task Sheet 3.7.1

NATIONAL SAFE TRACTOR AND MACHINERY OPERATION PROGRAM

Introduction

Understanding fires helps us to prevent and control them. Fires are often unexpected, but are usually predictable in their behavior. People, however, are unpredictable in their behavior with fire. People often panic when faced with a fire situation.

This task sheet discusses fire prevention and control as a means of helping the young agricultural worker deal calmly with unexpected fires. Task Sheet 3.7 describes the science of fire in detail.

Fire Prevention

A majority of fires can be prevented. Remember the science of the fire triangle? Fuel, air, and heat must react together for a fire to exist. Without any one of these factors, a fire is not possible A fire prevention program can be built around knowledge of the fire triangle.

Several steps will lead to a sound fire prevention program. Work-site analysis, maintenance, housekeeping, and fire prevention and control training are proven methods of reducing the risk of fire. Each of these items is discussed.

Work-site Analysis: Fire hazards should be surveyed at each farm. Combustible materials should be identified and stored properly. Fire extinguishers must be easily located and readily available. Fire extinguishers should be professionally inspected and/or recharged on an annual basis.

Maintenance and Housekeeping:

Equipment and facilities must be maintained and in working order. Regular maintenance schedules should be followed. For example, worn bearings on a motor shaft can overheat and ignite nearby flammable materials. A regular lubrication schedule can reduce that cause of fire. Good housekeeping helps prevent fires. Clean up oil-soaked rags to reduce the risk of sparks igniting the cloths.

Fire Prevention and Control Training: Everyone working on the farm must be a partner in the prevention and control of fires. All employees should have a job description which includes:

- Regular fire hazard inspection
- Training in fire extinguisher use
- Good housekeeping procedures

Each person is responsible to be knowledgeable in fire prevention and control procedures.

Figure 3.7.1.a. Sloppy housekeeping contributes to potential fire hazards in farm shops and other structures. What class of fire might exist in the area that this picture represents? Would it be Class A, B, or C? Review Task Sheet 3.7 for a clue.

People tend to panic in a fire. Such behavior causes further panic.

Learning Goals

- To be able to prevent fires
- To correctly select the proper fire extinguisher to use in a specific fire situation

Related Task Sheets:

Housekeeping	**2.6**
Ag Pesticides	**3.5**
Electrical Hazards	**3.6**
Fire Safety	**3.7**

Cooperation provided by The Ohio State University and National Safety Council.

Fire extinguishers should be regularly inspected for leakage!

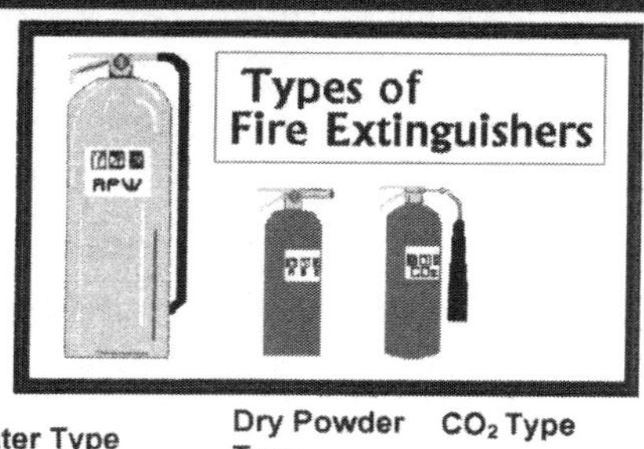

Water Type Dry Powder Type CO_2 Type

Figure 3.7.1.b. Fire extinguishers must be used for the class of fire for which they are rated.

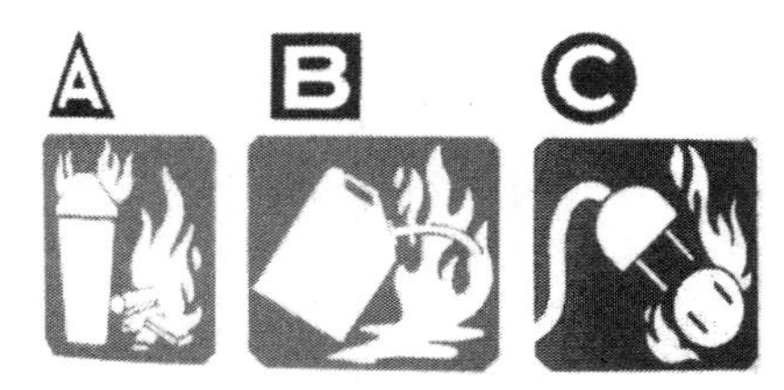

Figure 3.7.1.c. People do not have time to read directions in a fire emergency. Symbols attached to the fire extinguisher represent the type of fire. Can you identify the class of fire in each pictorial above? If you saw this pictorial on a fire extinguisher, what class or classes of fire would you be attempting to control?

Fire Extinguishers

Fire extinguishers are identified by a pictorial attached to the extinguisher body showing the type of fire for which they should be used. See Figure 3.7.1.c above. In an emergency, these standard graphics give us instant information about the extinguisher.

Water type extinguishers contain water under pressure. Use them for Class A fires only. Water spreads grease fires and conducts electricity. Water put on an electrical fire will conduct the electrical charge back to the user. Electrocution will result. The water cools the fire to extinguish it.

Water type extinguishers are made of stainless steel, have a pressure gauge, and long hose. See Figure 3.7.1.b above.

Chemical extinguishers contain a dry chemical powder. They can be used on class A, B, and C fires. The dry chemical suffocates the fire by eliminating the air. A small amount of material can extinguish an equipment or motor fire quickly. The dry chemical does leave a residue to clean up. The 10-pound, dry chemical extinguisher is recommended for use. See Figure 3.7.1.b above.

The dry chemical powder extinguisher is identified by its short, thick, red-colored container with a bright metal nozzle next to the pressure gauge.

Another chemical extinguisher is the Halon extinguisher. These extinguishers contain a gas that interrupts the chemical reaction that takes place when fuels burn. These types of extinguishers are often used to protect valuable electrical equipment since they leave no residue to clean up.

Carbon dioxide extinguishers contain CO_2 (carbon dioxide) gas. This extinguisher can be used on small class B and C fires. It leaves no residue. The pressurized CO_2 gas contacts the air and forms dry ice. The fire is cooled by the dry ice.

There are limits to the CO_2 extinguisher's use. Larger fires will require a greater capacity for control than what this extinguisher can provide. Also, the dry ice is so cold that it can burn the skin if a person touches the dry ice.

CO_2 extinguishers are identified by a red container with a larger black funnel-shaped nozzle which can pivot near the pressure gauge area. See Figure 3.7.1.b above.

DO NOT TREAT FIRE EXTINGUISHERS AS TOYS.

Squeezing the trigger to discharge the fire extinguisher just once will be enough to drain the pressure from the extinguisher. When it is actually needed, it will be worthless.

Using a Fire Extinguisher

To use a portable fire extinguisher, follow the steps called PASS. The steps include:

- Pull the pin
- Aim at the base of the fire
- Squeeze the trigger
- Sweep from side to side

Remember the acronym-PASS!

Important note: Always aim at the base of the fire. This is important for two reasons. First, a small fire extinguisher has limited material. It will be wasted aiming above the flame. Secondly, the fire extinguisher material will form a barrier above the fire. The flames can roll up under the barrier toward you.

See Figure 3.7.1.d for a graphic view of using a fire extinguisher.

Fire Preparedness

Being prepared to control a fire is different than prevention of fire hazards. There are a number of steps to take to be prepared for a fire emergency. Consider starting these practices in your home or place of employment.

- All family members/employees should be trained in fire prevention and control measures.
- Local fire company phone numbers should be accessible to all persons involved with the farm. Cell phones may be the best form of communication if phone lines are burned by fire.
- Written directions to the home or farm should be stored near each phone. In a panic, people commonly forget the simplest of directions or cannot state them clearly.
- Provide the local fire company with a detailed map of the farm including pesticide storage areas, fertilizer storage areas, manure pits and lagoons, and clean water pond sources. The fire company could have these on file, or they could be available in a weatherproof box at the farm lane.
- Install smoke alarms and carbon-monoxide detectors. Test the batteries regularly and replace them as needed.
- Schedule regular fire training and fire drills with the family and with the employees.
- Supply the correct fire extinguishers on all tractors.

Being prepared for a fire is good insurance that all persons involved will react in a focused and safe manner.

Figure 3.7.1.d. Review the PASS acronym as you look at these pictures. Can you repeat the steps to use a fire extinguisher?

Are you and your family and your employer fire prepared?

Figure 3.7.1.e. All tractors should have a dry, chemical- type fire extinguisher on board. Today's high-priced tractors and equipment should be fire control ready. What class(es) of fire will the dry chemical extinguisher control?

Safety Activities

1. What three factors make up the fire triangle?
2. Make a housekeeping inspection of the home shop, school shop, or a local farm shop to locate any hazards which could show a potential for fire. Make a list of those hazards. Ask for permission to eliminate the problem.
3. The kitchen stove catches fire while eggs are being fried. Should you throw water on the fire to control it? Why or why not?
4. How could you control a kitchen grease fire?
5. An electric motor is on fire. What fire extinguisher should you use and why?
6. Could a shovel full of soil be used to put out a small fire on the top of a farm machine? Explain your answer in terms of the fire triangle.
7. Does your computer room at school or at home have a Halon-type fire extinguisher available for use? Why is a Halon extinguisher a good idea in the computer area?
8. Recite the PASS process for using a fire extinguisher.
9. Conduct a survey of a local farm to determine how many fire extinguishers are found in the shop and on the tractors. Look for an inspection date. Are the extinguishers currently inspected?

References

1. Safety Management for Landscapers, Grounds-care Businesses and Golf Courses, John Deere Publishing, 2001.
2. Farm and Ranch Safety Management, John Deere Publishing, 2009.

Contact Information

National Safe Tractor and Machinery Operation Program
The Pennsylvania State University
Agricultural and Biological Engineering Department
246 Agricultural Engineering Building
University Park, PA 16802
Phone: 814-865-7685
Fax: 814-863-1031
Email: NSTMOP@psu.edu

Credits

Developed by WC Harshman, AM Yoder, JW Hilton and D J Murphy, The Pennsylvania State University. Reviewed by TL Bean and D Jepsen, The Ohio State University and S Steel, National Safety Council.
Revised 3/2013

This material is based upon work supported by the National Institute of Food and Agriculture, U.S. Department of Agriculture, under Agreement Nos. 2001-41521-01263 and 2010-41521-20839. Any opinions, findings, conclusions, or recommendations expressed in this publication are those of the author(s) and do not necessarily reflect the view of the U.S. Department of Agriculture.

HAY STORAGE FIRES

HOSTA Task Sheet 3.7.2

NATIONAL SAFE TRACTOR AND MACHINERY OPERATION PROGRAM

Introduction

Barn fires destroy property, stored crops, livestock, as well as cause a loss of revenue. Thousands of dollars can be lost as a result of barn fires. Investigations pinpoint many causes of these fires. Barn fires are a result of "spontaneous combustion," electrical malfunctions, poor housekeeping, and careless work habits.

Plant material (hay and straw) continues to respire (produce oxygen) for a short time after it is stored. Plant respiration and bacterial action creates heat as the plant oxygen is used up. Too much heat generated causes combustion.

This task sheet discusses recognizing hay fire risks and the proper handling of a hay crop as a means of preventing fires caused by spontaneous combustion.

The Chemistry of Hay Fires

Fresh cut forage crop cells continue to respire until the crop material dries or is cured. This chain of events occurring within the forage depends upon many factors. Moisture content is the most critical and is the only influence discussed from a fire safety standpoint.

Hay placed in storage should have a moisture content under 25%. Higher levels of moisture require an oxygen limiting storage system. The heat generated by the crop plus the presence of oxygen increases the risk of a fire.

Drying or curing of the forage takes several weeks, but the risk of fire in stored hay usually occurs within two to six weeks of storage. Stored hay of normal moisture levels undergoes some heating, but the heat is normally less than 125 degrees F. See Table 3.7.2.a. on page 2 of this task sheet.

Some hay growers apply chemical or biological additives and preservatives to the hay at harvest time to increase the rate of field drying or to bale and store the hay at higher moisture levels. The hay may still heat in storage.

Note: Stored cured hay can become damp due to a leaky barn roof, from ground moisture, or from high humidity and can still burn due to spontaneous combustion.

Figure 3.7.2.a. Hay storage fires result in devastating damage. The building, crop, and livestock losses can be tremendous, as well as the loss of income-producing facilities.

Store baled hay when the hay has less than 20% moisture.

Learning Goals

- To understand that improperly stored hay can ignite by spontaneous combustion
- To learn how to prevent hay storage fires
- To understand what to do if stored hay is getting too hot.

Related Task Sheets:

Fire Safety	**3.7**
Fire Prevention and Control	**3.7.1**

Cooperation provided by The Ohio State University and National Safety Council.

Table 3.7.2.a. Critical temperatures, conditions, and actions to take with hot hay according to the NRAES-18 publication. See the Reference Section of this task sheet.

Temperature	Condition and Action
125°F	No action needed.
150°F	Temperature will most likely continue to rise. Check temperature twice daily. Move hay to allow air circulation to cool the hay.
160°F	Check temperature every few hours. Move hay to allow air circulation to cool the hay.
175-190°F	Hot spots or fire pockets are likely. Alert fire service of a possible hay fire incident. Stop all air movement around the hay. Remove hot hay with assistance of fire service personnel.
200°F or above	Fire is present at or near the temperature probe. Inject water to cool hotspots before moving the hay. Fire service should be prepared for hay to burst into flame when contacting the air.

An experienced worker should monitor rising temperatures in hay storage, not a youth worker.

Hazards of Hay Fires

Three potential hazards exist from hay fires. They are:

- Sudden flareups of flame with exposure to fresh air
- Burned-out cavities in the hay that present a fall or entrapment hazard
- Toxic gases

Let us examine each of these in more detail.

Flareup of flame:

At temperatures between 150 and 170 degrees F the potential for spontaneous combustion of hay increases. Hay in this temperature range should be moved to allow for cooling. At the higher end of this temperature range, moving the hay exposes the heated material to oxygen and a sudden flareup can occur. Fire service officials should be notified if possible. Always have a charged water hose available.

Burned-out cavities in the hay:

Deep within the stored hay mass, temperatures may have reached levels where the hay has already burned. This burning has been a smoldering fire. Hollow cavities may have formed. These cavities can entrap a person who collapses the top of the hay pile by walking over it.

To prevent entrapment in burned-out cavities, place a wooden plank over the hay before walking over the area. A rope harness tied to a secure location is also recommended. Falls into a burned-out cavity may lead to broken bones, burns, and lung damage. Since the hay may have been chemically treated, a trained fire service person with a self-contained breathing apparatus (SCBA) should be called upon to provide the assistance needed in solving the potential fire problem.

Toxic gas exposure:

Smoldering and burning hay can be the source of toxic gases. Carbon monoxide can be concentrated within the smoldering fire and surrounding area. Chemically preserved hay crops may produce toxic gas vapors. Deadly gases add to the fire risk.

Crop preservative Material Safety Data Sheet (MSDS) information should be available to fire service personnel.

Note: The young farm worker should not be assigned to monitor temperatures of hay in storage. This poses an unnecessary risk to the inexperienced worker.

Monitoring Hot Hay

Smoldering hay gives off a strong, pungent odor. This odor indicates that a fire is occurring. At this point, stay off the hay, as a burned-out cavity may be found beneath where you would be walking.

The first reaction is to remove the heated hay. The temperature of the hay must be known before removal occurs. At lower temperatures, removing hay helps to move heat away from the hay by normal ventilation. When stored hay reaches 175 degrees F, any increased ventilation could result in rapid combustion.

Hay temperatures must be monitored. An experienced person should do this. Close coordination with a local fire service is of importance should the hay temperatures continue to rise.

Preventing Hay Fires

To prevent hay fires in storage areas, follow these approved practices to reduce the potential for forage crops to heat in storage.

Harvest Practices:

To reduce crop moisture levels rapidly, mow the forage early in the morning to allow one or more full days of drying time before baling. Storing dry hay reduces the risk of overheating.

Conditioning Practices:

Although it is difficult to achieve, the best weather conditions for hay curing is less than 50% relative humidity with some wind movement. Monitor the weather conditions and predictions to help schedule haymaking operations.

Hay mower conditioners, or crimpers, crush the forage stem and speeds the drying time of the crop. Windrow inverters, tedders, and hay rakes also speed the drying process. Each haying operation can shatter leaves from the stem and reduce the quality of the hay.

Chemical drying agents and preservatives may help to condition the forage crop. These materials can be used to speed up field drying rates. Most additives and preservatives increase the moisture level at which the forage can be safely preserved. Inoculant and acid-based preservatives increase the safe hay baling moisture levels to 25-30%. Spontaneous combustion ignition temperatures may be avoided when using these materials, but internal heating of the forage may cause heat-damaged protein. Heat-damaged protein reduces the nutritional value of the feed.

Baling Practices:

Bale the hay at 18-20% moisture to reduce the risk of conditions that support spontaneous combustion.

Storage Practices:

Store hay under cover to prevent rain damage and potential for heating. Leaky roofs and plumbing leaks can increase moisture levels of the stored forage to a point of reheating, which may lead to spontaneous combustion.

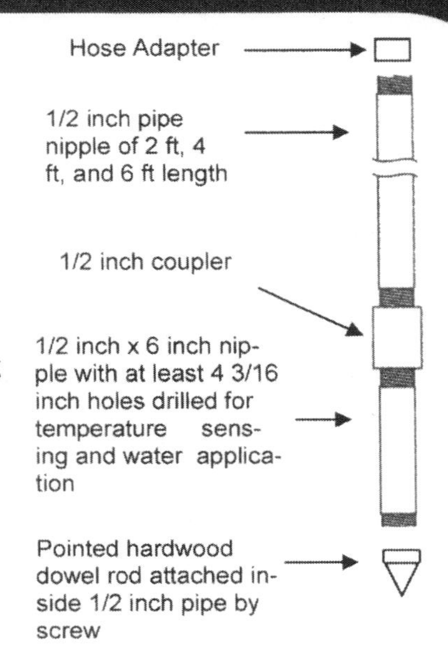

Figure 3.7.2.b. A homemade probe can be easily constructed.

Prevention of hay storage fires begins in the field with sound crop management.

Hay Loft—Top View

X X X

X X X

X X X

Figure 3.7.2.c. Use an organized pattern to monitor and record hay temperatures if overheating is suspected.

Safety Activities

1. Use a crop production reference to locate information about optimum moisture levels to harvest and store the major crops in your area. Make a chart to show what the moisture level should be for storage of those crops.

2. Contact your local agricultural chemical dealer to request brochures or labels for crop additives and preservatives. Write a report on these materials showing what they do.

3. Contact your local fire service personnel to ask about barn fires in your area. What were the causes? Were there hazardous chemicals involved? What special training do the fire service persons receive?

4. Develop a hay temperature monitoring kit of a probe, a thermometer and cord, and record sheet for use by farmers in your community.

5. Write a news release for your community farmers telling them about hay storage fire hazards.

6. Study silo fires, and write a report comparing a hay storage fire with a silo fire.

References

1. Extinguishing Fires in Silos and Hay Mows, NRAES-18, 2000, Cooperative Extension NRAES, 152 Riley-Robb Hall, Ithaca, NY 14853.
2. Visit www.cdc.gov/nasd/ Click on locate by topic/ Type in hay fires.

Contact Information

National Safe Tractor and Machinery Operation Program
The Pennsylvania State University
Agricultural and Biological Engineering Department
246 Agricultural Engineering Building
University Park, PA 16802
Phone: 814-865-7685
Fax: 814-863-1031
Email: NSTMOP@psu.edu

Credits

Developed by WC Harshman, AM Yoder, JW Hilton and D J Murphy, The Pennsylvania State University. Reviewed by TL Bean and D Jepsen, The Ohio State University and S Steel, National Safety Council. Revised 3/2013

This material is based upon work supported by the National Institute of Food and Agriculture, U.S. Department of Agriculture, under Agreement Nos. 2001-41521-01263 and 2010-41521-20839. Any opinions, findings, conclusions, or recommendations expressed in this publication are those of the author(s) and do not necessarily reflect the view of the U.S. Department of Agriculture.

CONFINED SPACES

HOSTA Task Sheet 3.8 **Core**

NATIONAL SAFE TRACTOR AND MACHINERY OPERATION PROGRAM

Introduction

Do you know what a confined space work area is? Farmers may think about silos, manure pits, and grain bins as the only confined spaces on their farms. Trenches, grain dryers, milk tanks, liquid manure spreaders, petroleum tanks, well shafts, and agricultural chemical tanks are other examples of confined spaces.

The Hazardous Occupations Order in Agriculture prohibits youth workers younger than age 16 from working inside confined spaces. See page 4 of this task sheet.

This task sheet discusses the hazards of confined space work areas. Young workers should not be assigned to work in these confined space areas.

Confined Space Definition

A confined space is defined by OSHA as:

- a space large enough and so configured that a person can enter and perform assigned work
- a space limited in openings for entry and exit purposes
- a space not intended for continuous human occupancy

Although specific standards for agricultural confined space work areas are not part of the OSHA regulations, the farm worksite contains confined space hazards for which every person associated with the farm should receive training. Silo, grain bin, manure storage, and farmstead chemicals are discussed in other task sheets in Section 3.

Think about the definition of these work areas. Have you been assigned to work in an area that meets the definition of a confined space?

- Do you have to enter an area to work by crawling, stooping, crouching or climbing into?
- Does the work area have an exit besides where you entered?
- Is there adequate, natural ventilation in the work space?
- Does that work space produce dangerous air contaminants as you do your work?
- Are there breathing hazards to be found in the confined space?
- Is the space capable of normal body movements for long time periods?

Youthful workers should discuss this type of work assignment with an adult before beginning the job.

Figure 3.8.a. A manure pit is a confined space work area. Many lives have been lost in manure pits due to toxic gases and lack of oxygen.

Youth younger than age 16 are prohibited from working in confined spaces.

Learning Goals

- To understand the hazards of confined space work areas

Related Task Sheets:

Cooperation provided by The Ohio State University and National Safety Council.

If you have to climb into a space to do work, you are confined. Know the risks.

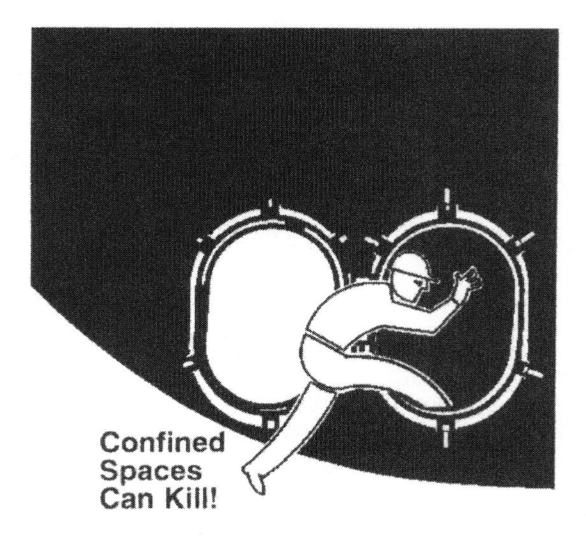

Figure 3.8.b. The OSHA Confined Space Entry Poster shows that confined space work areas pose a risk to the worker. You may not see this poster in farm confined space work areas, but remember the message shown.

Storage Tanks, Milk Tanks, and Oil Tanks

Some confined space work areas may appear to be safe for periodic inspection, cleanup, maintenance, or repair tasks. Storage tanks, milk tanks, and oil tanks may offer risk to health and safety. Consider these problems.

- an oxygen-deficient atmosphere
- a flammable atmosphere
- a toxic atmosphere

Oxygen-Deficient Atmosphere:

The air we breathe contains oxygen. At a minimum, the air should contain 19.5% oxygen. Oxygen levels may be normal when work begins inside a confined space, but the work being done can reduce the oxygen levels as the work proceeds.

Oxygen levels can be decreased by the presence of other gases and vapors. Welding inside a storage tank can deplete oxygen supplies. Cleaning rusty metal with a grinder will fill the atmosphere with particulates, which may reduce the available oxygen.

Toxic Atmosphere:

Depending upon the storage structure and its use, toxic material may be present when the worker enters the tank. The product stored in the tank may be toxic. Cleaning or scraping the tank can also release toxic chemicals.

The work being performed may cause chemical reactions. Cleaning a milk tank with degreasers and sanitizers must be done according to product directions. Some cleansing materials can harm the eyes and lungs if not handled properly. See Task Sheet 3.13.

Flammable Atmosphere:

Flammable materials can be gas, vapor, or dust in the proper mixture with oxygen. A source of ignition from welding or an electrical tool can ignite. An explosion inside the confined space can result.

Petroleum product storage tanks that must have repairs may contain highly flammable materials. These tanks may appear to be empty, but the residual vapors can be ignited. Vapors trapped in sludge-like material that must be scraped from tank walls are released and increase the risk of ignition.

It is recommended that welding on any storage tank should not begin until it is known what is inside the tank.

Working in Trenches

Trenches may be storage pits for silage or composting. The trench could be a ditch that is being dug for installation of electric utility or water lines. You may have been assigned to work in that trench. Is it a safe place to work?

Trench sidewalls can cave in and trap workers. Death by suffocation is possible. Trench cave-ins have trapped countless workers. Follow these safety plans for working in a trench.

- Do not enter a deep ditch that has sidewalls higher than your head unless it has steel retainer walls (trench box) to stabilize the trench.
- The trench should be cut so that "steps" or a sloping ramp are cut into the excavation to allow workers to exit easily.
- Use a hardhat and lifeline harness to protect yourself.
- While working in a trench, be within eyesight of another person who is not in the trench.

Reducing Confined Space Risks

Confined space work is usually done on a periodic basis rather than on a regular schedule. Safe work practices may not be remembered and repeated from one work period to another. To reduce the risks associated with working in a confined space, follow these approved practices.

Ventilation:

Ventilate confined space work areas before entering the area.

Isolate the confined space from entry:

Post signs at the confined space work area to warn of the hazard. Lockout/Tag out electric circuits to prevent start-up problems.

Test the Atmosphere:

If possible, monitor the atmosphere for oxygen deficiency. Most farms will not own this equipment, but fire service companies may have the equipment.

Self-Contained Breathing Apparatus:

Toxic atmosphere confined spaces should not be entered unless the worker is equipped with SCBA and has been trained in its use.

Safety Equipment:

Safety equipment needs are greater for confined space work. Respirators for a specific purpose are recommended. Hard hats and steel-toed shoes may be required. Communication equipment will be needed if direct contact with a helper cannot be made. Spark-proof tools will prevent ignition of flammable gases and dust. In addition, a safety harness and safety lines are advised.

Standby/Rescue:

Confined space work dictates that a helper or helpers must be available. Ladders, ropes, and lifts make immediate rescue possible. Do not work alone in confined spaces.

Figure 3.8.c. Trenches and pits in agriculture pose risk of sidewall cave-in and entrapment of the worker. Death by crushing can result. Suffocation can also occur.

Confined space work requires training and body harness equipment to be available.

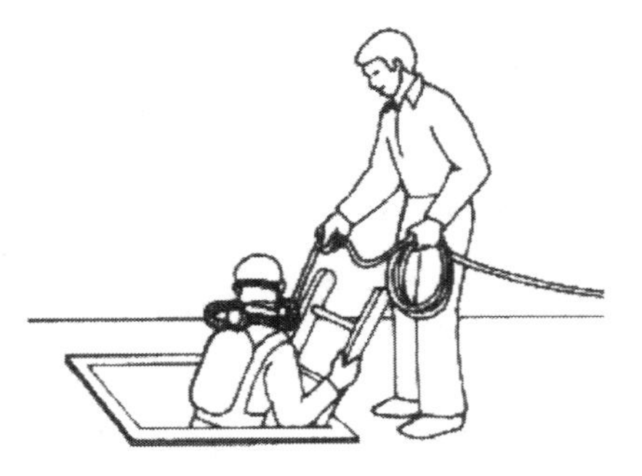

Figure 3.8.d. Confined space entry into oxygen-deficient areas requires SCBA. Training is necessary to use this equipment. Do not work alone in a confined space.

Cooperation provided by The Ohio State University and National Safety Council.

Hazardous Occupations Order in Agriculture Prohibitions

Some occupations in agriculture are considered to be particularly hazardous for the employment of youth younger than age 16. The Hazardous Occupations Order in Agriculture prohibits youth younger than age 16 from working inside the following areas. They include:

- fruit or grain storage designed to be oxygen-deficient or of a toxic atmosphere
- an upright silo within two weeks after silage has been added or when the unloading device is in operating position
- a manure pit
- a horizontal silo while operating a tractor for packing purposes

Other confined space work areas may be less well-defined. Many times a confined space work areas does not appear to be hazardous until an injury or fatality reminds us of the risks. Reread the section on page 3 of this task sheet about reducing confined space work risks.

Safety Activities

1. Interview a local fire service member to learn more about SCBA and its use.
2. Review the occupations that are considered hazardous for youth younger than age 16.
3. Visit OSHA's website (www.osha.gov), and search for the regulations regarding confined space work areas. Are there any points that farmers should consider for educating their employees and families? What are they?

References

1. www.cdc.gov/niosh/ Search for "A Guide to Safety in Confined Spaces," NIOSH Publication No 87-113, July 1987.
2. www.osha.gov/Type confined space in search box.

Contact Information

National Safe Tractor and Machinery Operation Program
The Pennsylvania State University
Agricultural and Biological Engineering Department
246 Agricultural Engineering Building
University Park, PA 16802
Phone: 814-865-7685
Fax: 814-863-1031
Email: NSTMOP@psu.edu

Credits

Developed by WC Harshman, AM Yoder, JW Hilton and D J Murphy, The Pennsylvania State University. Reviewed by TL Bean and D Jepsen, The Ohio State University and S Steel, National Safety Council. Revised 3/2013

This material is based upon work supported by the national Institute of Food and Agriculture, U.S. Department of Agriculture, under Agreement Nos. 2001-41521-01263 and 2010-41521-20839. Any opinions, findings, conclusions, or recommendations expressed in this publication are those of the author(s) and do not necessarily reflect the view of the U.S. Department of Agriculture.

SILOS

HOSTA Task Sheet 3.9 **Core**

NATIONAL SAFE TRACTOR AND MACHINERY OPERATION PROGRAM

Introduction

The farm silo serves the purpose of providing a storage space for finely chopped forages. These feeds ferment and become acidic. The low pH prevents bacteria from spoiling the silage.

Silos can be an upright tower or a trench, bunker, or stack or bag on the ground. Each has its own set of safety hazards. This task sheet discusses the safety considerations that a worker must understand when working with silos and ensiling of crops.

Silage Chemistry

Silage fermentation is the process of controlling bacterial actions that naturally break down the plant fibers of corn, hay, and other crops. Ideal silage is produced when silo oxygen is used up. Plant and bacterial respiration action will cause silage temperature to increase to 80-90 degrees F. During this stage, silage gas (see below) is produced. The silage becomes more acidic. This acid condition prevents further spoilage until oxygen enters the silo as the silage is fed.

Silo Gas

Silo gas is formed as the stored crop begins to ferment. Nitrogen dioxide and carbon dioxide are produced as the oxygen in the crop is depleted. During the first few days after filling the silo, the increase in these gases occurs.

The type of silo used determines which silo gas will predominate. In a sealed silo, carbon dioxide, an odorless, colorless, heavier-than-air gas is produced in large quantities. The carbon dioxide replaces the oxygen in the silo thus preventing the silage from spoiling. In a conventional silo nitrogen dioxide, a heavy, yellowish brown colored gas with a bleach-like odor is abundantly released. This heavier than air gas settles to low spots including feed rooms. Both of these gases cause death through asphyxiation (lack of oxygen).

Working Safely With the Chemistry of Silage

Understanding how silage is produced helps to prevent exposure to deadly silo gases. To prevent silage gas health problems, observe these precautions.

- Stay out of newly filled silos for at least two weeks. Use self contained breathing apparatus if the silo must be entered.
- Close the feed room door to the barn.
- If the silo must be entered, then:
    - Run the ventilation fan.
    - Get the help of an adult.
    - Wear a dust mask.

Figure 3.9.a. Silos may be 80-100 feet in height. A person overcome by silo gas will present a very difficult rescue problem.

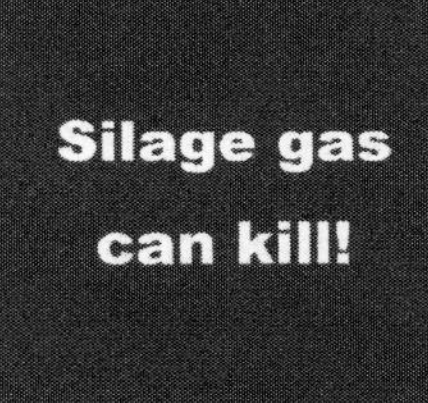

Learning Goals

- To understand how silo storage structures and equipment present hazards
- To develop safe work skills to use while working around silos

Related Task Sheets:

Cooperation provided by The Ohio State University and National Safety Council.

Shut down the tractor before attempting to unplug a silage blower or wagon.

Figure 3.9.b. While filling the silo, the work is done in close quarters. Two tractors are often involved with the PTO shafts operating the self-unloading wagon and the silage blower. Extreme caution is needed to do this work safely.

Working Safely with Silo Filling Equipment

Keep children away from silo filling operations.

Filling silo involves many tractors and many implements working together. Forage harvesters, self-unloading wagons, forage blowers, unloading platforms, bagging units, and silo distributors and augers are in constant use. The work area is crowded also. These machines are powered by PTOs or other moving shafts. An increased exposure to machine hazards occurs at silo filling time.

Silos produce the best silage when filled quickly and packed tightly. Much work occurs in a short time period. Corn silage harvest time can coincide with early fall and rainy weather. An increased need for safe work habits exists in changing work conditions. Let's examine each area that can pose a problem.

The Unloader

Before filling the conventional silo, the unloader must be raised by cable and pulleys to the top of the silo. One person at ground level can operate the electric control to do this job, but a second person observing the procedure from the blower pipe platform can signal if the cables become tangled. No one should be in the silo under the unloader as it is raised. Do not ride the unloader to the top of the silo in case the cables break.

Self-Unloading Wagons and Blowers

Self-unloading wagons contain moving aprons, beaters, conveyors or augers, and an assortment of chains and sprockets. PTO shafts are involved. The silage is moved by conveyor or auger to the PTO powered blower fan blades. These blades turn at high speeds to "blow" the silage to the top of the silo or further back into the trench. Silage bagging equipment also has numerous moving parts that pose risk.

Moving the moist, fine chopped crop can result in the equipment becoming plugged. Before attempting to unplug a clogged machine, follow these safety procedures.

- Disengage the power to the machine.
- Turn off the tractor engine.
- Wait for free-wheeling blower fan blades to come to a complete stop.
- Do not attempt to use your hands or feet to unplug a machine.

Falls

Falls account for a major source of injury to young agricultural workers.

Note: The Department of Labor Hazardous Occupations Order in Agriculture prohibits youth ages 14 and 15 from using a ladder higher than 20 feet from the ground.

Upright silos can be 80 to 100 foot tall. The silo's attached ladder may have a protective cage surrounding it. This cage offers some fall protection to the climber.

Trench or bunker silos often exceed 20 feet in height as well. Ladders may be placed against the silo walls for use when a plastic covering is installed.

Remember: Use three-point contact on the silo's ladder when climbing (two feet, one hand or two hands and one foot). Face the ladder while climbing. Stay inside the protective cage surrounding the silo's ladder.

Trenches, Bunkers, Stacks

Silos take many forms. Upright silos require expensive maintenance. Horizontal silos have capacity limited only by the location of the trench, bunker, or silage bag. Silage can even be stacked on a firm base. Each silo type has its own set of operation rules.

Horizontal silos like trenches, bunkers, and stacks must be packed tightly to exclude oxygen from the crop. Equipment rollover is a safety hazard as the silage pile is "packed." To avoid serious injury or death to the operator and to prevent costly equipment damage, use these practices.

- Use only tractors equipped with ROPS and seat belts.
- Use the seat belt when packing silage.
- Use low-clearance, wide front end tractors.
- Add weights to the front and back of the tractor to improve stability.
- Do not use wheel-type tractors on silage surfaces with slopes greater than 1 to 3 (1 foot of rise in 3 foot of run).
- Back up sloped silage surfaces, and drive down those areas.
- Distribute silage evenly in 6-inch layers for uniform packing.
- Front-wheel and assist-drive tractors provide extra traction and stability for packing and towing on silage.
- Mature, experienced operators should only be permitted to operate the packing tractor, unloading tractor, or forage wagon on the silage surface.

Trenches, bunkers, and stacks of silage are danger zones in crop harvest. Extra caution is needed to do this job safely and successfully.

Figure 3.9.c. When climbing silos, use the ladder, stay inside the enclosed protective cage, and maintain a three-point contact with the ladder.

Only mature, experienced operators should be assigned to pack silage in a horizontal silo.

Figure 3.9.d. Bunker silos must be packed tightly to limit the amount of oxygen available to bacterial and plant fermentation. Back and forth driving of a packing tractor can lead to tractor rollovers. Read the text on this page for safety pointers.

Safety Activities

1. Visit a local farm with upright silos to learn more about how the silo is loaded, unloaded, ventilated, and kept safe from youngsters or visitors. Develop warning signs that could advise operators or visitors about the dangers of the upright silo.

2. Visit a farm with horizontal silos to learn how they are filled, packed, and unloaded. Develop warning signs that could advise operators or visitors about the dangers of the horizontal storage areas.

3. Match the silo type with its description and related hazard.

_____A. Trench Silo	1. An upright silo with a roof and is accessible to workers. Presents silo gas hazard.
_____B. Bunker Silo	2. Can be a pit dug into the ground, which means an embankment collapse hazard is possible.
_____C. Stack Silo	3. Plastic wrapped silage where machinery operation by PTO is a safety risk.
_____D. Silo Bag	4. A horizontal silo with wooden timber or concrete sides. Packing this silo creates an increased risk of tractor rollover.
_____E. Oxygen-limiting silo	5. Tightly packed silage piled on the ground where the risk of tractor rollover is increased.
_____F. Tower silo	6. A lined, sealed steel or concrete silo with limited entry. Suffocation is likely if entered.

References

1. Farm and Ranch Safety Management, John Deere Publishing, 2009.
2. www.cdc.gov/nasd/ Type the keyword silo into the search box.

Contact Information

National Safe Tractor and Machinery Operation Program
The Pennsylvania State University
Agricultural and Biological Engineering Department
246 Agricultural Engineering Building
University Park, PA 16802
Phone: 814-865-7685
Fax: 814-863-1031
Email: NSTMOP@psu.edu

Credits

Developed by WC Harshman, AM Yoder, JW Hilton and D J Murphy, The Pennsylvania State University. Reviewed by TL Bean and D Jepsen, The Ohio State University and S Steel, National Safety Council. Revised 3/2013

This material is based upon work supported by the National Institute of Food and Agriculture, U.S. Department of Agriculture, under Agreement Nos. 2001-41521-01263 and 2010-41521-20839. Any opinions, findings, conclusions, or recommendations expressed in this publication are those of the author(s) and do not necessarily reflect the view of the U.S. Department of Agriculture.

HORIZONTAL SILO SAFETY

HOSTA Task Sheet 3.9.1

NATIONAL SAFE TRACTOR AND MACHINERY OPERATION PROGRAM

Introduction

Silage harvest time often becomes a hurry up time. Research shows filling the silo quickly and packing the forage tightly yields a higher quality silage. Increased opportunity for injury or fatality follows this haste. PTO and machinery entanglements, highway mishaps, dump truck incidents, silage baggers and horizontal silos have all played a role in injury and fatality during the busy silage harvest season. While corn is the most often harvested crop for silage, hay crops can also be placed into horizontal silos.

This task sheet discusses horizontal silo safety for those who may be required to work as part of a team during silo filling and feed-out.

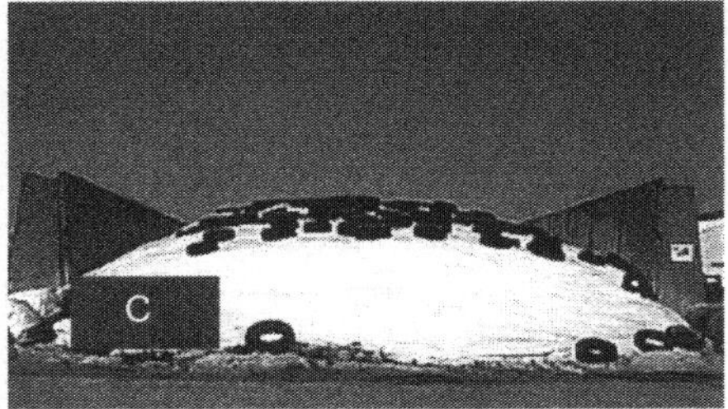

Figure 3.9.1.a. Three forms of horizontal silo are shown: a trench silo (A), a drive-over silage pile (B), and a concrete bunker silo (C). Horizontal silos can vary in height from only 4-5 feet high to over 40 feet.

Photo B, Courtesy of KJ Bolsen

Silo filling must be done rapidly; safe practices are necessary.

Horizontal silos

Horizontal silos vary in form. Whether it is a trench cut into the ground, forage piled on top of the soil, or a bunker-type silo with concrete or wooden sides, the process of filling and feeding from a horizontal silo is similar. Sizes of these trenches, bunks, and drive-over piles can vary according to the size of the livestock operation. Flexibility in expanding the storage area often makes the horizontal type storage more desirable compared to upright silos, but there are hazards involved in filling and feeding silage from horizontal silos.

Although wagons are often used to haul and dump forage, larger farms are increasingly using dump trucks for transport. As the forage depth and the side slope grows, the risk of tractor or dump truck overturn increases. As the forage depth becomes even greater, bunker sidewall capacity can be exceeded adding to the hazard potential. Dumping and packing equipment must operate away from these sidewalls to reduce stress on the structure, and to avoid overturning off the sidewall.

Learning Goals

- To understand the hazards involved with horizontal silos
- To safely fill horizontal silos
- To practice safe silage removal procedures

Related Task Sheets:

Silos	**3.9**
Packing Forage in a Horizontal Silo	**3.9.2**
Tractor Stability	**4.12**
Using the Tractor Safely	**4.13**
PTO	**5.4**
Dump Truck and Trailers	**6.5**
Silage Defacers	**6.9**

Cooperation provided by The Ohio State University and National Safety Council.

Harvest Pre-Inspection

Regardless of the style of horizontal silo, wear and tear occurs over time. Earthen trench sides can slip, earthen trench and drive-over pile site approaches can become muddy and rutted, and concrete or wooden-sided bunkers can become cracked over time. These should be repaired before use to maintain traction and stability for trucks and tractors.

Bunker silos should be equipped with iron pipe or steel sight rails. These sight rails give the operator a visual clue to the edge of the bunker while backing, unloading, and packing forage into the bunker. These rails are not intended to stop an overturning tractor or dump truck from toppling over the side of the bunker. They can also serve as fall protection for workers as they move around the sidewalls while working with plastic covering and weights used to seal the surface.

Figure 3.9.1.b. Bunker silo sidewalls are under great stress from silage weight and packing. A sidewall collapse could send a truck or tractor into an overturn. This situation must be repaired before harvest begins.

Figure 3.9.1.c. Sight rails offer a visual clue as to how close to the edge of the bunker you are operating the dump truck or packing tractor. These should be kept in good repair.

Fill and Pack Techniques

To help achieve the required amount of crop compaction a "progressive wedge" of forage is formed during the filling of the silo. The wedge provides a safe slope for the unloading and packing operations. A progressive wedge with a maximum slope of 1 to 3 minimizes the risk of roll-over. The resulting slope has a rise of 1 foot in every 3 feet of horizontal run. The operator must remember that while the surface is being packed it still can have ruts and soft areas that can lead to equipment roll-over. A rut or soft spot on the lower side of the truck or tractor can cause a sudden, unexpected shift of the vehicle to the side. There is seldom enough time to react to avoid this occurrence. Forage should be leveled before the next load is dumped and compacted.

Figure 3.9.1.d. Dump trucks are commonly used to fill horizontal silos. A truck equipped with an unloading, webbed-floor is more stable than a raised-bed truck. Operate the truck up and over the drive-over pile, not across the slope; and up then down on a trench fill silo.

Ideally, do not fill the trench or bunker higher than the sides of the retaining wall.

Figure 3.9.1.e. Use a ROPS equipped tractor fitted with dual wheels, extra weights, and a leveling blade to spread forages for packing. Be sure the seat belt is fastened when operating in a ROPS cab. Keep the pile level and rut-free before the next load arrives to be dumped and compacted.

Cooperation provided by The Ohio State University and National Safety Council.

Feeding silage safely

Work safely through the feed out process. Silage feed-out injury may occur from silage face collapsing due to undercutting, equipment roll-over, and entanglement. Observe where other workers and obstructions are located before beginning to work.

Silage face collapse may occur when equipment cannot reach the top of the feed-out face to remove an even amount of the feed. The silage that is removed from the bottom of the feed-out face allows heavy, unsecured silage from the top to break free. Workers, by-standers and even equipment operators can be buried beneath tons of silage. Numerous deaths have occurred with an avalanche of silage trapping persons and equipment. Using a silage defacer or equivalent accessory mounted on a material handler's boom to reach to the top of the silage face is a safe practice.

Equipment should not be operated from atop the silage. Edges of the feed face can be loosened allowing the silage face to collapse due to weight of equipment.

Using silage feeding equipment

Feed equipment to blend silage with other feedstuffs may be powered by a PTO shaft. The risk of entanglement in the turning shaft increases with use. PTO guards must be in place. Disengage the PTO and stop the tractor engine if adjustments or repairs must be made to feeding equipment. Never step across a turning PTO shaft for any reason.

Figure 3.9.1.f. Equipment that cannot reach to the top of the silage face will undercut the feed-out face at the bottom. A collapse of the silage can result. *Photo courtesy of KJ Bolsen*

Figure 3.9.1.g. A silage defacer mounted on a skid steer can reach the top of the silage face, but a material handler with boom may be needed to remove silage from higher reaches.

Other considerations

Silage can collapse. Fatalities have occurred while taking samples for nutritional analysis. Nutritionists, herds-persons, and students who must gather silage for forage quality evaluation should have an equipment operator scoop out and bring the silage sample to them. Avoid the feed-out face of the silage if it exceeds your own height.

Working near the top edge in a trench, bunker, or pile while removing the plastic cover or weights can cause the silage to collapse if it is weakened from undercutting. Freezing and thawing can weaken the face of the silage also. Do this job only if you can stay back away from the edge a safe distance.

When working around the silage face, use a fellow worker to assist should a collapse occur.

Remind children and by-standers of the dangers of machinery and silage face collapse.

Figure 3.9.1.h. This is not a safe place to be. What seems to be an activity in studying quality silage production or gathering feed samples could quickly become the scene of a tragedy should the silage feed-out face collapse.

Safety Activities

1. Conduct an Internet image search for horizontal silo topics. Without violating copyright laws use the pictures to make a scrapbook or poster display of the size and scope of corn or hay silage storage facilities. Label the pictures with important information found in this task sheet. For example: label the parts of the bunker silo, calculate and label the slope, or identify hazards.

2. Draw a map of your community and identify, with the help of your classmates or club members, where the silage trenches, bunkers, and drive-over piles are located. Estimate the total tonnage of silage that the largest bunker contains. Corn silage often weighs between 14 and 18 lbs/cubic foot as an average density.

3. Locate on the map from activity 2 where the local fire and rescue companies are to be found. How far from the furthest farm that has a horizontal silo are they?

4. With your class develop a 10-minute presentation about horizontal silo safety and the potential for silage face collapse and present the program to a local emergency rescue group or community farm or ranch group.

References

1. Horizontal Silo Safety, Fact Sheet E49, College of Agricultural Sciences, Department of Agricultural and Biological Engineering, Dennis J. Murphy and William C. Harshman

2. Farm Dump Truck and Trailer Safety, Fact Sheet E-44, College of Agricultural Sciences, Department of Agricultural and Biological Engineering, Dennis J. Murphy and William C. Harshman

Contact Information

National Safe Tractor and Machinery Operation Program
The Pennsylvania State University
Agricultural and Biological Engineering Department
246 Agricultural Engineering Building
University Park, PA 16802
Phone: 814-865-7685
Fax: 814-863-1031
Email: NSTMOP@psu.edu

Credits

Developed by WC Harshman, AM Yoder, JW Hilton and D J Murphy, The Pennsylvania State University and edited by KJ Bolsen, Professor Emeritus, Kansas State University
Version 7/2013

This material is based upon work supported by the National Institute of Food and Agriculture, U.S. Department of Agriculture, under Agreement No 2010-41521-20839. Any opinions, findings, conclusions, or recommendations expressed in this publication are those of the author(s) and do not necessarily reflect the view of the U.S. Department of Agriculture.

PACKING FORAGE IN A HORIZONTAL SILO

Hosta Task Sheet 3.9.2

NATIONAL SAFE TRACTOR AND MACHINERY OPERATION PROGRAM

Introduction

One task assigned to new employees may be packing freshly cut forage in a trench, bunker, or drive-over pile. The packing tractor is operated back and forth over the forage surface as the crop is harvested and dumped on a progressively, deeper wedge-shaped surface. As the forage depth grows and the sides become steeper, risks increase for a tractor roll-over.

This task sheet discusses forage packing safety.

Figure 3.9.2.a. Rapid and thorough packing of forage results in a high quality ensiled crop which when fed to livestock produces top yields of milk or meat. The packing tractor must be stable enough to operate on the forage surface.

The basics of top quality silage

Chopped forage must be packed tightly in trenches, bunkers, or drive-over piles. Rapid filling and packing limits air (oxygen) from the forage mass. Excessive air leads to a loss of plant sugars and undesirable fermentation by-products and potential for spoilage which limits the animal's consumption and utilization of the silage.

Figure 3.9.2.b. Steep sides should never be allowed to develop as the forage surface increases in depth. Steep slopes contribute to tractor rollover potential.

Tractor stability while packing forage may be a matter of life and death.

The packing equipment must be heavy enough (Figure 3.9.2.a) to achieve a dense pack. Heavier tractors equipped with dual wheels and a leveling blade attachment offer greater stability on the forage surface than smaller equipment. The ROPS enclosed cab and use of the seat belt offers a "zone of protection" to the operator. Smaller, older model tractors with single rear tires do not pack the chopped forage as densely and may cause ruts in the forage surface. The ruts in the surface coupled with the ever increasing side slope, (Figure 3.9.2.a and 3.9.2.b) increases the risk of rollover. Older tractors may not be fitted with a ROPS cab which further places the operator in danger.

Learning Goals

- To develop safe operating skills when packing forage in a horizontal silo.

Related Task Sheets:

Silos	3.9
Tractor Hazards	4.2
Tractor Stability	4.12

Cooperation provided by The Ohio State University and National Safety Council.

Safely packing forage in a horizontal silo

Since harvesting forage and filling silos must be done quickly it is important to do the job safely.

- Use a ROPS equipped tractor and fasten the seat belt.
- Recognize traffic flow of all vehicles involved in the operation.(Figure 3.9.2.c)
- No person on foot should be on the packing surface while equipment is operating.
- Develop a forage surface with side slopes not to exceed a 1 to 3 (1:3) slope. This means no more than a 1 foot rise in 3 feet of run.
- Do not exceed the side wall height of the bunker.

Figure 3.9.2.c. Coordination of dumping of forage with the packing operation is necessary. Movement of several trucks and more than one packing tractor means attention to their location is necessary. *Photo courtesy of KJ Bolsen*

- Sighting rails at the top of the bunker (Figure 3.9.2.d) provide a guide to help the operator stay back from the side.
- Do not let ruts develop. Ruts may throw the tractor to the side and cause an overturn.
- Dump wagons and dump trucks must be kept as level as possible when unloading. chopped forage.

Figure 3.9.12.d. Forage piles should not exceed the side wall height of the bunker. Sighting rails provide a guide to the operator. Do not operate the packing tractor close to the steep sides.

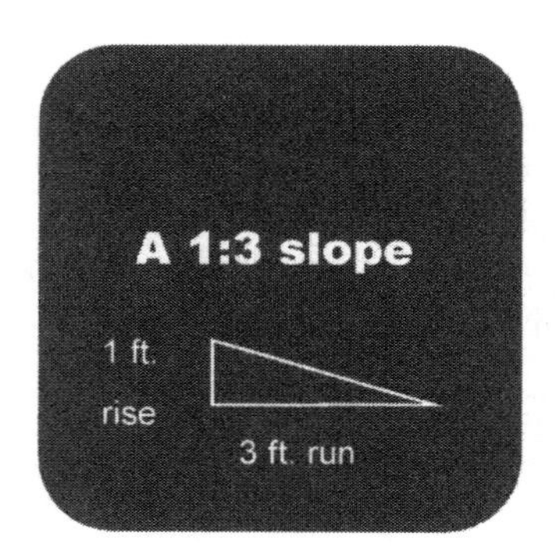

Safety Activities

1. Ask your instructor or club leader to schedule a field trip to observe the filling and packing of forage in a horizontal silo. If the field trip is not possible search the Internet using YouTube or a similar source to see if you can see a horizontal silo being filled.
2. Solve this problem. A corn grower has 1600 acres of corn silage to harvest. Yield is 24 tons per acre The average forage wagon hauls 7 tons and the average dump-equipped truck can haul 12 tons. How many loads must be hauled if only using the forage wagon? If only using the dump-equipped truck?
3. If a silage depth is 40 feet to the top-center, and a safe side slope of 1:3 ratio is recommended, how far out from the center does the silage extend? This is the distance for one side only so double the results to get the overall width of the storage area.

References

1. Horizontal Silo Safety, Fact Sheet E49, College of Agricultural Sciences, Department of Agricultural and Biological Engineering, Dennis J. Murphy and William C. Harshman
2. Internet search. Using any search engine, type in horizontal silage safety.

Contact Information

National Safe Tractor and Machinery Operation Program
The Pennsylvania State University
Agricultural and Biological Engineering Department
246 Agricultural Engineering Building
University Park, PA 16802
Phone: 814-865-7685
Fax: 814-863-1031
Email: NSTMOP@psu.edu

Credits

Developed by WC Harshman, AM Yoder, JW Hilton and D J Murphy, The Pennsylvania State University. Edits by KJ Bolsen, Professor Emeritus, Kansas State University.
Version 3/2013

This material is based upon work supported by the National Institute of Food and Agriculture, U.S. Department of Agriculture, under Agreement No.2010-41521-20839. Any opinions, findings, conclusions, or recommendations expressed in this publication are those of the author(s) and do not necessarily reflect the view of the U.S. Department of Agriculture.

GRAIN BINS

HOSTA Task Sheet 3.10 **Core**

NATIONAL SAFE TRACTOR AND MACHINERY OPERATION PROGRAM

Introduction

Unloading grain from storage bins and wagons exposes workers to the risk of being pulled into the flow of the grain and becoming entrapped. Moldy, damp grain creates a flow problem, often leading workers toward unseen hazards. Children playing in and around grain storage areas are often victims. Flowing grain entrapments cause an average of 12 deaths each year.

This task sheet discusses the hazards of flowing grain in storage bins, wagons, and trucks.

Flowing Grain

Grain harvest produces huge amounts of material to transport and store. Fortunately many labor-saving devices have been developed to make grain handling fast and efficient. Augers move grain rapidly. Gravity flow wagons and trucks make grain movement efficient. Flowing grain has many hazards that may go unnoticed.

Augers move grain from the bottom center of storage bins to the outer edge of the bin and into grain hauling vehicles or other storage bins. When the auger is running, grain flows out of the bin from directly above the outlet of the unloading auger in the center of the bin floor. A funnel-shaped flow on the top of the grain occurs with the grain flowing in a column below the surface toward the outlet (Figure 3.10.a.).This flow is like a moving conveyor belt or escalator.

With a large auger, a worker inside the bin can be pulled knee deep into the column of grain within a few seconds. Once your knees are covered by grain, it is almost impossible to free yourself without the assistance of others. If the knees are covered and the grain is still flowing, the flowing grain is similar to quicksand and can completely engulf a person very quickly. Figure 3.10.b and c. illustrate just how quickly a person will sink into flowing grain.

Note: Gravity unloading wagons have similar grain-flow patterns as grain bins. The grain flows in funnel-shaped form with a column of grain moving toward the unloading door of the wagon or truck.

Figure 3.10.a. The normal flow of grain from a bin is off the top and down a center column of grain flowing toward the unloading auger. The unloading auger is found at the bottom center of the grain bin.

Figure 3.10.b. Grain flowing out of storage causes a downward moving floor to move away from your feet. The victim is pulled waist deep in about 10 seconds. See Figure 3.10.c.

A 10-inch auger can move 85 cubic feet or 65 bushels of grain per minute.

Learning Goals

- To understand that flowing grain can be a deadly hazard
- To understand how to prevent flowing grain hazards while working with bins, wagons, and trucks

Related Task Sheets:

Common Respiratory Hazards	3.3
Respiratory Protection	3.3.1
Confined Spaces	3.8

Cooperation provided by The Ohio State University and National Safety Council.

Within seconds, a person can be helplessly trapped in flowing grain.

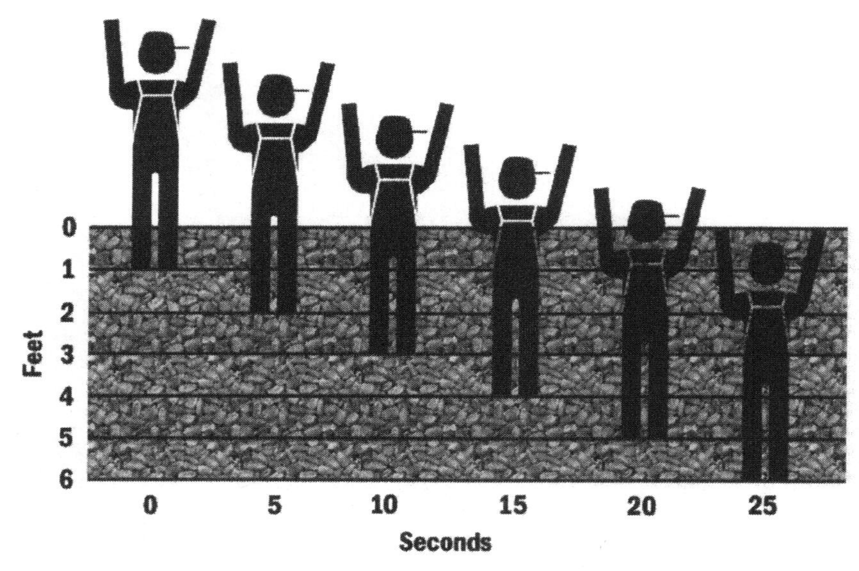

Figure 3.10.c. In a matter of a few seconds, a person standing in the grain bin can be helplessly trapped as the grain begins to flow. A person can be completely engulfed in the grain in about 25 seconds. Death from suffocation most often results.

Grain Bridging

Grain that is harvested before it has dried down adequately is damp and can mold quickly. This damp, moldy grain clumps together and hardens into a crusty mass. It gives the appearance of being a solid walking surface. This situation is often not recognized as a potential hazard.

Figure 3.10.d. A "grain bridge" cannot support the weight of the worker.

As poorly conditioned grain is unloaded from the bin, a cavity may develop. See Figure 3.10.d. Often the worker recognizes that the grain has stopped flowing but the bin appears full. The temptation is to enter the bin to break up the grain bridge. The "grain bridge" gives way as the worker walks over it (Figure 3.10.e), and the person is pulled into the flowing grain. Figure 3.10.c and d show the hazards of walking over the grain bridge.

Figure 3.10.e. As the grain bridge gives way, the worker is pulled into the pocket and is engulfed. The gain auger may have been left running and the flowing grain pulls the victim under the grain.

Wall of Grain Avalanches

In some cases, moldy grain will be found sticking to the walls of the bin. After removing the loose grain, the worker may be faced with a wall of crusted grain that must be broken free before it can be unloaded. If the wall of grain is higher than the height of the worker when the worker stands on the grain bin floor, an avalanche may occur as the worker tries to break up the crusted wall of grain. This avalanche could completely engulf the worker leading to injury and possible death (Figure 3.10.f). One foot of grain covering the engulfed worker would weigh approximately 300 pounds. This is normally too much weight for individuals to move to free themselves.

Preventing Flowing Grain Entrapment

The following steps can reduce the risk of flowing grain entrapment in storage bins, wagons, and trucks. These practices can save your life.

- Place entrapment warning decals on grain bins and grain transport vehicles.
- Prevent unauthorized entry to grain bins and grain transport vehicles, especially by children.
- Make sure all workers and children are aware of entrapment hazards.
- Keep grain in proper condition. This may include the use of mechanical stirrers to prevent the grain from molding. Out-of-condition grain is considered the leading cause of adult entrapments.
- Use inspection holes or grain bin level markers instead of entering a grain bin.
- Enter a grain bin or grain transport vehicle only if it is absolutely necessary. Use a body harness secured to the outside of the bin or vehicle.
- Use a pole to break up possible grain bridges from outside the bin.
- Lockout/tagout all power controls before entering a bin.
- Have at least two observers present during grain bin entry.
- Establish a form of nonverbal communication with observers (hand signals).
- Work from top to bottom when cleaning grain bin walls.

Special Notes:

Small children do not understand the hazards of agricultural work. Grain brought from the field to the farmstead has play appeal. Machinery that is moving grain draws their attention. The chances of a child being entrapped in flowing grain are very high. Most children do not survive grain storage entrapments.

Rescuing victims of grain bin entrapments calls for special tools and expertise from your local EMS groups (Figure 3.10.g.).

Figure 3.10.f. Damp, moldy grain can stick to the side of the grain bin. It can collapse on the worker who tries to dislodge it.

A 12-inch layer of corn covering a victim can weigh as much as 300 pounds.

Figure 3.10.g. It takes much force to remove a grain bin entrapment victim. Rather than removing the victim, it is easier to remove the grain. Special tools and skills are needed to cut through grain bins and remove the grain.

Special Note:

Grain vacuum equipment is becoming popular. The vacuum can quickly move grain from trucks to bins or can be used in more remote locations to empty wagons onto trucks. These vacuums can be moved over top of the grain in a side-to-side sweeping motion, and can remove thousands of bushels per hour. Hold the vacuum at an angle away from your body. If held close to the body, grain can rapidly be removed from under the operator's feet quickly pulling the operator down into the grain, possibly entrapping the person in the grain.

Fig.3.10.h The grain vacuum moves large quantities of grain from storage to truck, or from truck to grain bin.

Safety Activities

1. Arrange to visit a farm to observe grain being unloaded. Make a list of the hazards that can be found in this farm job.
2. Place a small doll in a grain-filled gravity unload wagon (above the grain unload door and on top of the grain). Open the unload door and describe what happens.
3. Use the Internet to search the Land Grant University College of Agriculture in your state to find information about grain moisture levels considered safe for preventing moldy grain. Fill in the blanks in the following chart.

Grain	Moisture Level Recommended for Safe Storage
Ear Corn	________________% Moisture
Shelled Corn	________________% Moisture
Wheat	________________% Moisture
Barley	________________% Moisture
Oats	________________% Moisture

References

1. Farm and Ranch Safety Management, John Deere Publishing, 2009.
2. Hazards of Flowing Grain, Task Sheet E43, Aaron M. Yoder, Dennis J. Murphy, and James W. Hilton, 2003, Agriculture and Biological Engineering, Penn State University, University Park, PA.

Contact Information

National Safe Tractor and Machinery Operation Program
The Pennsylvania State University
Agricultural and Biological Engineering Department
246 Agricultural Engineering Building
University Park, PA 16802
Phone: 814-865-7685
Fax: 814-863-1031
Email: NSTMOP@psu.edu

Credits

Developed by WC Harshman, AM Yoder, JW Hilton and D J Murphy, The Pennsylvania State University. Reviewed by TL Bean and D Jepsen, The Ohio State University and S Steel, National Safety Council. Revised 3/2013

This material is based upon work supported by the National Institute of Food and Agriculture, U.S. Department of Agriculture, under Agreement Nos. 2001-41521-01263 and 2010-41521-20839. Any opinions, findings, conclusions, or recommendations expressed in this publication are those of the author(s) and do not necessarily reflect the view of the U.S. Department of Agriculture.

MANURE STORAGE

HOSTA Task Sheet 3.11 **Core**

NATIONAL SAFE TRACTOR AND MACHINERY OPERATION PROGRAM

Introduction

The manure pit is full. It must be agitated and spread on the field. It is a routine in animal agriculture which must be done over and over again. The daily caution with machine hazards is coupled with exposure to manure gases.

Farm work exposes the worker to a variety of sights, sounds, and odors. Some of the odors, such as manure, are more than the strong smell. Some odors come from hazardous gases, which can also be harmful to us.

This task sheet discusses manure storage and the hazardous gases stored manure produces. Knowledge of manure gases is an important subject for those persons working in animal agriculture.

Manure Storage

Manure storage structures vary in size and type. The farm's animal numbers, the length of storage time needed, and the soil structure where the storage is built will influence what type of manure storage is used. Modern animal agricultural practices and environmental laws also make storage and management of manure a normal farming routine.

Manure storage is considered a confined space work area (Task Sheet 3.8).

Aboveground Storage

Manure sheds and aboveground storage tanks are used to store manure in many areas. The shed may have a roof covering and have open sides. Manure tanks are often open-top, silo-type structures. Semi-solid manure may be removed from sheds by tractor high-lifts. Liquid manure in tanks must be agitated and pumped to manure spreaders. In some cases, liquid manure is removed from storage by way of irrigation systems.

Belowground Storage

Manure storage pits may be separate structures from the barn or below the barn itself. Some manure pits are open. Manure is scraped into the pit. Other manure pits have slotted floors and storage lids or caps for covers. Animal foot traffic and gravity fill the pit. Pump-out pits are usually of smaller capacity, serve as temporary storage structures, and are pumped to larger storage structures.

Manure storage pits directly beneath animals, pits under the farm building, and closed or covered pump-out pits pose the most risk of manure storage gas hazards. Fatalities to humans and to livestock have been documented.

While odor may be a tell-tale sign indicating the presence of manure gas, several toxic gases are odorless and colorless when

Figure 3.11.a. Manure storage pits are often found below ground level. They are covered until time to pump them out. The agitation and pumping releases toxic gases.

A manure pit lacks the oxygen needed to keep you alive.

Learning Goals

- To understand the hazards of liquid and semi-solid manure storage

Related Task Sheets:

First Aid and Rescue	**2.11**
Respiratory Hazards	**3.3**
Respiratory Protection	**3.3.1**
Confined Spaces	**3.8**

Cooperation provided by The Ohio State University and National Safety Council.

Manure gases can be so concentrated that they can suffocate you instantly!

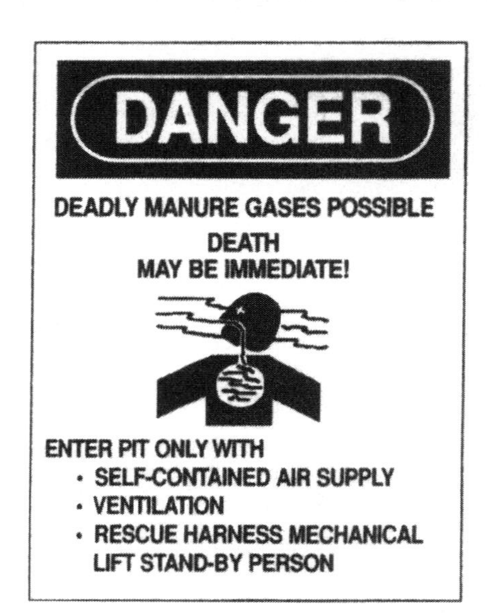

Figure 3.11.b. A danger sign posted near a manure storage structure provides a clear warning that immediate death is possible from manure gases.

Manure Gases

Manure is the product of digestion. Undigested feed materials, body cells and tissues, and minerals pass through the animal and are excreted. This material is in the beginning stages of decomposition, rot or fermentation. Fermentation or the rotting process produces manure gases.

Manure gases are poisonous. Low-level exposure produces lung and eye irritations, dizziness, drowsiness, and headaches. Additionally, some manure gases are heavier than air and deplete or displace the oxygen in the storage area. High levels of manure gases can quickly render a person unconscious. Death from suffocation can occur.

Four hazardous gases can be found in stored manure. They are:

- Hydrogen sulfide
- Ammonia
- Carbon dioxide
- Methane

Each of these gases is discussed further.

Hydrogen Sulfide - Hydrogen sulfide has a foul odor similar to rotten eggs. It is rapidly released from agitated manure. It can cause headache, dizziness, and nausea in as low a concentration as 0.5%. At a concentration of 1% in the atmosphere, hydrogen sulfide can cause death. It is heavier than air and settles to the lower level of the manure storage or on top of the manure level.

Ammonia - Ammonia is a colorless, pungent gas with a bleach-like odor. It is soluble in water and irritates the eyes, nostrils, lungs, and throat. The burning effect on the eyes and nose is reduced with breathing fresh air. It is lighter than air and rises out of the storage area rapidly.

Carbon Dioxide - Carbon dioxide is an odorless and colorless gas. It exists in low levels in the air we breathe, but in high concentration causes difficult breathing, headaches, and even death. It is heavier than air and concentrates in low areas of the storage.

Methane - Methane is a nontoxic, colorless, odorless gas. This gas is lighter than air and rises from storage areas. Headaches may be experienced in methane concentrations of 50% of the atmosphere. Methane in manure gas is just as explosive as the methane gas found in a coal mine.

All of these gases are released into the atmosphere when the manure is agitated and pumped prior to spreading. The gases can also remain in the manure pit or tank even after the manure is removed.

Manure Gases Can Kill

A 31-year-old male dairy farmer and his 33-year-old brother died after entering a 25 square foot, 4 1/2 feet deep manure pit inside a building on their farm. A pump intake pipe in the pit had clogged, and the farmer descended into the pit to clear the obstruction. While in the pit, he was overcome and collapsed. The victim's brother was standing at the entrance of the pit and apparently saw the victim collapse. He entered the pit in an attempt to rescue him. The brother was overcome and collapsed inside the pit. Four hours later, another family member discovered the two victims inside the pit and called the local fire department to rescue them. The victims were pronounced dead at the scene by the coroner. The coroner's report attributed the cause of death in both cases to methane asphyxiation.

See the NIOSH reference for other case examples.

Manure Storage Precautions

Safe work practices can be applied to manure storage areas. The following approved practices will reduce the risk of exposure to deadly manure gases and drowning hazards. They include:

- Keep people and animals out of confinement buildings during manure storage agitation and pumping.
- Ventilate the area for several hours following pumping activities. A back-up ventilation system and emergency power source should be considered in the event that the power should fail.
- Allow one to two feet of air space above the manure surface for gases.
- Eliminate or prohibit smoking or any source of ignition near manure storage facilities.
- Keep manure agitators below the liquid manure's surface to reduce the volume of gas released.
- Remove temporary access ladders leaning against aboveground manure tanks.
- Lock access to permanent ladders on the aboveground manure tanks.
- Do not drive on crusted manure surfaces of aboveground, open-air manure storage tanks, as the crust is not uniformly solid and can break.
- Warn visitors and guests of the hazards of manure storages.
- Provide signs at the manure storage area, and give verbal instruction to all visitors and guests.

Figure 3.11.c. Open manure storage areas pose a less deadly gas hazard than belowground pits. The major hazard of the open manure storage becomes drowning. Fencing and warning signs alert people of the liquid manure hazard.

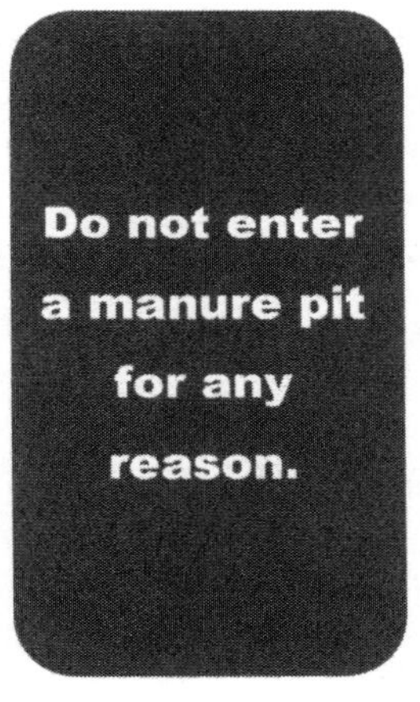

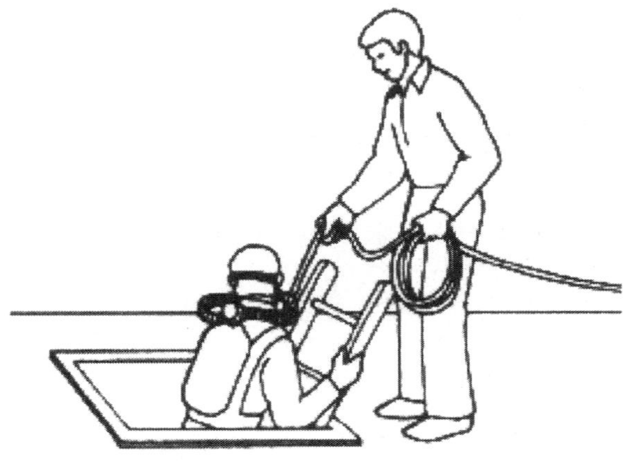

Figure 3.11.d. The only safe way to enter a manure pit is by using a self-contained breathing apparatus (SCBA). Only trained persons should use the SCBA. A lifeline is also a part of safe entry. Do not work alone.

Safety Activities

1. Conduct a survey of the farms in your area. Make a chart comparing how many aboveground manure storage facilities exist compared with the number of belowground manure storage structures.

2. One mature cow produces approximately 1 cubic foot of waste per day. For a herd of 500 cows, how many cubic feet of storage space would be necessary to store the waste for 180 days?

3. Using farm magazines, newspapers, the Internet, or any other source, make a collection of news articles which tell about manure storage injuries or fatalities.

4. Contact your local fire and emergency response company to learn more about self-contained breathing apparatus. Write a report for your group or employer.

5. Invite local firefighters to visit a farm to learn more about the hazards associated with manure storage.

6. Research the topic, "positive ventilation systems". Determine which is better at ventilating a manure pit, a positive ventilation system or a negative ventilation system. Write 2-3 paragraphs with your answer or explain your answer to your instructor or leader.

References

1. www.cdc.gov/NIOSH/Topics/Agriculture/Type manure storage in search request.
2. Farm and Ranch Safety Management, John Deere Publishing, 2009.
3. American Society of Agricultural and Biological Engineers (ASABE), ANSI/ASABE S607, Ventilating Manure Storage to Reduce Entry Risk.

Contact Information

National Safe Tractor and Machinery Operation Program
The Pennsylvania State University
Agricultural and Biological Engineering Department
246 Agricultural Engineering Building
University Park, PA 16802
Phone: 814-865-7685
Fax: 814-863-1031
Email: NSTMOP@psu.edu

Credits

Developed by WC Harshman, AM Yoder, JW Hilton and D J Murphy, The Pennsylvania State University. Reviewed by TL Bean and D Jepsen, The Ohio State University and S Steel, National Safety Council. Revised 3/2013

This material is based upon work supported by the National Institute of Food and Agriculture, U.S. Department of Agriculture, under Agreement Nos. 2001-41521-01263 and 2010-41521-20839. Any opinions, findings, conclusions, or recommendations expressed in this publication are those of the author(s) and do not necessarily reflect the view of the U.S. Department of Agriculture.

ANHYDROUS AMMONIA

HOSTA Task Sheet 3.12 **Core**

NATIONAL SAFE TRACTOR AND MACHINERY OPERATION PROGRAM

Introduction

Plant growth is improved with fertilizer application. Nitrogen is one plant food element. Nitrogen is responsible for green, healthy, productive leaves. Soils usually lack nitrogen so this element must be added to the soil.

Anhydrous ammonia is a powerful source of nitrogen containing 82% nitrogen. Nitrogen solutions are caustic. Caustic chemicals can burn plant and human tissues.

This task sheet discusses the hazards of anhydrous ammonia.

Youth younger than age 16 are forbidden by the Hazardous Occupations Order in Agriculture regulations from handling or using anhydrous ammonia. There are no exceptions to these regulations based upon a supplemental training program. If assigned to the task of working with anhydrous ammonia, tell your employer that you are not permitted to do so.

Even so, youth may be working around anhydrous ammonia and should understand its hazards.

Use of Anhydrous Ammonia

Anhydrous ammonia (NH_3) is a powerful ammonia nitrogen fertilizer. Stored under pressure, anhydrous ammonia exists in liquid form. In the air, anhydrous ammonia becomes a gas.

Pressurized tanks (nurse tanks) are used to store and deliver this form of fertilizer to application tanks used on the farm. Field application tanks apply the anhydrous ammonia by injection into the soil. Soil moisture then attracts and holds the nitrogen.

Anhydrous means "without water." Anhydrous ammonia is quickly attracted to any form of moisture. Soil moisture absorbs the fertilizer rapidly.

Just as soil moisture reacts quickly with anhydrous ammonia, so does the human body. Moist skin, eye, and lung tissues react with NH_3 by severe burning of those body areas. Severe health problems will result by improper handling and application of anhydrous ammonia. **Anhydrous ammonia can result in permanent damage to your lungs.**

Using anhydrous ammonia is more complex than applying dry, granular fertilizer. Pressurized tanks, control valves, and pressure hoses must be in working order and used properly. The operator must follow several specific procedures exactly. Safety equipment must be nearby and not stored away from the job site.

Important: The danger of using anhydrous ammonia comes through the risks of handling the material. Youth workers younger than age 16 are not permitted to handle anhydrous ammonia.

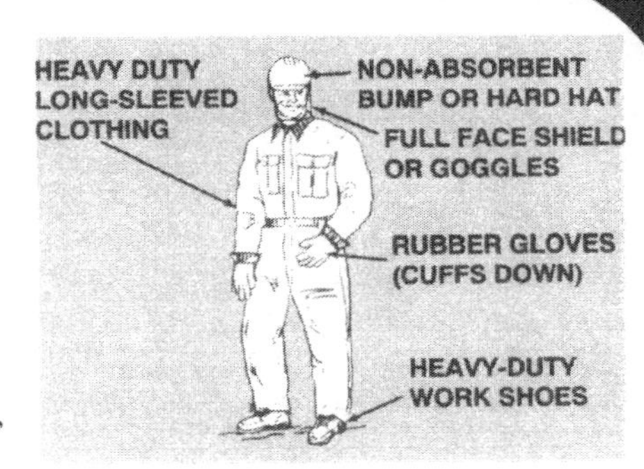

Figure 3.12.a. Personal protective equipment for working with anhydrous ammonia lowers the risk of being exposed to the burning effects of the NH_3 on the skin, the eyes, and the lungs. *Farm and Ranch Safety Management, John Deere Publishing, 1994. Illustrations reproduced by permission. All rights reserved.*

NH_3 is also added to corn silage at the silo to increase protein levels

Learning Goals

- To understand anhydrous ammonia uses and the risks that this material can pose

Related Task Sheets:

Hazardous Occupations Order I in Agriculture	**1.2.1**
Occupational Safety and Health Act	**1.2.2**
Personal Protective Equipment	**2.10**
First Aid and Rescue	**2.11**
Respiratory Hazards	**3.3**
Respiratory Protection	**3.3.1**
Silos	**3.9**

 Cooperation provided by The Ohio State University and National Safety Council.

An estimated 80% of NH_3 injuries and fatalities are the result of a lack of knowledge or training.

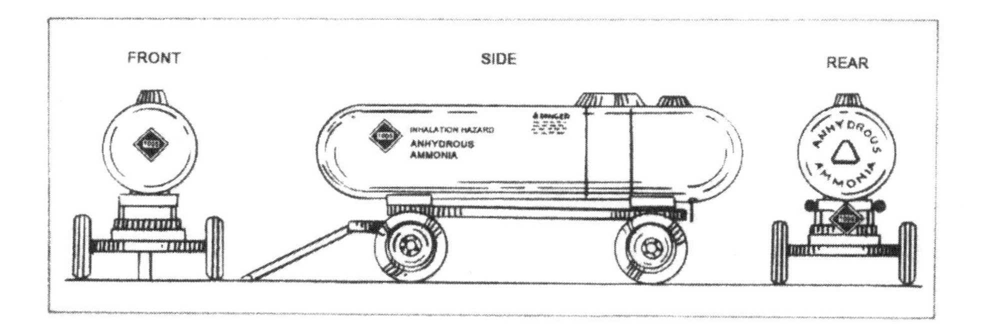

Figure 3.12.b. Anhydrous ammonia tanks must be plainly marked on all surfaces as containing an inhalation hazard. Such markings provide the message that this material is deadly. *Farm and Ranch Safety Management, John Deere Publishing, 1994. Illustrations reproduced by permission. All rights reserved.*

Anhydrous Ammonia Systems and Safety

The anhydrous ammonia system is made of several components. Each component operates under a pressurized condition. System components include:

- the nurse tank (the delivery tank)
- control valves for withdrawal, fill, pressure relief, and return lines
- pressure gauges
- transfer hoses
- the applicator tank (for field application)

Anhydrous ammonia system components must meet rigorous safety standards. Anhydrous ammonia is corrosive, therefore system parts must be of high strength steel or other suitable materials. Fittings should be made of black iron. All parts and surfaces must withstand a minimum of 250 pounds per square inch of pressure (psi). Containers used to store anhydrous ammonia must be painted white or silver to reflect away the heat of the sun to control tank temperatures and pressure.

Daily system checks and routine maintenance are a must. A regular, scheduled replacement program of valves and hoses is recommended. Leaks in the system must receive immediate attention. Dents, gouges and cracks must be repaired by qualified service representatives. Certified welders must be utilized for repairs requiring welding.

Equipment markings must warn users and bystanders of the hazards of anhydrous ammonia. The labels, markings, and safety signs include:

- anhydrous ammonia labeling in 4-inch letters on the side and rear of the tank
- inhalation hazard labeling required by the federal Department of Labor must appear as 3-inch high lettering on both sides of the tank
- nonflammable gas placard with the numbers 1005 (identification number for anhydrous ammonia) must appear on both sides and both ends of the tank
- SMV emblem must be displayed on the rear of the tank
- valves must be labeled by color or legend as vapor valves (Safety Yellow color) or liquid valve (Omaha Orange color). Lettering must be at least 2 inches in height and within 12 inches of the valves.

Anhydrous Ammonia Safety Precautions

Anhydrous ammonia is a deadly material. It can kill or cripple a person quickly. Constant attention to safety must be part of working with this material. Follow these safe practices.

- Use the correct personal protective equipment (a face shield or splash-proof goggles, rubber gloves and heavy-duty, long-sleeved shirts and pants are recommended).
- At least 5 gallons of clean, fresh water is required to be carried with each vehicle transporting anhydrous ammonia (exposure from spills or splashes will require at least a 15-minute flushing with water to dilute the anhydrous ammonia).
- Operators who are working directly with the NH_3 should carry a squeeze bottle of water in their immediate possession to treat exposure.
- Remove contaminated clothing which can become frozen to the skin (NH_3 works as a cooling gas in the air).
- The operator should be trained in system components and how they operate.
- Daily safety inspections are necessary.
- All labels, markings, and safety signs must be in place and clear for visibility.
- Highway towing speeds should be reduced to less than 25 mph to decrease the risk of upsets or damage.
- Safety chains must be used for highway transport.
- Use a qualified service person to repair the tank, valves, fittings, and hoses.
- Keep untrained persons away from the anhydrous ammonia tanks and equipment.

The same safe practices are to be followed if anhydrous ammonia is to be injected into corn silage as it is blown into the silo. Anhydrous ammonia is a valuable crop nutrient and feed additive if handled safely.

Figure 3.12.c. Transfer of anhydrous ammonia from nurse tanks to field application tanks is a critical time period for using personal protective equipment. Standard operating procedures like having a source of water nearby to flush away spills is important. *Farm and Ranch Safety Management, John Deere Publishing, 1994. Illustrations reproduced by permission. All rights reserved.*

A small squirt bottle of water should be carried with the worker as they work with anhydrous ammonia.

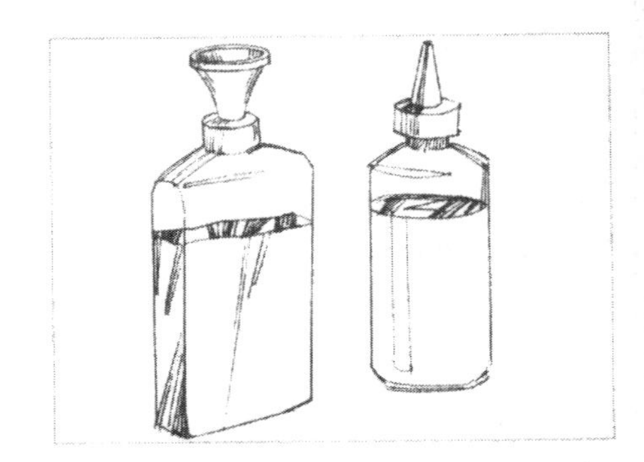

Figure 3.12.d. A small amount of water carried on the person working with anhydrous ammonia may be the needed protection to flush the eyes should a spill or splash occur. This small amount of water will help until the victim can get to a larger supply of water. *Farm and Ranch Safety Management, John Deere Publishing, 1994. Illustrations reproduced by permission. All rights reserved.*

Safety Activities

1. Draw a sketch of the parts of an anhydrous ammonia fertilizer system. Label the parts by name and function. The information can be found by using the website www.cdc.gov/nasd.

2. Use the website of the Department of Labor (www.dol.gov) or your own state's Department of Transportation website to locate information on hazardous materials placards. Print a copy of the various placards that are found on trucks hauling materials through your community.

3. Practice flushing the eyes with water for 15 minutes to prepare yourself for spills or splashes of any chemicals which could contact your eyes. Is there a water temperature that is best recommended? What source of water is best recommended?

4. Conduct a survey of local farmers to determine how many use anhydrous ammonia. Present the results at your 4-H club, FFA meeting, or to your mentor.

5. Research the possibility of purchasing small squeeze water bottles to use for eye flushing. Make these bottles available to local farmers along with a brochure on anhydrous ammonia safety.

6. Write a letter to local fire service groups informing them of the dangers of anhydrous ammonia. Ask them if they have the necessary equipment to work with local farmers who may need their emergency services.

References

1. Visit www.cdc.gov/nasd/ Click on search by topic/ Type anhydrous ammonia in search box.
2. Farm and Ranch Safety Management, John Deere Publishing, 2009. Illustrations reproduced by permission. All rights reserved.

Contact Information

National Safe Tractor and Machinery Operation Program
The Pennsylvania State University
Agricultural and Biological Engineering Department
246 Agricultural Engineering Building
University Park, PA 16802
Phone: 814-865-7685
Fax: 814-863-1031
Email: NSTMOP@psu.edu

Credits

Developed by WC Harshman, AM Yoder, JW Hilton and D J Murphy, The Pennsylvania State University. Reviewed by TL Bean and D Jepsen, The Ohio State University and S Steel, National Safety Council. Revised 3/2013

This material is based upon work supported by the National Institute of Food and Agriculture, U.S. Department of Agriculture, under Agreement Nos. 2001-41521-01263 and 2010-41521-20839. Any opinions, findings, conclusions, or recommendations expressed in this publication are those of the author(s) and do not necessarily reflect the view of the U.S. Department of Agriculture.

FARMSTEAD CHEMICALS

HOSTA Task Sheet 3.13 **Core**

NATIONAL SAFE TRACTOR AND MACHINERY OPERATION PROGRAM

Introduction

Not all chemicals used on a farm are toxic pesticides. Pesticides are those chemicals which kill pests. Workers younger than age 16 are not permitted to work with restricted use pesticides in any manner.

There are many other chemicals used on the farm which are not pesticides. Chances are high that you will be exposed to some chemicals which are not regulated under pesticide laws.

This task sheet discusses farmstead chemicals and working with these chemicals safely.

Farmstead Chemicals

The beginning farm worker may be assigned to the milking parlor of a dairy farm, the animal treatment area, a livestock center, the field crop area, or to the farm shop. The milking process involves working with cattle, cleaning facilities, and equipment including milk pipelines. The animal treatment area may expose the worker to disinfectants and medicinals. Livestock center chores may range from baby pig care to feeding and care of beef steers. Field crop work involves handling fertilizer and lime. Farm shop work finds a young worker cleaning parts and servicing equipment.

Dairy farm work involves using cleaners and sanitizers. Acid rinses, alkaline compounds, chlorine, and iodine materials are commonly found on farms. These can damage skin and produce toxic fumes.

The animal treatment area of a dairy farm has potentially hazardous materials. Animal medications may be applied externally or by injection. Young persons are often trained to administer vaccinations. The needles can expose workers to vaccines or puncture wounds.

Livestock center work parallels the work of the dairy industry. Animal medications mixed into the animals drinking water are used. Foot bath chemicals are mixed to treat foot health problems.

Field crop work with the exception of pesticide application will be assigned to most young workers. Hauling fertilizer and lime is a dusty chore. Those particulates can create respiratory health risks and skin irritation.

Farm equipment becomes greasy and dirt-covered. Degreasers and solvents may be needed to clean the parts. Hydrosulfuric acid will be encountered while servicing a battery (Task Sheet 4.6.2). These materials are also hazardous.

There are many types of chemical materials used on the farm. They are so numerous that the list would be endless. Every year new products are added to the list. It is impossible to discuss all farmstead chemicals in this task sheet.

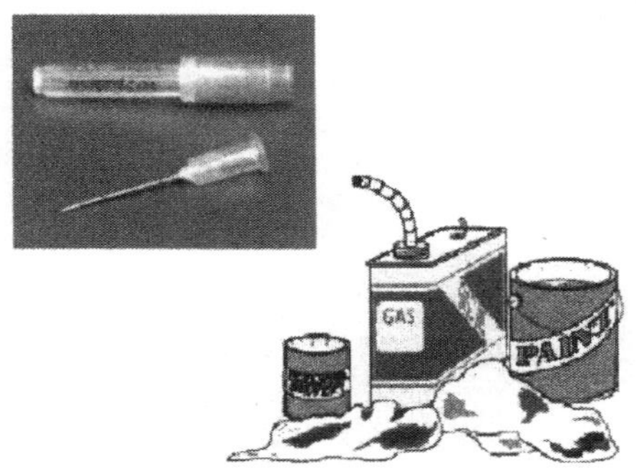

Figure 3.13.a. All farmstead chemicals are not pesticides. Animal medicines, fuels, paints, and solvents are also chemicals. The young farm worker may be exposed to many such products.

Do you know what the chemicals you are handling can do to your body?

Learning Goals

- To understand that farmstead chemicals are used in agriculture and that each one of these must be handled in a safe manner

Related Task Sheets:

Hazardous Occupations Order in Agriculture	**1.2.1**
Worker Protection Standards	**1.2.4**
Injuries Involving Youth	**2.1**
Age-Appropriate Tasks	**2.4**
Respiratory Hazards	**3.3**
Respiratory Protection	**3.3.1**
Working with Livestock	**3.4**
Agricultural Pesticides	**3.5**
Fire Safety	**3.7**
Fire Prevention and Control	**3.7.1**

 Cooperation provided by The Ohio State University and National Safety Council.

Young farm workers are often assigned to work with animal cleaning and sanitation products.

Figure 3.13.b. A young farm worker should not be surprised that beginning level jobs will involve cleaning equipment. Dairy facilities, for example, require use of a variety of cleaners, degreasers, and sanitizers to meet milk inspection standards.

Animals and Chemicals

Working with dairy, livestock, and small animals often requires the use of a variety of chemical products. Animals must be kept clean and healthy. Equipment used with animals must be disinfected. Unhealthy animals must be treated.

A variety of chemical products may be handled by young farm workers. Disinfectants are used with livestock to reduce infectious organisms. These products may be applied to the animal directly by the worker. The material may be diluted with water and applied by way of foot baths.

Direct application of chemical formulations to the animal can be done by sanitary wipes or dust application. Udders and teats of the dairy cow are disinfected with individual sanitary wipes. Teat dips are used before and after milking to reduce bacterial infection. Foot baths contain copper sulfate solutions to control and prevent foot rot organisms from destroying hoof tissues of cattle, horses, and sheep.

Milking equipment, milk pipelines, and bulk tanks must be cleaned and sanitized. Butterfat and protein particles must be removed by degreasing chemicals. The milking equipment components must also be sanitized to prevent growth of harmful microorganisms.

Livestock equipment must be disinfected to prevent spread of disease from one group of animals or from one farm to another. Weigh scales and head locks are treated with disinfectants and may be applied by pressure-washing equipment. Livestock tools, such as dehorning and castration equipment, must be sterilized after each use.

Many animal medicinals or pharmaceuticals are also agricultural chemicals.

Dairy and livestock must be treated for disease or vaccinated to prevent disease. Injections supplement nutritional needs of the animal as well. The young farm worker will often be trained to assist with these injections.

Safe work habits will prevent you from unnecessary exposure to the active ingredients in these products. Follow these safety points:

- Read product labels to understand the safety requirements of the product.
- Do not mix chemical solutions without adult supervision.
- Use proper personal protective equipment to protect eyes, skin, and lungs.

Note: The maturity and strength of a young worker must be considered when accepting animal care tasks.

Lime and Fertilizer

Fertilizer and lime are necessary for plant growth. Fertilizer provides the plant food elements like nitrogen, phosphorous, and potash. Lime neutralizes soil acidity to make fertilizer elements more available to the plant. Fertilizer materials are applied in dry, gas, or liquid form. Lime is applied in a dry powder or liquid form.

Fertilizer is a hygroscopic material. This means that it attracts moisture. As it pulls moisture from the skin, eyes, nose, or mouth, tissues can blister and burn. Exposure occurs when fertilizer is being handled. Operator exposure is increased when you are unprotected.

Lime in the hydrated form is also a hygroscopic material. Hydrated lime is often used to treat barn alleyways as a disinfectant and as a fast-acting soil amendment.

Wear long-sleeved shirts, long pants, and eye protection while handling and applying these materials. A toxic particle dust mask is also recommended.

Machinery and Chemicals

Farm machinery must be maintained and repaired. There are many chemicals used for maintenance and repair tasks. The chemicals include but are not limited to:

- fuel
- oils and lubricants
- degreasers
- antifreeze
- battery acid
- solvents

Each of these materials can be toxic, caustic, or flammable.

Toxic materials poison a person if they are ingested, spilled on the skin or in the eyes, or inhaled. Petroleum products can be fatal if swallowed. Antifreeze poisons a person who has swallowed it.

Caustic materials burn skin tissues quickly. Battery acid burns skins and clothes. Solvents can dry the skin and cause irritation.

Flammable materials can explode or ignite and burn violently. Petroleum products and cleaning solvents are class B fuels for fire sources (See Task Sheets 3.7 and 3.7.1).

Safe work habits should be practiced in all areas of the farm. Shop safety with chemicals should include:

- Use of personal protective equipment, such as goggles, chemical gloves, and aprons
- Understanding label directions for the material's use in mixing and application
- Adult guidance for those areas of confusion

Special note: Shop rags pose a hazard as well. The rags may be soaked in toxic material from wiping up an area. The rags can be soaked in caustic material, such as battery acid, or the rags could contain flammable materials. Rags can expose the worker to hazardous materials and should be disposed of after use to prevent fires.

Figure 3.13.c. . Exercise care when using all farmstead chemicals. Spills can pose hazards such as slips and falls. Follow cleanup and proper disposal procedures on the product's label.

Young farm workers are often assigned to clean equipment and to move crop supplies.

Figure 3.13.d. Machinery cleanup may require a solvent. What would you use to clean the grease away from the hydraulic fittings?

Safety Activities

1. Visit a dairy farm, a horse farm, a beef farm, or a swine facility. With the owner's permission, make a list of all the farmstead chemicals that you can find. Do not include pesticides.

2. If you are studying this material in a group, have the group make a list of farmstead chemicals that they have used on their farm or a farm where they are working.

3. Are dairy cleansers, sanitizers, and medicines covered by the Worker Protection Standards Act? You will have to refer to Task Sheet 1.2.4, or use the Internet to search for the subject of Worker Protection Standards.

4. Research foot rot in livestock and how it's controlled.

5. Find out what procedures a local farmer would use to clean up an oil, antifreeze, or fuel spill. Write the procedures in outline form.

6. Define these terms:
   a. sanitize
   b. acid compound
   c. alkaline compound
   d. hydrated lime

References

1. Farm and Ranch Safety Management, John Deere Publishing, 2009.

Contact Information

National Safe Tractor and Machinery Operation Program
The Pennsylvania State University
Agricultural and Biological Engineering Department
246 Agricultural Engineering Building
University Park, PA 16802
Phone: 814-865-7685
Fax: 814-863-1031
Email: NSTMOP@psu.edu

Credits

Developed by WC Harshman, AM Yoder, JW Hilton and D J Murphy, The Pennsylvania State University. Reviewed by TL Bean and D Jepsen, The Ohio State University and S Steel, National Safety Council. Revised 3/2013

This material is based upon work supported by the National Institute of Food and agriculture, U.S. Department of Agriculture, under Agreement Nos. 2001-41521-01263 and 2010-41521-20839. Any opinions, findings, conclusions, or recommendations expressed in this publication are those of the author(s) and do not necessarily reflect the view of the U.S. Department of Agriculture.

ANIMAL, WILDLIFE, AND INSECT RELATED HAZARDS

HOSTA Task Sheet 3.14

NATIONAL SAFE TRACTOR AND MACHINERY OPERATION PROGRAM

Introduction

Farm work may bring you into contact with animals on the farm, as well as, wildlife that may occupy the same area. Sometimes these contacts can be hazardous.

Understanding the risks of these exposures is important. Some animal health problems can be transferred to humans. Farm workers may unexpectedly encounter potentially hazardous animals, snakes and insects.

This task sheet discusses animal, wildlife, and insect related hazards.

Zoonoses

Definition: Zoonoses is the term that denotes diseases that can be transmitted between vertebrate animals and humans. These diseases can be transferred in several ways.

Direct Animal Contact

Animal manure, urine, bedding, and products (raw meat, unprocessed milk, hides, hair, etc.) can serve as a source of human infection. Disease causing organisms and disease carrying insects can be found in and on these products.

Animal manure contains bacteria from the animal's digestive system. E. coli, a bacteria, is found in manure. This bacteria can cause intestinal disease, with nausea and general feelings of ill health.

Animal products such as meat and milk can carry microorganisms that can cause disease. Meat can be a source of Salmonella or Listeria, both of which are bacterial organisms. These organisms can cause fever, nausea, vomiting, and diarrhea. Processing or pasteurization is used to control and eliminate these micro-organisms.

Animal hides and hair may harbor insects that can carry disease, bite, or sting a person. Workers who must handle raw animal products are placed at risk for exposure to insects and ticks (See Page 3).

Infections of the animal's reproductive tract can be transmitted to people who assists with the birthing of calves, piglets, lambs, and foals. Sterile, disposable gloves should be worn to protect against harmful organisms. Such organisms can enter the body through cuts and scratches. Just as importantly infection from a person's hands can enter the animal's reproductive tract and cause disease to the animal.

Indirect Animal Contact

Soil, plants, and water can be contaminated by animal wastes. Surface water (streams and ponds), as well as water wells and reservoirs, can be contaminated with animal waste. Avoid drinking such water to reduce your exposure to potential health risks..

Figure 3.14.a. Cattle can transmit ringworm, rabies, and other micro-organisms to humans.

Ringworm is an example of a zoonotic disease

Learning Goals

- To understand the hazards of zoonotic diseases, wildlife, and insects to the worker.

Related Task Sheets:

The Work Environment	1.1
First Aid and Rescue	2.11
Working with Livestock	3.4

Insects and snakes are found in the fields and barns where farm employees work.

Figure 3.14.b. Stinging insects and poisonous reptiles are found throughout the United States. Each geographic area may have its own set of insect and snake species which may be hazardous.

Stinging Insects and Poisonous Snakes

Stinging Insects

Wasps, hornets, bees, and other stinging and biting insects, as well as, spiders and tarantulas are found throughout America. Many a farm worker has been stung by one or more of these pests with various reactions.

Insect bites create health problems for some people. Allergic reaction to the sting or bite is one such reaction. *Anaphylactic shock* is caused by insect venom and is a serious medical emergency.

Anaphylactic shock is characterized by swelling of the throat which can cause suffocation and a sudden decline in blood pressure. Both of these can cause death. A person who has such a reaction must be taken immediately to emergency medical care.

Poisonous Snakes

Various species of poisonous snakes are found throughout the United States. Rattlesnakes, copperhead snakes, and others pose little danger to most people if they are left alone in their surroundings. They are generally found away from human populations, so most workers will not often encounter a snake.

Occasionally a farm worker may encounter a snake that may strike. Farm work in seldom used barns, along fences, and near woodlots can bring the worker into a surprise encounter with a snake. Quick identification of the snake as poisonous or harmless is necessary.

Poisonous snakes have a angular head with a pit in front of the eyes. If such a snake is encountered these are recommended actions:

- Slowly back away from the snake.
- Make no sudden or threatening moves.
- Report the incident to others who may have to work in the same area.

If a snake bite occurs, the following ideas can prevent the wound from become more serious than it need be:

- Allow bite to bleed freely for 15-30 seconds.
- Clean and disinfect the area.
- Stay calm.
- Get assistance to travel to emergency medical care.

Be aware of snake habitats and watch your movements carefully.

 Cooperation provided by The Ohio State University and National Safety Council.

Rabies

Rabies is a viral disease of mammals. It is transmitted through the bite of an infected animal. Most cases of rabies come from wild animals such as raccoons, skunks, bats, and foxes. Cats, cattle, and dogs can also become infected.

Rabid animals appear to be confused, paralyzed, excitable, and frothing from the mouth.

The best way to prevent rabies is to avoid animals that show strange behavior. Report such animals to your employer or parents.

If bitten by an animal that is suspected of having rabies, kill the animal if need be, handle the animal carcass with disposable gloves, and submit the animal for post-mortem testing. A person who has been exposed to rabies will need medical treatment quickly.

Lyme Disease

Ticks often attach themselves to warm-blooded animals and feed on their blood. Their blood filled bodies are commonly found on dogs (dog tick) and deer (deer tick). These same ticks can also attach and feed on human blood.

Ticks are often found on people who have been walking in tick infected areas. Adult ticks wait on host weed species and pass on to warm-blooded hosts as they pass by.

Deer ticks are common in the northeast United States. Deer ticks can be found on deer hunters who are processing the animals. Deer ticks may carry Lyme disease and must be removed immediately.

Lyme disease, first reported in the Lyme, Connecticut, has spread nationwide. It affects people who have been bitten by a deer tick, but failed to notice the insect attached to their bodies. At least 48 hours of infectious contact will result in the onset of the disease. Lyme disease left untreated can cause a rash and flu-like symptoms followed by loss of coordination, memory loss, irregular heartbeat, and arthritis. Lyme disease is rarely fatal however.

Lyme disease is preventable. These considerations will reduce the risk of Lyme disease exposure.

- Wear light colored clothing when in infested areas (to be able to see the tick)
- Tuck pants into socks to keep ticks out
- Use an insect repellant approved for tick control to treat clothing before going into woods or fields.
- Avoid weedy, brushy areas that may harbor ticks
- Check your body for ticks when returning home

Lyme disease presents a concern, but should not keep anyone from enjoying walking or working in the fields and woodlands and from hunting or fishing.

If you suspect that you are infected with Lyme disease consult a physician immediately. A second opinion may be needed as Lyme disease can be diagnosed as one of many other nervous system problems. Antibiotics are used to treat Lyme disease.

Figure 3.14.c. Raccoons are common carriers of rabies. If you find these animals acting abnormally around humans be alert for the danger of a rabid animal bite.

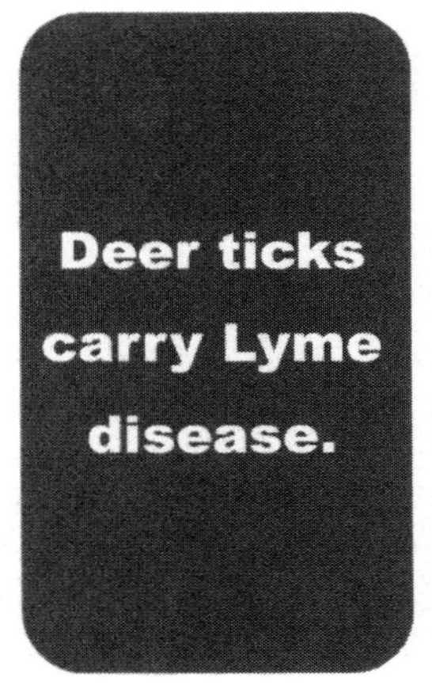

Figure 3.14.d. Deer harbor a tick that can carry Lyme disease. Hunters who bag a deer should take precautions to avoid becoming infested. Inspect your body after handling infected deer during processing.

Safety Activities

1. Use the Internet website of your state Land Grant University's Entomology Department to locate pictures of stinging insects. Make a collage of the insects that you have seen, or that have stung you. Place a label on the insect picture to identify it.

2. Interview 25 persons to determine how many have had an allergic reaction to an insect sting. From the percentage of persons calculated to be allergic, determine how many people that may be in the United States if the total population is estimated to be a total of 300,000,000 people. How many people may have this allergic reaction. (This is not a scientific study.)

3. Word Search. Draw a line through as many words about zoonotic diseases and their carriers as you can find. Use the word list. Words may be horizontal, vertical, diagonal, frontwards, and backwards.

R	A	B	C	B	T	A	M	O
I	R	D	R	A	T	I	A	Z
N	A	E	B	T	T	R	L	X
G	B	F	G	T	I	E	A	X
W	I	N	S	E	C	T	R	S
O	E	H	I	R	K	C	I	I
R	S	U	R	I	V	A	A	T
M	J	K	L	A	M	B	N	E
P	A	R	A	S	I	T	E	S

Use these words: parasites, bacteria, ringworm, rabies, malaria, virus, snake, tick, insect, rat, bat.

References

1. Safety and Health for Production Agriculture, ASAE Textbook Number 5, Dennis J. Murphy, American Society of Agricultural Engineers, St. Joseph, MI.
2. The Internet. Type a key word on animal health, wildlife, insect, or disease into the search box and scroll for the sites you wish to visit.

Contact Information

National Safe Tractor and Machinery Operation Program
The Pennsylvania State University
Agricultural and Biological Engineering Department
246 Agricultural Engineering Building
University Park, PA 16802
Phone: 814-865-7685
Fax: 814-863-1031
Email: NSTMOP@psu.edu

Credits

Developed by WC Harshman, AM Yoder, JW Hilton and D J Murphy, The Pennsylvania State University. Reviewed by TL Bean and D Jepsen, The Ohio State University and S Steel, National Safety Council. Revised 3/2013

This material is based upon work supported by the National Institute of Food and Agriculture, U.S. Department of Agriculture, under Agreement Nos. 2001-41521-01263 and 2010-41521-20839. Any opinions, findings, conclusions, or recommendations expressed in this publication are those of the author(s) and do not necessarily reflect the view of the U.S. Department of Agriculture.

AGRICULTURAL TRACTORS

HOSTA Task Sheet 4.1 **Core**

NATIONAL SAFE TRACTOR AND MACHINERY OPERATION PROGRAM

Introduction

In 1892 a man named John Froelich developed a successful tractor to power a grain thresher. By 1918 a PTO shaft was used to power equipment drawn behind the tractor. Before these time periods, farm work was done by hand, by horse, or by huge stationary steam engines.

You will be operating a tractor designed to accomplish greater amounts of work than ever thought possible in the early 1900s. The speed, power, flexibility, adaptability, and handling ease of modern tractors is what makes them valuable and indispensable for modern day farming. This task sheet describes agricultural tractors, with an emphasis on what tractors are designed to do.

Tractor Types/Sizes

Tractors have both narrow and wide front ends, use both wheels and tracks, and can be two-wheel drive, four-wheel drive, or articulated. A narrow front end ("tricycle") will be an older tractor, as they have not been produced this way since the 1960s. Articulated tractors are usually very large (at least 250 hp) and are usually operated only by very experienced farmers. Young and inexperienced tractor drivers usually operate tractors ranging from lawn and garden-size (~ 20 hp) to large two- and four-wheeled drive tractors (around 150 hp). Many older and smaller tractors will not have a rollover protective structure (ROPS), while most new tractors will have a ROPS and seat belt.

Figure 4.1.a. Tractors come in all shapes and sizes.

Tractor Purposes

Farm tractors were designed for four primary purposes:

1. Load Mover (High Lift)
2. Remote Power Source (PTO)
3. Implement Carrier (3 Pt. Hitch)
4. Transport Unit (Drawbar Unit)

Understanding that ordinary farm tractors are not recreational vehicles is very important. Farm tractors are not to be used for fun, play, or for mud-bogging or racing, unless specifically modified for that purpose. You must use the tractor only for work purposes. Other uses can increase the chance of injury to you or others, or to the tractor, implements, and other property.

Figure 4.1.b. Tractors should be used for their designed purpose.

Tractors are work horses, not race horses.

Learning Goals

- To describe how tractors vary in size, shape and age
- To describe how tractors are designed for work

Related Task Sheet:

Tractor Hazards **4.2**

Cooperation provided by The Ohio State University and National Safety Council.

Tractor Characteristics

Here are some design elements of a tractor.

- Rear wheels adjustable for width
- "Turn-on-a-dime" steering
- High-powered engine with many gear ranges for relatively low speeds
- Great clearance beneath the tractor
- More weight over traction wheels
- Individual brakes for each rear wheel
- Adjustable drawbar hitch
- Power controls to increase pulling power
- Potential to add or subtract weights for ballast
- Hydraulic system for added power source
- PTO shaft to transfer power to towed machine
- Differential lock for added traction
- Adapted to carry or pull equipment
- Fitted with a Rollover Protective Structure (ROPS) or a Falling Object Protective Structure (FOPS)

A tractor is designed to do work. Use the tractor only for this purpose!

Safety Activities

1. Take photos or video camera footage of tractors being used for the four intended purposes. Make a display for your club or classroom or employee lunch room where you work.
2. Collect newspaper and magazine articles on farm tractor safety. Share the main points of the articles with classmates.
3. Locate a farmer in your community who has been injured with a tractor or farm machine and see if they will discuss the incident with you.
4. Use the Internet to find information on tractor safety. Find articles that describe people injured by a tractor because they were not using it for its designed purpose.
5. Do a survey of tractors at area farms or at an equipment dealer's lot and record how many tractors: a) have a tricycle or wide front end; b) have a ROPS with seat belt; c) have wheels or a track; if it has wheels, is it a two-wheel, four-wheel, or an articulated tractor? Also record the engine horsepower and tractor age.

References

1. Farm and Ranch Safety Management, John Deere Publishing, 2009.

Contact Information

National Safe Tractor and Machinery Operation Program
The Pennsylvania State University
Agricultural and Biological Engineering Department
246 Agricultural Engineering Building
University Park, PA 16802
Phone: 814-865-7685
Fax: 814-863-1031
Email: NSTMOP@psu.edu

Credits

Developed by WC Harshman, AM Yoder, JW Hilton and D J Murphy, The Pennsylvania State University. Reviewed by TL Bean and D Jepsen, The Ohio State University and S Steel, National Safety Council. Revised 3/2013

This material is based upon work supported by the National Institute of Food and Agriculture, U.S. Department of Agriculture, under Agreement Nos. 2001-41521-01263 and 2010-41521-20839. Any opinions, findings, conclusions, or recommendations expressed in this publication are those of the author(s) and do not necessarily reflect the view of the U.S. Department of Agriculture.

TRACTOR HAZARDS

HOSTA Task Sheet 4.2 **Core**

NATIONAL SAFE TRACTOR AND MACHINERY OPERATION PROGRAM

Introduction

Tractors are a primary source of work-related injury on farms, however, not all of the injuries happen while the tractor is being used for work.

Nationally, nearly one-third of all farm work fatalities are tractor-related. Injuries occur for a variety of reasons and in a number of different ways. This task sheet will describe types of tractor hazards and the nature and severity of injuries associated with using farm tractors.

Hazard Groups

There are several hazards associated with tractor operation. Tractor hazards are grouped into the following four categories:

1. Overturns
2. Runovers
3. Power Take-Off Entanglements
4. Older Tractors

Each of these is discussed briefly in this task sheet. Other task sheets will cover some of these topics in more detail.

Overturn

Tractor ***overturns*** is one major hazard group and accounts for the most farm-work fatalities. Approximately 50% of tractor fatalities come from tractors turning over either sideways or backward. There are dozens of examples of tractor turnover situations. Most are preventable if operators follow good safe tractor operation practices. Some common examples of tractor overturns include:

- Turning or driving too close to the edge of a bank or ditch
- Driving too fast on rough roads and lanes and running or bouncing off the road or lane
- Hitching somewhere other than the drawbar when pulling or towing objects
- Driving a tractor straight up a slope that is too steep
- Turning a tractor sharply with a front-end loader raised high

A rollover protective structure (ROPS), a structural steel cage designed to surround the operator—particularly one that is built into an enclosed cab—can protect the operator from being killed when a tractor overturns. This is especially true if the operator has fastened the seat belt. Remember, though, that a ROPS can protect you from injury but cannot keep the tractor from overturning in the first place. This explains the importance of operating a tractor safely even if the tractor has a ROPS.

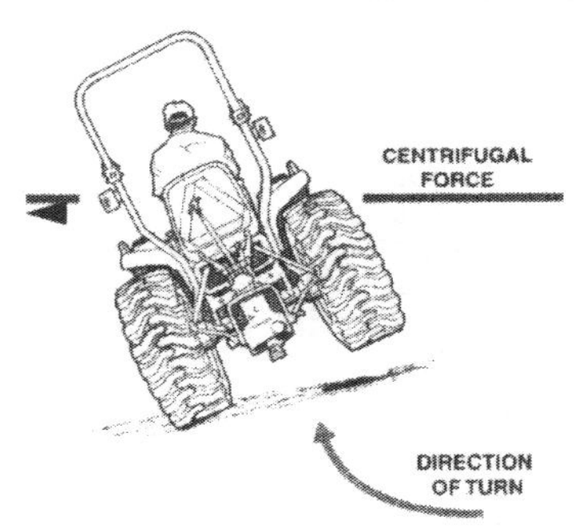

Figure 4.2.a. Tractor overturns can occur with high speed sharp turns. Avoid sudden sharp movements in all tractor work. *Safety Management for Landscapers, Grounds-Care Businesses, and Golf Courses, John Deere Publishing, 2001. Illustrations reproduced by permission. All rights reserved.*

Top-heavy, powerful tractors can upset if used improperly.

Learning Goals

- To recognize and avoid those hazardous situations which can result in exposure to overturns, runovers, PTO entanglements, and older tractor safety deficiencies.

Related Task Sheets:

Reaction Time	2.3
Mechanical Hazards	3.1
Agricultural Tractors	4.1
Tractor Stability	4.12
Using the Tractor Safely	4.13
Using PTO Implements	5.41

 Cooperation provided by The Ohio State University and National Safety Council.

Follow this rule! One seat on a tractor means one rider only– the operator. Keep all others away.

Figure 4.2.b. Tractor runovers have claimed many lives. Extra riders can slip from the tractor and be crushed before the operator can stop. Say no to your friends who want to hitch a ride.

Runover

There are three basic types of tractor *runover* incidents. One is when a passenger (extra rider) on the tractor falls off. Extra rider incidents happen because there is only one safe place for a person to be on a tractor, and that is in the operator's seat. *Some new, larger tractors have an extra seat for temporary instructional purposes, but only if the tractor has an enclosed ROPS cab.* The tractors that most young and inexperienced operators drive will have only one seat—the operator's seat. Standing on the tractor drawbar, axle housing, side links of three-point hitches, rear-wheel fenders, and the area immediately around the operator's seat are common locations unsafely occupied by extra riders. Extra riders rarely keep a tight handgrip on the tractor. Thus they can be easily thrown from the tractor.

Another runover incident involves the tractor operator either falling off the tractor as it is operating or being knocked out of the seat by a low-hanging tree branch or other obstacle. This most often happens on older tractors that do not have a ROPS and have an older seat that has no arm or back rest (often called pan seats). A person can more easily lose his or her balance and be knocked off or bounced out of a pan seat. An operator can also be run over while trying to mount or dismount a moving tractor. This type of incident can occur when the operator leaves the tractor seat without first shutting off the tractor and setting the brake or placing it in PARK, and the tractor moves unexpectedly. This may happen during the hitching and unhitching of equipment. Shut off the tractor before dismounting for any reason.

The third type of runover incident involves a person who is on the ground near a tractor. This may include the tractor operator who tries to start a tractor from the ground while the tractor is in gear. This usually involves an older tractor that can be started in gear or a newer tractor when an operator attempts to bypass a newer tractor's safe start-up design. Bypass starting hazards are discussed in more detail in Task Sheet 4.8.

Small children, often under the age of 5, are sometimes run over by a tractor (and equipment) as it is moved around the farmstead. Often, the tractor operator is unaware that the child is near the tractor. A loud noise, such as the start up of a tractor, is often attractive to a young child, and he or she may run toward it as it starts or begins to move.

Power Take-Off (PTO) Entanglement

The tractor *power take-off (PTO) stub* is another major hazard. The PTO stub transfers power from the tractor to PTO-powered machinery. The PTO stub normally turns between 540 and 1,000 revolutions per minute. At this rate, the stub is turning from 9 to 17 times per second. This is much faster than a human being can react if he or she is caught and pulled into or around the PTO stub or shaft. A person can have an arm or leg wrapped around a PTO stub shaft before they know they are in danger. A PTO master shield protects a person from the PTO stub. Some tractors have PTO stub guards that fasten to the PTO stub. All tractors should have a PTO master shield to protect the tractor operator and helpers.

Older Tractors

Older tractors should always be included when talking about tractor hazards. Many farm tractors still used for work may be 30 to 40 years old or older. These older tractors are often less safe to operate because they do not have modern safety features, and because some parts of the older tractor may not have been maintained in good working condition. A list of reasons why older tractors may be less safe to operate includes:

- Lack of ROPS and seat belt
- A seat without arm and back rests (pan seat)
- Seat does not adjust easily or at all
- Absence of a safety start system
- No bypass starting protection
- Rear brakes and brake pedals do not operate properly
- Front wheels do not turn as quickly as the steering wheel turns
- Tractor has no warning flashers or the flashers do not work
- PTO master shield is missing or does not offer adequate protection

Young and inexperienced workers may be given older tractors to operate in many cases. The older tractor is best suited for the types of jobs a young or inexperienced operator is hired to do. These tractors are best suited for raking hay, hauling wagons, and mowing fields or pastures. Young and inexperienced operators should be given newer tractors to operate when possible.

Figure 4.2.c. Power take-off stub and PTO shaft must be properly guarded to prevent entanglements. Locate the PTO area on every tractor you operate. Check whether or not that area is safely guarded.

PTO shafts kill or cripple countless victims. Some of these victims most likely live in your community.

Figure 4.2.d. Older tractors are often assigned to younger drivers to do less heavy chores. Raking hay, pulling wagons, and hauling feed to livestock does not require the most powerful tractor. Older tractors may have safety deficiencies due to age and missing safety features. This tractor does not have a ROPS or seat belt.

Safety Activities

1. Match the tractor hazard with the safety situation. (Some choices may be used more than once.)

___A. Overturn
___B. Runover
___C. PTO entanglement
___D. Older tractor deficiency

1. High lift carried in raised position in transit
2. Pet dog was tied to wagon
3. Bypass starting
4. PTO stub shaft missing
5. Driving too close to ditch embankment
6. A friend is helping to drop the hitch pin

2. Write a letter to your best friend explaining why you won't let him/her ride on the fender of the tractor to go to the field to help you make hay.

3. Explain how people are run over when they choose to bypass the ignition switch to start the tractor engine.

4. Learn more about the hazards of bypass starting a tractor engine by contacting a tractor salesperson or mechanic.

References

1. Safety Management for Landscapers, Grounds-Care Businesses, and Golf Courses, John Deere Publishing, 2001. Illustrations reproduced by permission. All rights reserved.
2. Farm and Ranch Safety Management, John Deere Publishing, 2009.

Contact Information

National Safe Tractor and Machinery Operation Program
The Pennsylvania State University
Agricultural and Biological Engineering Department
246 Agricultural Engineering Building
University Park, PA 16802
Phone: 814-865-7685
Fax: 814-863-1031
Email: NSTMOP@psu.edu

Credits

Developed by WC Harshman, AM Yoder, JW Hilton and D J Murphy, The Pennsylvania State University. Reviewed by TL Bean and D Jepsen, The Ohio State University and S Steel, National Safety Council. Revised 3/2013

This material is based upon work supported by the National Institute of Food and Agriculture, U.S Department of Agriculture, under Agreement Nos. 2001-41521-0126 and 2010-41521-20839. Any opinions, findings, conclusions, or recommendations expressed in this publication are those of the author(s) and do not necessarily reflect the view of the U.S. Department of Agriculture.

NAGCAT TRACTOR OPERATION CHART

HOSTA Task Sheet 4.3 **Core**

NATIONAL SAFE TRACTOR AND MACHINERY OPERATION PROGRAM

Introduction

Farm families often provide much of the labor for the operation of the farm. Farm work may start early in a child's life as a means of learning responsibility and contributing to the productivity of the farm. Tractor operation can come at an early age for many farm youth because tractors are a large part of how farm work is done. Tractor work can range from the simple to the complex.

This task sheet presents a Tractor Operation Chart as a guide to appropriate tractor work for young tractor operators.

Youth and Tractors

Examples of common jobs performed by youth operating tractors include:

- Mowing pastures, fields, yards and lanes
- Raking and baling hay and straw
- Towing hay and grain wagons between fields and storage
- Picking rocks and other obstacles from fields using a front-end loader
- Scraping manure from barn floors with a tractor-mounted blade
- Using the tractor to power augers and elevators during unloading operations
- Pulling old fence posts and tree stumps out of the ground with log chains

Several hazards can arise during the course of these and other jobs that involve tractor use. Many times the larger the tractor, the more complex the operation of that tractor becomes. Additionally, large and complex equipment may be attached to and powered by the tractor.

Young tractor operators do not usually have the experience needed to skillfully and safely operate large and complex combinations of tractors and machinery.

North American Guidelines for Children's Agricultural Tasks (NAGCAT) Tractor Operation Chart

Farm injury prevention specialists from the U.S. and Canada have developed consensus opinion that a guide to tractor operations by age groups is a way of matching youthful capabilities with tractor operation jobs. The NAGCAT chart is presented on the reverse side of this task sheet.

You can use this chart:

- To see if you have been doing jobs with the size tractor that matches your age
- To guide an employer in determining what they can reasonably expect a person of your age to do with various types and sizes of tractors

It is common for youths to be over confident in their ability to react safely to new or unexpected hazard situations with tractors.

Figure 4.3.a. Oftentimes the youthful farm worker is willing to do more work than his or her mental and physical maturity will safely permit. Youth should never operate a tractor without a ROPS and seat belt.

Learning Goal

- To review safety recommendations in matching tractor size and tasks with the age of the tractor operator

Related Task Sheets:

Safety and Health Regulations	**1.2**
Injuries Involving Youth	**2.1**
Age-Appropriate Tasks	**2.4**

Cooperation provided by The Ohio State University and National Safety Council.

Refer to the specific guideline for recommended supervision	Size of Tractor →			
Increased Complexity of Job →	LAWN & GARDEN less than 20hp	SMALL 20hp to 70hp	MEDIUM/LARGE more than 70hp	ARTICULATED
OPERATING A FARM TRACTOR (no equipment attached)	12-13 years	12-13 years	14-15 years	16+ years
TRAILED IMPLEMENT fieldwork	12-13 years	12-13 years	14-15 years	16+ years
3-POINT IMPLEMENTS fieldwork	12-13 years	14-15 years	14-15 years	16+ years
REMOTE HYDRAULICS fieldwork	14-15 years	14-15 years	14-15 years	16+ years
PTO-POWERED IMPLEMENTS fieldwork	14-15 years	14-15 years	14-15 years	16+ years
TRACTOR-MOUNTED FRONT-END LOADER	14-15 years	16+ years	16+ years	16+ years
WORKING IN AN ORCHARD	14-15 years	16+ years	16+ years	16+ years
WORKING INSIDE BUILDINGS	14-15 years	16+ years	16+ years	16+ years
DRIVING ON PUBLIC ROADS*	N/A	16+ years	16+ years	16+ years
PULLING OVERSIZED OR OVERWEIGHT LOAD	Due to increased hazard and complexity, these jobs should **not** be assigned to children.			
HITCHING TRACTOR TO MOVE STUCK OR IMMOVABLE OBJECTS				
SIMULTANEOUS USE OF MULTIPLE VEHICLES				
ADDITIONAL PERSONS WORKING ON A TRAILING IMPLEMENT				
PESTICIDE OR ANHYDROUS AMMONIA APPLICATION*				

*Follow State/Providence Laws

References

1. www. nagcat.org/Click on Guidelines/Select T1 Tractor Operation Chart, February 2013.
2. Cooperative Extension Service of your State's Land Grant University.

Contact Information

National Safe Tractor and Machinery Operation Program
The Pennsylvania State University
Agricultural and Biological Engineering Department
246 Agricultural Engineering Building
University Park, PA 16802
Phone: 814-865-7685
Fax: 814-863-1031
Email: NSTMOP@psu.edu

Credits

Developed by WC Harshman, AM Yoder, JW Hilton and D J Murphy, The Pennsylvania State University. Reviewed by TL Bean and D Jepsen, The Ohio State University and S Steel, National Safety Council. Revised 3/2013

This material is based upon work supported by the National Institute of Food and Agriculture, U.S. Department of Agriculture, under Agreement Nos. 2001-41521-01263 and 2010-41521-20839. Any opinions, findings, conclusions, or recommendations expressed in this publication are those of the author(s) and do not necessarily reflect the view of the U.S. Department of Agriculture.

Cooperation provided by The Ohio State University and National Safety Council.

TRACTOR INSTRUMENT PANEL

HOSTA Task Sheet 4.4 **Core**

SAFE TRACTOR AND MACHINERY OPERATION PROGRAM

Introduction

Instruments, or gauges, on the tractor control panel tell the driver about the operating conditions within and around the tractor. All tractor drivers should know what instruments are available to indicate that the tractor is operating properly.

When tractor systems are not working properly, continued operation may cause costly repairs and possible injury.

This task sheet will identify and explain instruments and gauges commonly found on tractors. Using tractor Owners' Manuals and obtaining the help of an experienced tractor operator will help you to learn the information in this task sheet.

Instruments and Gauges

Instruments can be warning lights, analog gauges, computer digital displays, buzzers, or standard gauges.

It is important for the beginning operator to develop the habit of regularly checking the instrument panel. Check the gauges:

- At start up
- At regular intervals during operation
- When changes occur in the normal sounds of operation

Abnormal gauge readings, plus changes in operating sounds, indicate that there is a problem. You should immediately **stop the engine in a safe place**, and seek help from the owner or an experienced operator.

Instruments you will use may include the following (there may be many more):

- Engine speed indicator (Tachometer)
- Oil Pressure Indicator
- Engine Temperature Indicator
- Fuel Gauge
- Air Filter Condition Indicator
- Transmission Temperature Indicator
- Hydraulic System Oil Level Indicator
- Hour meter
- Charge Indicator

Each of these instruments is important to the safe tractor operation as well as avoiding damage to the tractor. Other gauges may be found on the tractor you operate. Be sure to understand the meaning of all instruments, gauges and warnings before operating a tractor.

Figure 4.4.a. The modern tractor instrument panel may appear as complex as the cockpit controls of a jet airliner. The operator must know what each instrument or gauge is telling him/her about operating conditions.

Learn which warning lights, gauges, and digital displays are on your tractor.

Learning Goals

- To understand the instruments and gauges used to monitor the tractor's operation and performance
- To be able to make operating decisions based upon the information and gauges provide to the operator

Related Task Sheets:

Preventative Maintenance and Pre-operation Checks	**4.6**
Fuel, Oil, and Coolant Levels	**4.6.1**
Lead Acid Batteries	**4.6.2**

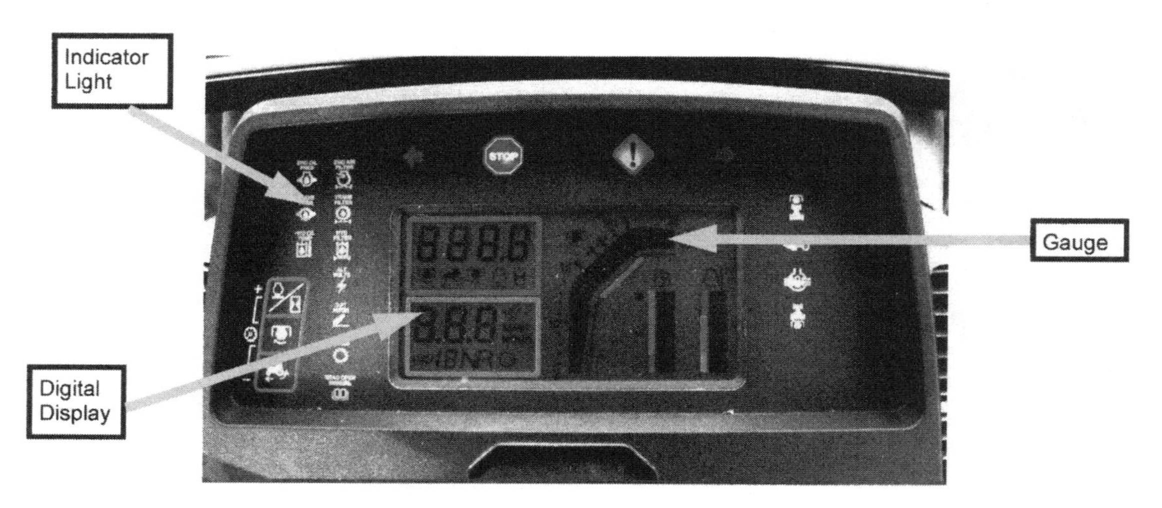

Figure 4.4.b. You may find indicator lights, standard gauges, computerized digital displays, and buzzers as instruments to show operating conditions.

Engine speed must match the work being done to be safe and to avoid engine and driveline damage.

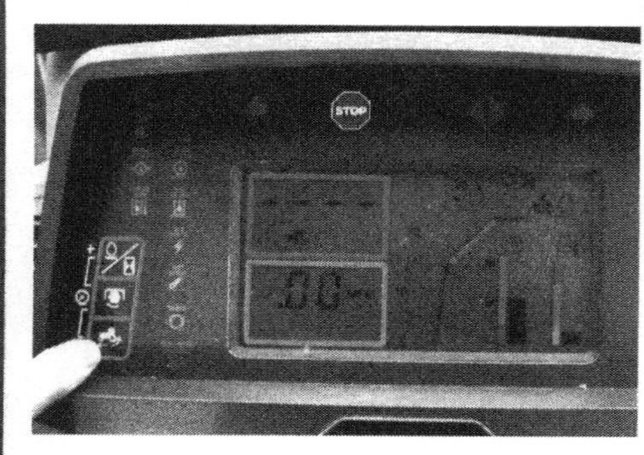

Figure 4.4.c. Check the manufacturer's RPM recommendations for various jobs to be done. Tachometers may be a gauge type or a digital display as shown above.

Tachometer (Engine Speed Indicator)

Tachometers show revolutions per minute **(RPM)**. Engine RPM must be matched to the job being done.

Incorrect RPM can lead to:

- Engine damage
- Driveline and PTO damage
- Hazardous situations

Low engine speed while in a higher gear and beginning to pull a heavy load can stall the engine.

High-engine speed with a low gear while attached to a heavy load can also create enough torque (rotational force) to tip the tractor backward. Accelerating quickly with a heavy load going up a slope can cause the tractor to rear up and tip backward.

Engine RPMs must also match PTO-driven machine requirements. Speed up the engine before engaging the PTO to operate an implement. **Low-engine speed** could stall the tractor. **High-engine speed** could shear off the implements safety shear pin if the pin was already under load. (Example: a plugged hay baler).

Follow the manufacturer's recommendations for engine speed selection.

Charge Indicator and Oil Pressure Indicator

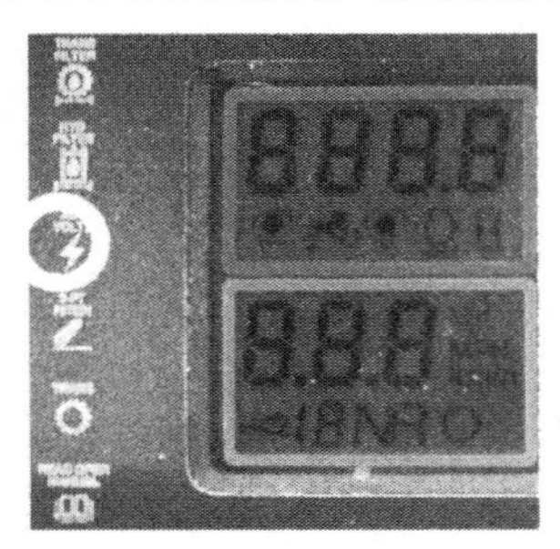

Figure 4.4.d. If low or zero oil pressure is indicated, shut down the tractor engine immediately to avoid costly engine rebuilds.

Charge Indicator

The charge indicator, or ammeter, shows whether the alternator or generator is charging the battery properly. Each time the tractor is started, the battery is discharged. During operation, the battery is recharged. Gauges will indicate + or - charge. Lights will show red at low charge. If the battery is discharging, find out the problem. The engine may not start the next time due to a low battery.

Oil Pressure Indicator (Oil Light or Gauge)

This indicator is important to the long life of an engine. If oil pressure falls because of an oil leak or low oil levels, the light or gauge shows you must stop the engine immediately. Never operate the engine with low oil pressure or oil levels. Oil lubricates the internal parts of the engine and prevents major repair expense.

Engine Temperature and Other Gauges

Engine Temperature Indicator

The engine must be cooled to prevent damage. Water-cooled engines can overheat if coolant is lost, radiators become clogged with debris, or the radiator leaks.

If the engine overheats, stop the engine, allow it to cool, then check for the problem.

Never open the radiator cap while the engine is hot. Scalding from extremely hot water can result.

Never open the radiator cap when the tractor engine is hot. Scald injury can occur.

Fuel Gauge

Check the fuel gauge before leaving for the field. Running out of fuel is inconvenient. On some tractors, running out of fuel (diesel) means time-consuming bleeding of air from the fuel lines in order to be able to start the tractor again.

Other Gauges

Tractors may come equipped with instruments to monitor air filter conditions, transmission temperatures, hydraulic system oil levels, and of course hours of work (hour meter). Become familiar with all instruments before operating the tractor.

Figure 4.4.e. Wait until the engine is cool to remove the radiator cap. *Safety Management for Landscapers, Grounds-Care Businesses, and Golf Courses, John Deere Publishing, 2001. Illustrations reproduced by permission. All rights reserved.*

Safety Activities

Answer these questions

1. If you are operating the tractor in the field and the oil light comes on, what should you do?

   a. drive to the shop　　b. stop and let the engine idle

   c. shut down immediately　　c. shut off the engine until it cools and then restart

2. What can happen if you remove a radiator cap from an overheated tractor's coolant system?

   a. nothing　　b. explosive pressure can hurt you

   c. a fire may start　　d. you can be scalded by hot steam

3. When pulling a heavy load of hay up a hill, which gear/RPM (engine speed) combination should you use?

   a. 5th gear/high RPM　　b. lower gear with medium RPM

   c. highest gear with lowest RPM

4. The letters RPM represent:

   a. ground speed measurement　　b. oil pressure measurement

   c. engine speed measurement

Activities:

1. Demonstrate to your teacher how many hours of use have been placed on the tractor by showing the hour-meter reading for that tractor.

2. Demonstrate to your teacher how to scroll through the various computer digital read-outs to show engine RPM, engine temperature, and hours of use information on that tractor.

References

1. Safety Management for Landscapers, Grounds-Care Businesses, and Golf Courses, John Deere Publishing, 2001. Illustrations reproduced by permission. All rights reserved.
2. Operator's Manuals from various tractor manufacturers

Contact Information

National Safe Tractor and Machinery Operation Program
The Pennsylvania State University
Agricultural and Biological Engineering Department
246 Agricultural Engineering Building
University Park, PA 16802
Phone: 814-865-7685
Fax: 814-863-1031
Email: NSTMOP@psu.edu

Credits

Developed by WC Harshman, AM Yoder, JW Hilton and D J Murphy, The Pennsylvania State University. Reviewed by TL Bean and D Jepsen, The Ohio State University and S Steel, National Safety Council. Revised 3/2013

This material is based upon work supported by National Institute of Food and Agriculture, U.S. Department of Agriculture, under Agreement Nos. 2001-41521-01263 and 2010-41521-20839. Any opinions, findings, conclusions, or recommendations expressed in this publication are those of the author(s) and do not necessarily reflect the view of the U.S. Department of Agriculture.

TRACTOR CONTROLS

HOSTA Task Sheet 4.5 **Core**

NATIONAL SAFE TRACTOR AND MACHINERY OPERATION PROGRAM

Introduction

To help tractor drivers identify controls and use them correctly, many tractor manufacturers use the same color code for specific tractor controls. The direction that you move controls have also become standardized.

Many older tractors do not have controls with uniform color coding. Sometimes those colors wear off or a control is replaced with a irregular color control knob. Moving a control that is not color-coded may not result in the expected operation.

This task sheet will identify the four main groups of tractor controls, their colors, and their direction of movement. Each group of controls will be discussed in more detail in their own task sheet.

Controls and Colors

The American Society of Agricultural Engineers (ASAE) has published standards for tractor controls (standards are widely accepted rules set in place by experts). The four main groups of color-coded controls are discussed below.

*Commit this color code to memory***.** You will use this information to operate a modern tractor.

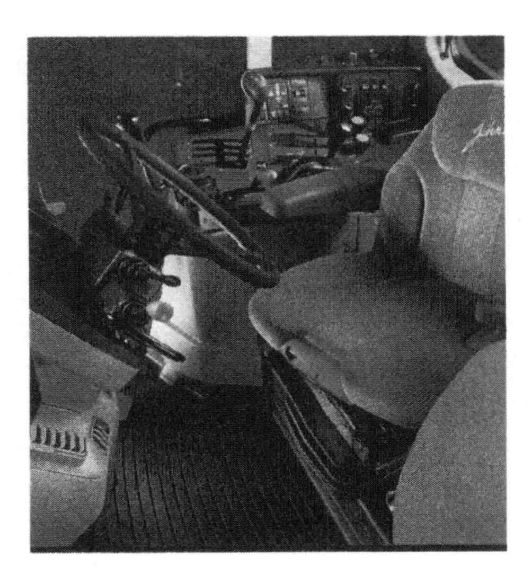

Fig 4.5.a. Know where each control is located and what it controls. Color codes will help you learn the function of each control.

Color Coding for Controls -

STOP ENGINE—**RED** color

GROUND MOTION- **ORANGE** color (engine speed, PARK-Lock, transmission)

POWER ENGAGEMENT – **YELLOW** color (engage PTO or remote power sources)

POSITIONING and ADJUSTING– **BLACK** color (choke the engine, turn lights on)

Remember that older tractors may not use these colors, or you may not be able to see them. If the tractor you need to use does not have color controls, take time to learn about the controls on that tractor.

The same control on an older tractor may not produce the same result as on a newer tractor.

Learning Goals

- To identify tractor controls by their color coding
- To identify what action will result when a control is moved in a particular direction

Related Task Sheets:

Cooperation provided by The Ohio State University and National Safety Council.

Moving Controls

As a general rule, controls will function in the following way:

- To engage a foot brake, push in. To set a hand brake, pull up.
- A foot clutch is disengaged when it is pushed in and engaged when let up.
- A hand-operated engine speed control (throttle) increases the engine speed if the throttle is moved upward or forward. A foot-operated throttle increases speed as it is pushed forward or downward by toe pressure.
- The direction the tractor travels is controlled by specific forward and reverse gears or by directional controls. If a hand-operated directional control is used, the tractor moves in the same direction as the control is moved.
- The engine stop control is by key and by mechanical push-pull control. A key is always turned counterclockwise to stop an engine. A push-pull lever is always pulled out to stop the engine.
- Controls that lift or lower attachments or implements are generally pushed forward, down, or away for lowering, and pulled back, up, or toward you for lifting.
- A PTO is usually engaged when pulled up or pushed forward.

Figure. 4.5.b. What does this yellow control knob do?

You are responsible for many controls. Know the use of each one.

Safety Activities

1. Matching color with function. (Place the letter of the correct color next to the control function.)

____Engage PTO	A. Red
____Lift a High-Lift Bucket	B. Orange
____Throttle Up	C. Yellow
____Stop the Diesel Engine	D. Black

2. Identify as many specific controls as you can on one or more tractors, and group them by control function.
3. What will happen if you pull an orange-colored control in order to stop the tractor engine?

References

1. American Society of Agricultural and Biological Engineers, ANSI/ASABE, EP443.1 Color Coding of Hand Controls, St. Joseph, MI.
2. Owners' Manuals for Specific Tractors.
3. Farm and Ranch Safety Management, John Deere Publishing, 2009.

Contact Information

National Safe Tractor and Machinery Operation Program
The Pennsylvania State University
Agricultural and Biological Engineering Department
246 Agricultural Engineering Building
University Park, PA 16802
Phone: 814-865-7685
Fax: 814-863-1031
Email: NSTMOP@psu.edu

Credits

Developed by WC Harshman, AM Yoder, JW Hilton and D J Murphy, The Pennsylvania State University. Reviewed by TL Bean and D Jepsen, The Ohio State University and S Steel, National Safety Council. Revised 3/2013

This material is based upon work supported by the National Institute of Food and Agriculture, U.S. Department of Agriculture, under Agreement Nos. 2001-41521-01263 and 21010-41521-20839. Any opinions, findings, conclusions, or recommendations expressed in this publication are those of the author(s) and do not necessarily reflect the view of the U.S. Department of Agriculture.

ENGINE STOP CONTROLS

HOSTA Task Sheet 4.5.1 **Core**

NATIONAL SAFE TRACTOR AND MACHINERY OPERATION PROGRAM

Introduction

"How do I stop the engine?" What is a routine operation on one tractor can be a little confusing on a different tractor.

For many years, tractor manufacturers have used the same color for certain controls to help drivers identify controls and use them correctly. This task sheet discusses the "Stop Engine" control.

The Color Red

Red is the color code for the single purpose of "Stop Engine" control. Whether it is a gasoline engine tractor, a diesel engine tractor, or an alternative fuel engine, the color red indicates a stop engine function.

Gasoline Engine—Red letters on key switch.

Diesel Engine—Red fuel shut-off switch (Remember, most diesel engines are shut off with the fuel shut-off switch, not by the ignition key.)

Some Rules for "Red"

Here are a few more points to remember for the red engine stop control. If a mechanical push-pull fuel switch is used, it must:

- Be within 6 inches of the key switch
- Be pulled to stop
- Be labeled "Pull to Stop Engine"
- Remain in the stop position without continued effort

Key switch controls turn counterclockwise to stop the engine.

Some newer diesel engines are also stopped simply by turning the key counterclockwise to the off position.

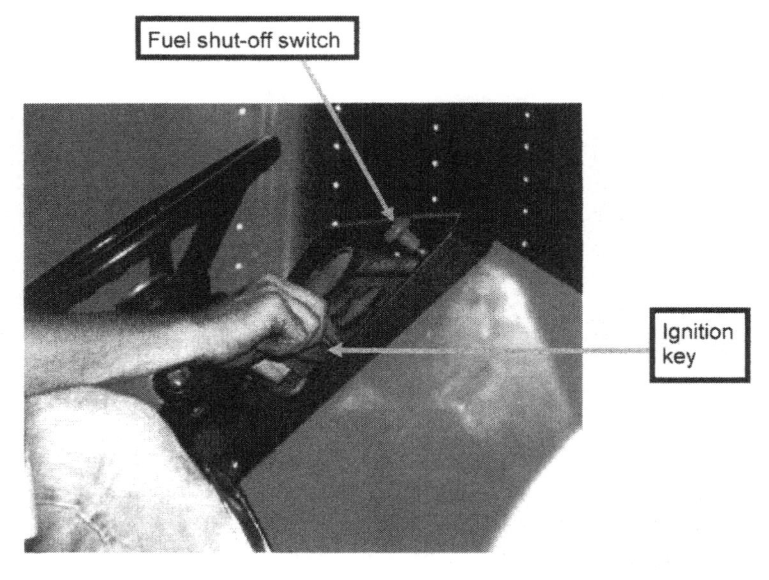

Figure 4.5.1.a. Remember, diesel engines are often shut off with the fuel shut-off switch, not by the ignition key. Newer diesel engines may be shut off by the key only.

A red control knob means "Stop the Engine."

Learning Goals

- To identify tractor engine stop controls used on modern tractors by their color
- To identify the results when an engine stop control is moved in a particular direction

Other related sheets:

Tractor Controls **4.5**

Cooperation provided by The Ohio State University and National Safety Council.

Pictorial Study

Figure 4.5.1.b. Diesel engines are often stopped by shutting off the fuel flow from the fuel pump.

Figure 4.5.1.c. Key switch on lower left of older tractor. Push-pull switches may be found also.

Figure 4.5.1.d. Quiz time. What if the red color is missing on older tractors?

A similar colored control on an older tractor may not have the same result as the control on a newer tractor.

Safety Activities

1. Compare the ignition switch and stop engine control methods of diesel and gasoline engine tractors by tracing the wiring of each.
2. Find the oldest tractor model you can in your community, and determine if color-coding would indicate how to stop the engine. Record the following information:

Tractor Model	Approximate Age of Tractor	Color-Coded Stop Control Y/N
__________	_______	________________
__________	_______	________________

References

1. American Society of Agricultural and Biological Engineers, ANSI/ASABE, EP443.1 Color Coding of Hand Controls, St. Joseph, MI.
2. Owners' Manuals for Specific Tractors.
3. Farm and Ranch Safety Management, John Deere Publishing, 2009.

Contact Information

National Safe Tractor and Machinery Operation Program
The Pennsylvania State University
Agricultural and Biological Engineering Department
246 Agricultural Engineering Building
University Park, PA 16802
Phone: 814-865-7685
Fax: 814-863-1031
Email: NSTMOP@psu.edu

Credits

Developed by WC Harshman, AM Yoder, JW Hilton and D J Murphy, The Pennsylvania State University. Reviewed by TL Bean and D Jepsen, The Ohio State University and S Steel, National Safety Council. Revised 3/2013

This material is based upon work supported by the National Institute of Food and Agriculture, U.S. Department of Agriculture, under Agreement Nos. 2001-41521-01263 and 2010-41521-20839. Any opinions, findings, conclusions, or recommendations expressed in this publication are those of the author(s) and do not necessarily reflect the view of the U.S. Department of Agriculture.

GROUND MOTION CONTROLS

HOSTA Task Sheet 4.5.2 **Core**

NATIONAL SAFE TRACTOR AND MACHINERY OPERATION PROGRAM

Introduction

"How do I get this tractor to move?" "How do I stop this operation?" For many years, tractor manufacturers have used the same color for certain controls to help drivers identify and use them correctly. This task sheet discusses "Ground Motion" controls.

Figure 4.5.2.a. Ground motion controls include transmission controls, park-lock, and gear-shift levers.

The Color Orange

Orange is the color code for tractor ground motion controls. Ground motion controls include:

- Engine Speed
- Transmission Controls
- Parking Brake or Park-Lock
- Independent Emergency Brakes
- Differential Lock

You can easily become confused if you are not familiar with the tractor. Do not hesitate to ask for a demonstration of the controls and job you will be doing.

An orange control knob shows you where to control ground motion.

Some Rules for "Orange"

Here are more important points to remember for orange ground motion controls.

- Engine speed controls are operated with the right hand and/or right foot.
- Transmission gearshift patterns must be clearly and permanently identified.
- Differential lock controls are engaged with a forward or downward motion.
- Brake locks may be a mechanical lock on the drive train versus a lock on the axle.

Learning Goals

- To identify tractor ground motion controls by the orange color coding
- To identify what action results when a ground motion control is moved in a particular direction

Other related sheets:

Tractor Controls **4.5**

Cooperation provided by The Ohio State University and National Safety Council.

Pictorial Study

Figure 4.5.2.b. The foot throttle on the tractor is orange in color. Orange is the color code for ground motion controls.

Figure 4.5.2.c. Brakes are locked together and the orange lever is for setting the brakes on this tractor.

Figure 4.5.2.d. Quiz time
What if the orange color is missing on older tractors?

A similar colored control on an older tractor may not produce the same result as the control of a newer tractor.

Safety Activities

1. Ask the farmer/owner if you can inspect all the tractors on a farm. Note the orange color-coded controls. What does each control do? Make a comparison of how older model tractor controls are identified for ease of recognition compared with newer model tractors.
2. Identify as many ground motion controls as you can on several different tractors. Compare their locations and the direction in which they are moved.

References

1. American Society of Agricultural and Biological Engineers, ANSI/ASABE, EP443.1 Color Coding of Hand Controls, St. Joseph, MI.
2. Owners' Manuals for Specific Tractors.
3. Farm and Ranch Safety Management, John Deere Publishing, 2009.

Contact Information

National Safe Tractor and Machinery Operation Program
The Pennsylvania State University
Agricultural and Biological Engineering Department
246 Agricultural Engineering Building
University Park, PA 16802
Phone: 814-865-7685
Fax: 814-863-1031
Email: NSTMOP@psu.edu

Credits

Developed by WC Harshman, AM Yoder, JW Hilton and D J Murphy, The Pennsylvania State University. Reviewed by TL Bean and D Jepsen, The Ohio State University and S Steel, National Safety Council. Revised 3/2013

This material is based upon work supported by the National Institute of Food and Agriculture, U.S. Department of Agriculture, under Agreement Nos. 2001-41521-01263 and 2010-41521-20839. Any opinions, findings, conclusions, or recommendations expressed in this publication are those of the author(s) and do not necessarily reflect the view of the U.S. Department of Agriculture.

POWER ENGAGEMENT CONTROLS

HOSTA Task Sheet 4.5.3 **Core**

NATIONAL SAFE TRACTOR AND MACHINERY OPERATION PROGRAM

Introduction

"How do I get this implement to run? How can I stop this machine? Where is the PTO control for this tractor?"

For many years, tractor manufacturers have used the same color for certain controls to help drivers identify and use them correctly. This task sheet discusses the "Power Engagement" control.

The Yellow Color

Yellow is the color code for the controls which engage mechanisms using the tractor as a remote power source. The same color coding is used for self-propelled machines. Here are a few of the power engagement-type controls:

- PTO
- Cutterheads
- Feed Rolls
- Elevators
- Winches
- Unloading Augers

  You can easily become confused if you are unfamiliar with a tractor. A quick review of the Owner's Manual will help identify controls and their function. Do not hesitate to ask for a demonstration of the job you will be doing.

Figure 4.5.3.a. Yellow color-coded controls engage accessories. This is often done through the PTO.

Some Rules for "Yellow"

Here are a few more points to remember for yellow power-engagement controls. These controls can be knobs, toggle or rocker switches, levers, or pedals.

1. PTO controls are designed to move to the rear or downward to disengage the PTO.
2. Horizontal-mounted rocker switches use the right side to begin normal machine operation.
3. Vertical-mounted rocker switches use the upper side of the switch to begin normal machine operation.

A yellow-colored control knob means "engage remote power" to a machine.

Learning Goals

- To identify tractor power-engagement controls on modern tractors by their color coding
- To identify what action results when a power-engagement control is moved in a particular direction

Related Task Sheets:

Cooperation provided by The Ohio State University and National Safety Council.

Pictorial Study

Figure 4.5.3.b. Most control levers are right-side mounted.

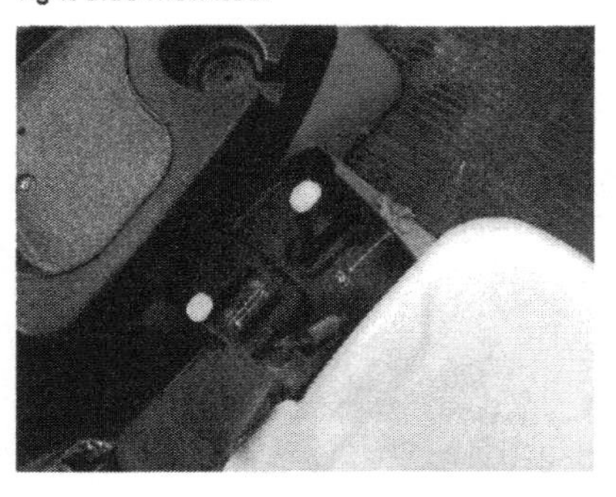

Figure 4.5.3.c. Some control levers may be left-side mounted.

Figure 4.5.3.d. Rocker arm switches may be used. If you find a control feature with which you are unfamiliar, ask for instructions before operating costly equipment.

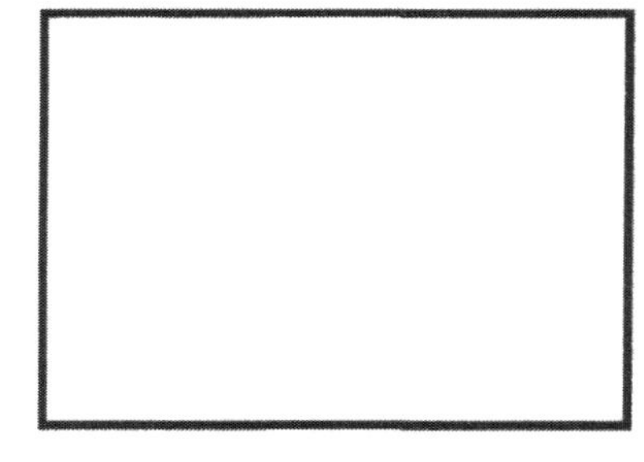

Quiz time? What if the yellow color is missing on older tractors? How would you find the PTO control? Make a sketch here of an older PTO control.

A similar colored control on an older tractor may not produce the same result as the control on a newer tractor.

Safety Activities

1. Ask the farmer/owner if you can inspect all the tractors on the farm. Note the yellow color-coded controls. What does each control do? Make a comparison of how older model tractor controls are identified for ease of recognition when compared with newer model tractors.
2. Identify as many power-engagement controls as you can on several different tractors, and compare their locations and the directions in which they move.

References

1. American Society of Agricultural and Biological Engineers, ANSI/ASABE, EP443.1 Color Coding of Hand Controls, St. Joseph, MI.
2. Owners' Manuals for Specific Tractors.
3. Farm and Ranch Safety Management, John Deere Publishing, 2009.

Contact Information

National Safe Tractor and Machinery Operation Program
The Pennsylvania State University
Agricultural and Biological Engineering Department
246 Agricultural Engineering Building
University Park, PA 16802
Phone: 814-865-7685
Fax: 814-863-1031
Email: NSTMOP@psu.edu

Credits

Developed by WC Harshman, AM Yoder, JW Hilton and D J Murphy, The Pennsylvania State University. Reviewed by TL Bean and D Jepsen, The Ohio State University and S Steel, National Safety Council. Revised 3/2013

This material is based upon work supported by the National Institute of Food and Agriculture, U.S. Department of Agriculture, under Agreement Nos. 2001-41521-01263 and 2010-41521-20839. Any opinions, findings, conclusions, or recommendations expressed in this publication are those of the author(s) and do not necessarily reflect the view of the U.S. Department of Agriculture.

POSITIONING AND ADJUSTING CONTROLS

HOSTA Task Sheet 4.5.4 **Core**

NATIONAL SAFE TRACTOR AND MACHINERY OPERATION PROGRAM

Introduction

"Every control knob seems to be black in color except that red, orange, and yellow one. I want to lift the scraper blade to clean the free stall alley like the owner told me to do. Let's see…which one of these levers will I use?"

For many years, tractor manufacturers have used the same color for certain controls to help drivers identify and use them correctly. This task sheet discusses "Positioning and Adjusting" controls.

Figure 4.5.4.a. Black controls adjust accessory position and control electrical components. They also show choke location as pictured here.
Graphic courtesy of Hobart Publications.

The Black Color

Black is the color code for the many controls which position or adjust tractor work accessories. A few of the positioning/adjusting controls are:

- Remote hydraulic control
- Implement hitches
- Unloading components on self-propelled equipment
- Engine chokes and steering column position
- Lights, flashers, and signals
- Cab comforts (fans, radio, etc.)

You can easily become confused if you are unfamiliar with a tractor. Do not hesitate to ask for a demonstration of the controls to use for the job you will be doing.

Some Rules for "Black"

Here are a few more rules to help you use the black color coded controls. These controls can be knobs, toggle or rocker switches, levers, or pedals.

1. Lift controls operated from the tractor seat must be clearly identified and are found on the right side of the cab.
2. Front-end loader controls must be located on the right side of the operator.
3. Foot controls must be pushed forward to lower equipment.

A black control knob means "position or adjust."

Learning Goals

- To identify tractor positioning and adjusting controls on modern tractors by their color coding
- To identify what action will result when a position/adjustment control is moved in a particular direction

Related Task Sheets:

Tractor Controls **4.5**

Cooperation provided by The Ohio State University and National Safety Council.

Pictorial Study

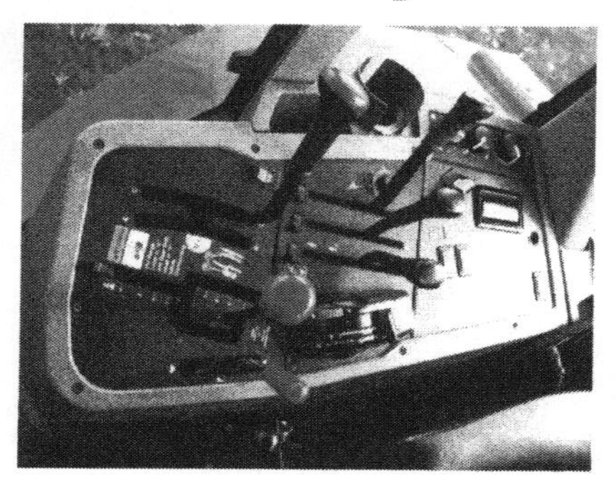

Figure 4.5.4.b. High lift controls are color-coded black.

Figure 4.5.4.c. The light control switch is a black rocker switch.

Figure 4.5.4.d. There are many seat adjustments shown here. Seat positioning and adjusting is coded with black knobs.

Figure 4.5.4.e. What if the black color is missing on older tractors? Where is the light switch?

A similar colored control on an older tractor may not produce the same result as the control on a newer tractor.

Safety Activities

1. Ask a farmer/owner if you can inspect all the tractors on the farm. Note the black color-coded controls. What does each control do? Make a comparison of how older model tractor controls are identified for ease of recognition compared with newer model tractors.
2. Obtain a tractor's Operator's Manual and read the instructions for setting the 3-point hitch for depth control of plows or scraper blades.

References

1. American Society of Agricultural and Biological Engineers, ANSI/ASABE, EP443.1 Color Coding of Hand Controls, St. Joseph, MI.
2. Owners' Manuals for Specific Tractors.
3. Farm and Ranch Safety Management, John Deere Publishing, 2009.

Contact Information

National Safe Tractor and Machinery Operation Program
The Pennsylvania State University
Agricultural and Biological Engineering Department
246 Agricultural Engineering Building
University Park, PA 16802
Phone: 814-865-7685
Fax: 814-863-1031
Email: NSTMOP@psu.edu

Credits

Developed by WC Harshman, AM Yoder, JW Hilton and D J Murphy, The Pennsylvania State University. Reviewed by TL Bean and D Jepsen, The Ohio State University and S Steel, National Safety Council. Revised 3/2013.

This material is based upon work supported by the National Instiute of Food and Agriculture, U.S. Department of Agriculture, under Agreement Nos. 2001-41521-01263 and 2010-41521-20839. Any opinions, findings, conclusions, or recommendations expressed in this publication are those of the author(s) and do not necessarily reflect the view of the U.S. Department of Agriculture.

LOCATION AND MOVEMENT OF TRACTOR CONTROLS

HOSTA Task Sheet 4.5.5 **Core**

NATIONAL SAFE TRACTOR AND MACHINERY OPERATION PROGRAM

Introduction

Tractors are designed for multiple tasking (doing many jobs at once). Several functions may occur at the same time. A safe operator will be able to maintain control of each function. For example, you are mowing hay with a 12-ft. wide mower-conditioner. As you approach an uphill grade, you must downshift (ground motion control). A huge rock in the field also means you must raise the mower head to avoid damage to the knife guards and knife sections (machine positioning control). You are steering, shifting, and using remote hydraulic controls simultaneously.

This task sheet will identify several important tractor controls and their direction of movement.

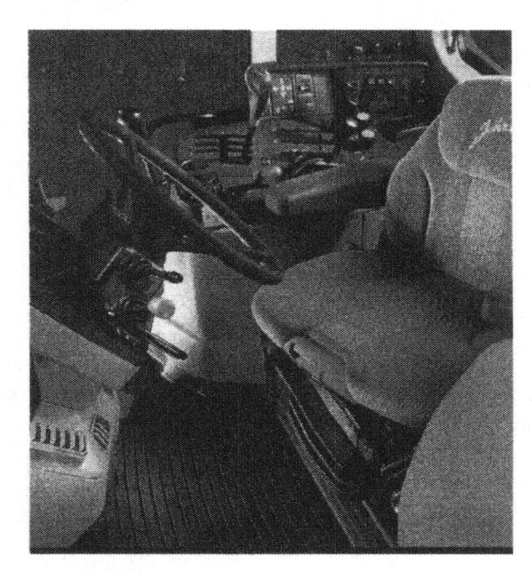

Figure 4.5.5.a. Tractor controls are placed in specific locations so that operators do not have to search for them. Older tractors may have controls placed in various locations. How the control is used may be entirely different from tractor to tractor.

Tractor controls allow you to multi-task.

Control Devices and Functions

Tractor manufacturers have tried to help tractor drivers identify controls and use them correctly for many years. For example, specific controls are located on the same side of the operator's seat and move in the same direction to obtain a desired effect. Similar to the color-coding of main groups of controls, many older tractors may have controls or directions of movements that are not the same as newer tractors.

Three common types of control devices are used on a tractor.

They are:

1. **Foot Controls**—Pedals
2. **Hand Controls**—Levers, toggles, switches, knobs, and buttons
3. **Combination Hand and Foot**—Engine throttles

These controls apply brakes, operate the clutch, speed the engine, change gears, lock the differential, steer, stop the engine, lift implements, engage the PTO, and control electrical and hydraulic flow. Computer functions are also part of the control panel on modern tractors.

Learning Goals

- To identify the location of major operating controls on tractors
- To move a tractor's major operating controls to obtain a desired function

Related Task Sheets:

Tractor Controls	**4.5**
Engine Stop Controls	**4.5.1**
Ground Motion Controls	**4.5.2**
Power Engagement Controls	**4.5.3**
Positioning and Adjusting Controls	**4.5.4**

Cooperation provided by The Ohio State University and National Safety Council.

Movement and Location of Controls

The same location and direction of motion for controls makes it easier to operate the tractor safely and efficiently. The ASABE standard for location and direction of motion for tractor controls is listed in the reference section. Below are the most common rules for the location and direction of motion for tractor controls, including some combinations of control functions. There are several exceptions to these rules. Study the Owner's Manual for all the tractors you operate. Consult the tractor owner to be sure you know where a control is located and what happens when you move a control. Do this before operating the tractor.

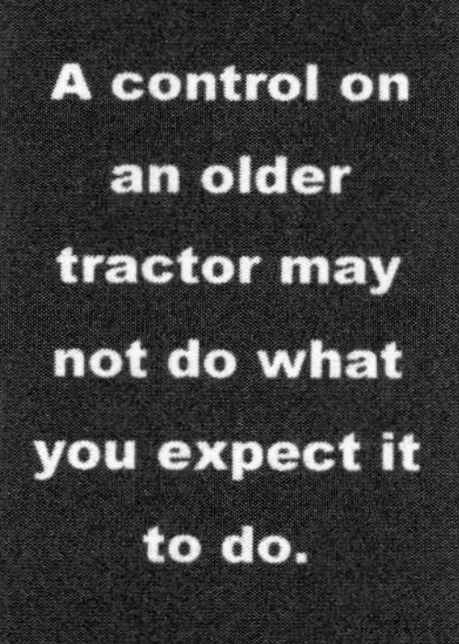

Figure 4.5.5.b. Brake control—Foot brake pedals must be located on the right side. Push the brake forward and/or downward to engage. If a hand brake is provided, it can be on either side and must be pulled to be set. Brake locks may be lifted to be set.

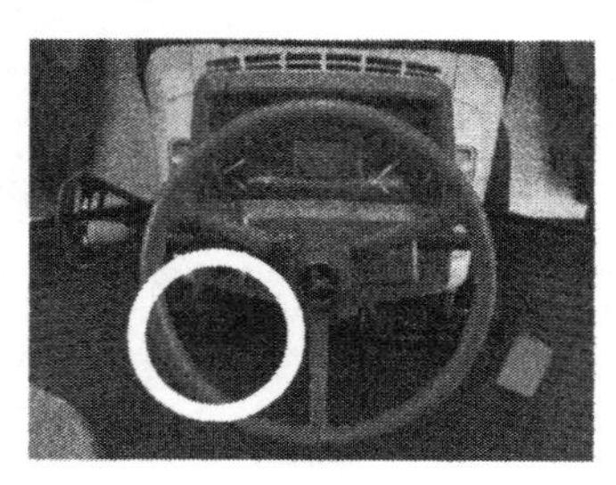

Figure 4.5.5.c. Clutch control– A foot clutch pedal must be located on the left side. The pedal is moved forward or downward for disengagement. A hand-operated clutch can be located on either side and must be moved toward the operator to be disengaged.

Combination clutch and PTO control—A foot-operated combination will be on the left side and moved forward and/or downward to cause clutch and PTO disengagement.

Combination clutch and brake—A foot-operated combination will be found on the left side and moved forward and/or downward to cause clutch disengagement and brake engagement.

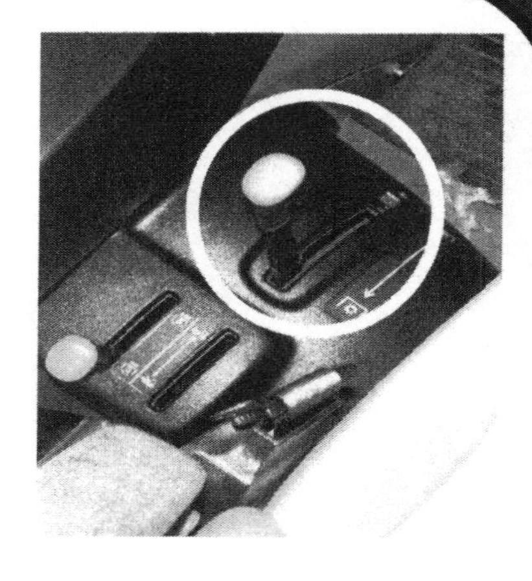

Figure 4.5.5.d. Power Take-Off (PTO) control—A hand-operated PTO control can be located on either side and will be moved upward or forward for engagement and rearward or downward for disengagement.

Figure 4.5.5.e. Engine speed control—The control is located on the right side. If the hand-operated control is located next to the tractor seat, the direction of motion must be forward or upward to increase engine speed and rearward or downward to slow engine speed.

Cooperation provided by The Ohio State University and National Safety Council.

If the hand-operated speed control is located near the steering wheel, the direction of motion must be rearward and/or downward to increase speed and forward and/or upward to slow engine speed.

If a foot-operated control is provided, it must be on the right side and moved forward and/or downward to increase speed.

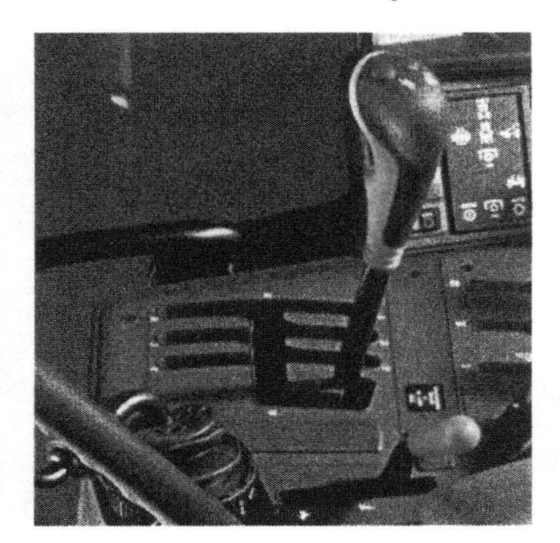

Figure 4.5.5.f. *Ground speed control*—A hand-operated forward-reverse (non-variable speed) directional control must be moved forward for forward travel and rearward for reverse. A hand-operated variable speed control must be moved forward and/or upward to increase speed and rearward and/or downward to decrease speed.

A hand-operated combination direction and variable speed control must be moved forward or away from the operator—from the neutral position—for forward travel and increasing speed. To reverse and to increase reverse speed, the control is moved rearward or toward the operator, from a neutral position.

A foot-operated combination direction and variable speed control(s) must be on the right side. If a single pedal is used, it must produce forward motion with a forward or downward toe motion, and move in reverse with a rearward or downward heel motion. If two pedals are used, the inner pedal must be moved forward or downward for forward motion, and the outer pedal must be moved forward or downward for backing up. Also, the forward or downward pressure on both pedals must increase speed and automatically return to a neutral position when a foot is taken off the pedal.

Figure 4.5.5.g. *Differential lock control*—A differential lock must be moved forward or downward for engagement.

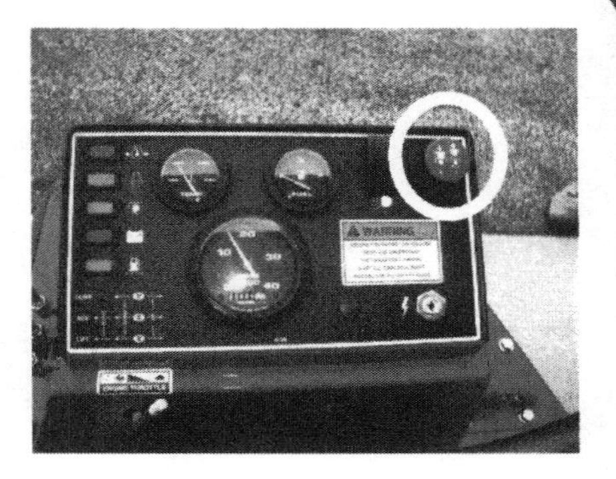

Figure 4.5.5.h. *Engine stop control*—A key switch must be rotated counterclockwise to stop the engine. A mechanical pull-push control must be within 6 inches of the key switch and pulled to stop the engine. Engine stop and ground speed controls that are combined into a single lever must move in the same direction to first slow ground speed and then stop the engine.

Figure 4.5.55.i. *Lift controls for implements or attachments*—Lift controls must be located on the right side. A hand-operated control must be moved forward, downward, or away from the operator for lowering, and backward, upward or toward the operator for lifting.

Cooperation provided by The Ohio State University and National Safety Council.

Safety Activities

1. Visit area farms or equipment dealers and review with the farmer or dealers how the major controls operate. Make a record of which ones follow ASAE standards for location and direction of motion.
2. Solve this word search puzzle on tractor controls and color coding.

Tractor Controls

```
Y E L L O W C O N T R O L D Y
Z T L X A Y L A Y A A O I T B
V T O O U D Y A Y Q R F L G K
D J R J U O P U D T F N O I N
P O T S E N I G N E D E R X E
C A N Q G N G O R E P L T K F
D W O R I C C E L N W T N M B
I K C W L E N H A B M T O H E
V V K I G T W W P R C O C O W
D W C N I C L C G A S R T B F
G D A A P P F B J K I H F F X
E R L H J D S I C E C T I E R
O M B P X F Q C K S Q W L F P
Y E L P G K G B W D P A P J T
```

Words to Use:

Black Control	Gearshift	Red Engine Stop
Brakes	Lift Control	Throttle
Differential	Orange Control	Yellow Control
Foot Pedal	PTO	

References

1. American Society of Agricultural and Biological Engineers, ANSI/ASABE, S335.4 Operator Controls on Agricultural Equipment, St. Joseph, MI.
2. Various Owners' Manuals for Specific Tractors
3. Farm and Ranch Safety Management, John Deere Publishing, 2009.

Contact Information

National Safe Tractor and Machinery Operation Program
The Pennsylvania State University
Agricultural and Biological Engineering Department
246 Agricultural Engineering Building
University Park, PA 16802
Phone: 814-865-7685
Fax: 814-863-1031
Email: NSTMOP@psu.edu

Credits

Developed by WC Harshman, AM Yoder, JW Hilton and D J Murphy, The Pennsylvania State University. Reviewed by TL Bean and D Jepsen, The Ohio State University and S Steel, National Safety Council. Revised 3/2013.

This material is based upon work supported by the National Institute of Food and Agriculture, U.S. Department of Agriculture, under Agreement Nos. 2001-41521-01263 and 2010-41521-20839. Any opinions, findings, conclusions, or recommendations expressed in this publication are those of the author(s) and do not necessarily reflect the view of the U.S. Department of Agriculture.

TRACTOR OPERATION SYMBOLS

HOSTA Task Sheet 4.5.6 **Core**

NATIONAL SAFE TRACTOR AND MACHINERY OPERATION PROGRAM

Introduction

Operational symbols were designed to promote and improve tractor and equipment use and safety in the agricultural workplace. Operational symbols were developed to show tractor and equipment operating functions. Operation symbols are pictures used to transmit information with minimal use of words and are displayed in a standard way.

This task sheet discusses uniform, tractor operation symbols that workers on farms should recognize and understand. Use Owners' Manuals to learn more about these symbols.

Tractor Operation Symbols

Symbols are designed to draw your attention to operating functions and alert you to malfunctions. These symbols may be found on agricultural, construction, and industrial equipment. Owners' Manuals detail operating symbols of particular importance to your tractor or machine.

Symbols quickly help a person to recognize a function or malfunction. *Learn what each symbol communicates.* This information can help you prepare for work or respond to a malfunction. Use the reference section to find a complete exhibit of tractor and equipment operation symbols.

This symbol represents diesel fuel. Be sure of which fuel you are putting into the tank. From this pictorial, can you identify the type of fuel pump and the type of fuel supplied?

This symbol serves as a reminder to use the seat belt. A tractor equipped with a ROPS can save your life when used with the seat belt.

This symbol is an ALERT for a malfunction. Alert symbols usually are found in conjunction with another symbol.

Figure 4.5.6.a. The symbol for oil should draw your attention to checking the oil or the oil fill area. You will see this symbol with engine lubricant and hydraulic systems.

Tractor operation symbols provide quick information regarding operating functions and malfunctions.

Learning Goals

- To recognize the messages that tractor operation symbols are conveying in normal tractor use
- To recognize the messages that tractor operation symbols are conveying in order to react to possible malfunctions

Other related sheets:

Hazard Warning Signs	**2.8**
Tractor Instrument Panel	**4.4**
Tractor Controls	**4.5**

Cooperation provided by The Ohio State University and National Safety Council.

Oil Lubricant
Type & Frequency

Grease Lubricant
Frequency

Figure 4.5.6.b. Universal symbols provide operating information. The oil can symbol may be used to indicate frequency of oil changes and the SAE number of oil to use. This picture would represent a SAE 20 oil changed at 50 hours. The grease gun shaped object shows a grease point and how often to apply the lubricant.

Operating Symbols

During tractor operation, these symbols will indicate what to do or what is happening.

Engine R.P.M.

This symbol will show you what the engine speed is.

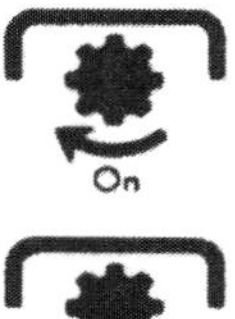

Power Take-off

The PTO symbol indicates an engaged/disengaged function.

Speed Range

The throttle symbol reminds us of the slow turtle and fast rabbit story, or speed control.

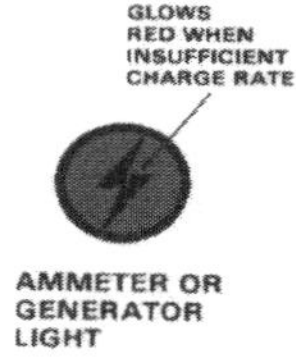

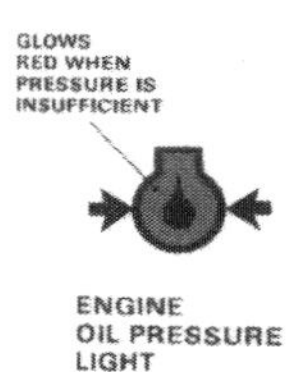

The engine oil pressure and ammeter symbols are used to draw your attention to malfunctions during operations. An oil pressure gauge that begins to show red lighting is an indication to stop the engine immediately. A red glowing ammeter light display indicates the battery is not charging properly. The operator could still use the tractor with a low battery, but the problem must be fixed soon.

An oil light or gauge that indicates low oil pressure is a message to stop the engine immediately. Major repairs will occur if you do not react.

Figure 4.5.6.c. This symbol shows the only recommended lift point to attach a chain for moving a heavy weight. Damage or injury can occur if any other lift point is used.

Cooperation provided by The Ohio State University and National Safety Council.

Other Operation Symbols

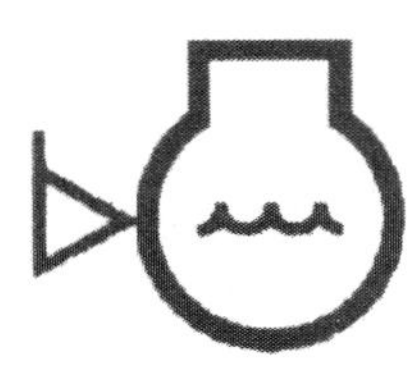

This symbol shows engine coolant level. If an oil drop was shown in the center of the engine block form, what would this symbol represent?

This symbol represents the clutch. If you do not know how the clutch engages the transmission to the engine, find someone who can explain this operation to you.

More Operation Symbols

You may need to use the accessories on a tractor. Operation symbols will be found on the equipment as well as on the tractor.

This symbol indicates that the resulting operation will raise the high-lift bucket. This is a positioning and adjusting control symbol.

This symbol indicates that the resulting operation will tilt the high-lift bucket to the rear.

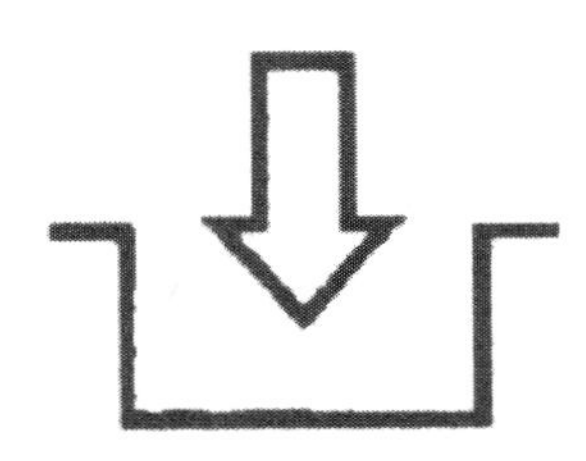

Some symbols may be more difficult to understand. This sign tells you that this is an engage control function. Recall that engagement controls are yellow in color. Remote power operation occurs.

This sign tells you this is a disengage control function. Engagement controls are yellow in color.

You may not encounter all the symbols used, but you should study them for future reference.

Figure 4.5.6.d. Older tractors will not have operation symbols. What will you find on this tractor to tell you the information you need about oil pressure and engine temperature?

Make an effort to learn all the symbols. Owners' Manuals can be helpful.

In the space above, draw an operation symbol that would show someone that the engine oil filter needs to be checked. Check the asae.org website to compare the standard to what you have drawn.

Safety Activities

1. List the top 5 operating symbols you would locate and respond to before you start a tractor. Tell why you think these 5 symbols are important. (There is no wrong answer for this discussion.)

2. You are assigned to rake hay in a field one mile from the farm shop. The engine oil pressure light comes on. Draw the symbol that shows this malfunction.

3. In problem 2, what should be done with the tractor when the problem is observed?

   a. drive it back to the farm shop
   b. continue to rake hay
   c. shut down immediately
   d. let the tractor idle while you use the cell phone to notify the owner of the tractor

4. A tractor you are using begins to show a low-battery charge problem. What should you do?

   a. return to the shop area without finishing to rake the hay
   b. shut down immediately
   c. return to the shop area after finishing to rake the hay
   d. none of these

5. Use the Internet website shown in the reference section to locate the ASABE Graphic Symbols for Operator Controls and Displays on Agricultural Equipment section. Print out this information to share with your class, group, or club. (There are 32 pages to print. Ask your leader or instructor for permission to print first.)

References

1. American Society of Agricultural and Biological Engineers, ANSI/ASABE, S304 Graphic Symbols for Operator Controls and Displays on Agricultural Equipment, St. Joseph, MI.
2. Farm and Ranch Safety Management, John Deere Publishing, 2009.

Contact Information

National Safe Tractor and Machinery Operation Program
The Pennsylvania State University
Agricultural and Biological Engineering Department
246 Agricultural Engineering Building
University Park, PA 16802
Phone: 814-865-7685
Fax: 814-863-1031
Email: NSTMOP@psu.edu

Credits

Developed by WC Harshman, AM Yoder, JW Hilton and D J Murphy, The Pennsylvania State University. Reviewed by TL Bean and D Jepsen, The Ohio State University and S Steel, National Safety Council. Revised 3/2013.

This material is based upon work supported by the National Institute of Food and Agriculture, U.S. Department of Agriculture, under Agreement Nos. 2001-41521-01263 and 2010-41521-20839. Any opinions, findings, conclusions, or recommendations expressed in this publication are those of the author(s) and do not necessarily reflect the view of the U.S. Department of Agriculture.

PREVENTIVE MAINTENANCE AND PRE-OPERATION CHECKS

HOSTA Task Sheet 4.6 **Core**

NATIONAL SAFE TRACTOR AND MACHINERY OPERATION PROGRAM

Introduction

John is a part-time farmer. Two years ago he purchased a small utility tractor with backhoe and scraper blade for $12,000. He wanted to push snow, clean the barn, and do odd jobs on his property. While driving his tractor down the road, the engine overheated, began to make noise, lost power, and shut down. A neighbor stopped by and John asked, "What could be the problem?" He was already pouring water in the radiator. "Could it be the hydrostatic transmission?" he asked as he checked that dipstick.

The neighbor suggested the engine oil, but John didn't know where to find that dipstick, which turned out to be hidden by the high-lift arms. The dipstick registered no oil at all.

Performing tractor maintenance is a critical task for every tractor operator. This task sheet discusses the proper way to maintain a tractor to avoid costly and unnecessary repairs.

Pre-Operation Checks

A good operator uses a daily checklist of items and systems to inspect before starting the tractor. This is often called a pre-operation checklist. Many drivers write down what needs to be inspected and then check off the list as they examine each item.

Things to check include:

- Fuel level
- Coolant level
- Engine oil level
- Hydraulic oil level
- Battery condition
- Lug nuts and wheels
- Tire condition
- Loose or defective parts
- SMV emblem
- Fluid leaks
- Operators platform/steps
- Seat/Adjustment
- Seat belt
- Fire extinguisher
- Lighting/Flashers
- Visibility from operator's seat

Some Practical Hints

Here are several things to look for as you perform a pre-operation check:

- Low tires and leakage from the valve stem
- Oil or hydraulic leaks on the ground beneath the tractor
- A frayed or worn fan belt
- Corroded battery terminals
- Loose bolts or lug nuts on wheels
- Dirty cab windows that obstruct your vision
- Headlights or warning lights with broken bulbs or glass
- An SMV emblem that is faded or distorted in either color or shape
- A fire extinguisher with a pressure gauge in the "recharge" range
- Several tools or supplies on the operator platform

If you were to buy a new, expensive tractor, what would you want your friends to check before they started the engine?

Learning Goals

- To conduct pre-operation checks on a daily basis to reduce repair costs and downtime

Related Task Sheets:

Fuel, Oil, Coolant Levels	**4.6.1**
Lead Acid Batteries	**4.6.2**
Tire and Wheel Condition	**4.6.5**
The Operator Platform	**4.6.6**

Cooperation provided by The Ohio State University and National Safety Council.

Don't start the engine until you have completed the "walk-around" inspection and are sure all systems are ready to work for you.

Safe Starts

Some newer utility or lawn tractors may have safety start systems. If so, the owner should also have in good working order one or both of the following items:

Seat Switch/Safety Interlock that prevents starting the tractor if the operator is not in the seat

Neutral-Start Safety Switch that prevents the tractor from starting if the tractor is in gear

A good operator takes responsibility for the tractor he or she operates.

Safety Activities

1. Make a chart of maintenance items to be done on your tractor. Use the following format, or develop your own chart. If you have a computer, make a spreadsheet or database project to help with maintenance records.

Tractor Maintenance Log

Date	Item Checked	Problem Found	Corrective Action

2. Help someone change the oil and oil filter on a tractor.
3. Help someone change an air filter on a tractor.
4. Call a tractor dealer/service center, and ask for any maintenance charts or record forms that they can send to you.
5. Memorize the "pre-op" checklist, and recite this list as you conduct a pre-operation inspection for your class or an interested adult.
6. Math Problem: You forgot to check the engine oil in the tractor before starting. When the oil light came on, you continued working. Now the engine must be rebuilt to the amount of $5000. This is the only tractor that can pull the forage harvester and chop 40 acres per day for the next 5 days. An estimated nutrient loss value of $10 per acre will occur due to the delay in harvest. Calculate the dollar loss to the producer.

References

1. Farm and Ranch Safety Management, John Deere Publishing, 2009.
2. Owners' Manuals for specific tractors.

Contact Information

National Safe Tractor and Machinery Operation Program
The Pennsylvania State University
Agricultural and Biological Engineering Department
246 Agricultural Engineering Building
University Park, PA 16802
Phone: 814-865-7685
Fax: 814-863-1031
Email: NSTMOP@psu.edu

Credits

Developed by WC Harshman, AM Yoder, JW Hilton and D J Murphy, The Pennsylvania State University. Reviewed by TL Bean and D Jepsen, The Ohio State University and S Steel, National Safety Council. Revised 3/2013.

This material is based upon work supported by the National Institute of Food and Agriculture, U.S. Department of Agriculture, under Agreement Nos. 2001-41521-01263 and 2010-41521-20839. Any opinions, findings, conclusions, or recommendations expressed in this publication are those of the author(s) and do not necessarily reflect the view of the U.S. Department of Agriculture.

FUEL, OIL, AND COOLANT LEVELS

HOSTA Task Sheet 4.6.1

NATIONAL SAFE TRACTOR AND MACHINERY OPERATION PROGRAM

Introduction

A tractor is a huge investment to make farm work more efficient. Even a mid-size tractor may cost $40,000 or more.

The tractor must be kept in top operating condition. Downtime for engine and tractor repairs are costly. An engine rebuild may cost over $5000 in parts and labor. A crop in the field may be lost because of harvest delays. Crop losses can lead to increased costs to purchase replacement feeds or protein supplements.

Therefore, tractor and equipment pre-operation checks are an economic necessity. A damaged engine or an empty fuel tank at the farthest field from the barn is no excuse for the skilled operator.

This task sheet discusses the importance of checking the fluid levels of the

- fuel
- coolant, and
- oils

before you touch the tractor ignition switch. Developing this habit will help you to understand that the tractor engine is ready for field work.

Figure 4.6.1.a. Before driving the tractor to the field, check for the possibility of an empty fuel tank. If you run out of fuel during a workday, you are causing downtime losses.

What to Do

Fig. 4.6.1.b. Check the fuel level.

Fig. 4.6.1.d. Check the coolant level with the engine cold.

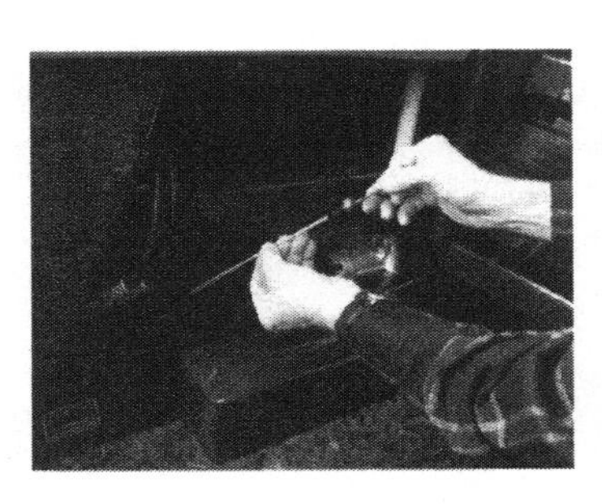

Fig. 4.6.1.c. Check the oil level.

Save an engine from costly repairs; check the fuel, coolant, and oil levels before starting the engine.

Learning Goals

- To understand how to check fuel levels of common engines (alternative fuels excluded here)
- To safely check coolant levels of liquid cooled engines
- To correctly check oil levels of any engine

Related Task Sheets:

Tractor Instrument Panel	**4.4**

Cooperation provided by The Ohio State University and National Safety Council.

Why You Should Check Fuel, Coolant and Oil Levels

Fuel

Check the fuel level before leaving the barnyard or shop area. You cannot assume that someone else has done this job. Failure to check the fuel level may result in lost field time. Or it may result in the need to mechanically bleed air from diesel fuel lines in some older tractors.

Be sure you do not fill diesel fuel tanks with gasoline and vice versa.

Oil

Oil bathes metal surfaces to prevent the heat of friction from damaging the moving parts. Low engine oil allows engine parts to overheat, expands them, and "seizes" the engine. Overfilling the engine oil results in oil seal damage.

Use the oil dipstick daily to prevent engine damage.

Coolant

Coolant fluid (water and antifreeze) carries engine heat away from the engine. Air flowing across the radiator then reduces the coolant temperature. Lack of coolant causes overheating of the engine. Water used as a coolant by itself will cause rust in the water pump.

Check coolant levels while the engine is cold to prevent severe scalds.

Figure 4.6.1.e. Never remove a radiator cap from a hot engine. Steam and hot water from the radiator can scald your skin. *Safety Management for Landscapers, Grounds-Care Businesses, and Golf Courses, John Deere Publishing, 2001. Illustrations reproduced by permission. All rights reserved.*

If the engine oil light comes on while you are operating the tractor, shut down immediately.

Safety Activities

1. Park the tractor at the farthest field from the barn, and time your walk back to the farm shop or fuel area. This is wasted time or downtime when cropping work could be completed.
2. Call a tractor dealer's service department to ask about the cost to rebuild a tractor engine damaged from lack of oil. Provide this information to your class and instructor.
3. Using a hydrometer (device to measure specific gravity of coolant or antifreeze for level at which the liquid would freeze), test engine coolant for level of temperature protection that coolant would provide.
4. Explain the meaning of the term "oil viscosity."
5. Describe the difference between diesel fuel and gasoline. How does the storage of these fuels differ?

References

1. Safety Management for Landscapers, Grounds-Care Businesses, and Golf Courses, John Deere Publishing, 2001. Illustrations reproduced by permission. All rights reserved.
2. Farm and Ranch Safety Management, John Deere Publishing, 2009.
3. Owners' Manuals of Several Tractors.

Contact Information

National Safe Tractor and Machinery Operation Program
The Pennsylvania State University
Agricultural and Biological Engineering Department
246 Agricultural Engineering Building
University Park, PA 16802
Phone: 814-865-7685
Fax: 814-863-1031
Email: NSTMOP@psu.edu

Credits

Developed by WC Harshman, AM Yoder, JW Hilton and D J Murphy, The Pennsylvania State University. Reviewed by TL Bean and D Jepsen, The Ohio State University and S Steel, National Safety Council. Revised 3/2013.

This material is based upon work supported by the national Institute of Food and Agriculture, U.S. Department of Agriculture, under Agreement Nos. 2001-41521-01263 and 2010-41521-20839. Any opinions, findings, conclusions, or recommendations expressed in this publication are those of the author(s) and do not necessarily reflect the view of the U.S. Department of Agriculture.

LEAD ACID BATTERIES

HOSTA Task Sheet 4.6.2

NATIONAL SAFE TRACTOR AND MACHINERY OPERATION PROGRAM

Introduction

Lead acid batteries provide a source of electrical current to start an engine and power tractor accessories, such as lights, emergency flashers, instrument panel gauges and meters, computerized digital read-outs, and other machine functions. Tractor electrical power may be used to operate and monitor functions of towed equipment.

Battery electrical current results from a chemical reaction produced by sulfuric acid and water mixture. This chemical solution, called electrolyte, can burn your skin and eyes. The energy produced is stored as positive (+) and negative (-) electrical charges on the battery plates. An explosive gas is produced by this reaction as the battery charges and discharges.

Modern tractors may have one or two batteries to provide current to the starting motor (starter).

Correct battery care and use will provide countless starts of the tractor engine in a safe manner.

This task sheet discusses battery construction, battery hazards, and battery care and safety.

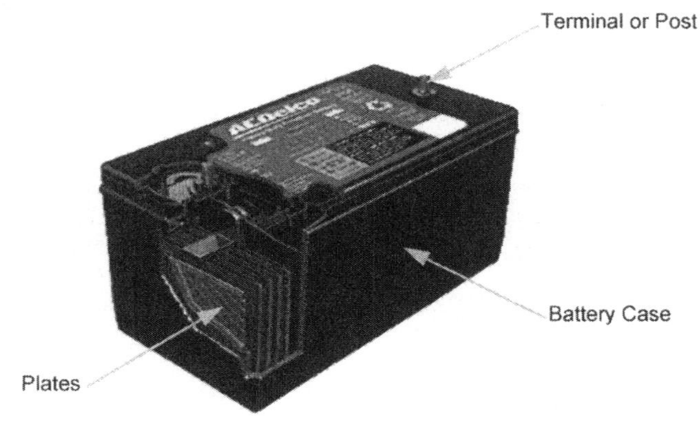

Fig. 4.6.2.a. Battery Construction. Inside the battery case are plates, which hold the electrical charges (+) and (-) and sulfuric acid-water mix. Also in the battery case are the connections to the battery terminals (posts). These terminals also extend outside the battery case. Battery cables connect the posts to the starter motor and a grounding surface.

Parts of a Battery

Battery Part	What it Does
Battery Case	A container to hold the battery acid solution and electrical storage plates
Battery Plate	Holds electrical charges (+) and (-)
Terminals	Connected to the storage plates and become the connecting points for battery cables leading to the starter (+) and the ground (-)

Use of safety goggles and protective clothing is a must when working with a lead acid battery.

Learning Goals

- To identify battery parts and functions
- To become familiar with hazards of lead acid batteries
- To use safe practices in working with and caring for batteries

Related Task Sheets:

Using a Battery Charger	**4.6.3**
Using Jumper Cables	**4.6.4**

Cooperation provided by The Ohio State University and National Safety Council.

Battery Hazards

Hazard	Definition	Safety Precautions
EXPLOSIONS	Battery acid produces hydrogen gas, which is explosive. A spark can lead to fire (dust, chaff, etc., around the battery) or explosion of hydrogen gas from the battery itself.	Check fluid level often to prevent gas buildup. Maintenance of fluid levels reduces the space in a battery where gases can accumulate.
CHEMICAL BURNS	The electrolyte solution in a battery is caustic to the skin and eyes and can burn holes through clothing.	**Use splashproof safety goggles and rubber gloves**. Keep the battery posts clean of corrosion.
ELECTRICAL SHOCK	The electrical charge of a battery may be only 12-26 volts, but with the effects of the ignition coil on spark ignition engines may produce voltages in the range of 100,000 volts. You can receive a severe shock. Wiring and electrical parts can be damaged.	Keep tools and parts away from the positive (+) terminal. It is best to remove the ground cable first when removing a battery or working on any part of the electrical system. When replacing the battery, connect the ground cable last.

Battery Safety Practices

1. Check battery fluid levels often. Low electrolyte levels increase the space where hydrogen gas can accumulate.
2. Prevent electrical sparks by keeping tools and parts away from the positive (+) terminal. The battery cable leading to the starter is usually the positive, or "hot" wire. Cap it with an insulating material when working near it.
3. When removing a battery for replacement or bench work, remove the ground cable first.
4. When replacing a battery, install the ground cable last.
5. Use safety goggles, long sleeves, and rubber gloves when refilling battery liquid. Distilled water is recommended for the refill. Any clean water can be used in an emergency if the battery is nearly dry.
6. Keep battery terminals clean of corrosion for best electrical contact. Prevent the corroded material from getting on your skin or in your eyes.
7. If you spill battery acid on your skin, flush it off with water immediately.
8. If you splash battery acid in your eyes, flush with warm water for at least 15 minutes. Seek medical attention.

Safety Activities

1. Check the fluid (electrolyte) level in your family's car, truck, riding mower, or tractor if it has fluid fill caps. If there are no fill caps, observe how the battery is checked for electrolyte. Use eye and skin protection.
2. With the help of an adult supervisor, clean the battery terminals of a corroded battery by removing the battery cables (ground cable first and positive or "hot" cable last). Use a battery terminal cleaner or mixture of baking soda and water. Re-attach battery cables with the "hot" or positive first and the ground cable last.
3. Search the Internet to learn more about batteries. One source is www.ACDelco.com. You can also use www.ask.com to ask questions about the batteries, their construction and operation.

References

1. www.ACDelco.com
2. Farm and Ranch Safety Management, John Deere Publishing, 2009.
3. www.ask.com

Contact Information

National Safe Tractor and Machinery Operation Program
The Pennsylvania State University
Agricultural and Biological Engineering Department
246 Agricultural Engineering Building
University Park, PA 16802
Phone: 814-865-7685
Fax: 814-863-1031
Email: NSTMOP@psu.edu

Credits

Developed by WC Harshman, AM Yoder, JW Hilton and D J Murphy, The Pennsylvania State University. Reviewed by TL Bean and D Jepsen, The Ohio State University and S Steel, National Safety Council. Revised 3/2013.

This material is based upon work supported by the National Institute of Food and Agriculture, U.S. Department of Agriculture, under Agreement Nos. 2001-41521-01263 and 2010-41521-20839. Any opinions, findings, conclusions, or recommendations expressed in this publication are those of the author(s) and do not necessarily reflect the view of the U.S. Department of Agriculture.

USING A BATTERY CHARGER

HOSTA Task Sheet 4.6.3

NATIONAL SAFE TRACTOR AND MACHINERY OPERATION PROGRAM

Introduction

If batteries are not cared for properly (see Task Sheet 4.6.2.), or if they are nearing the end of their useful life expectancy, (a 60-month guaranteed battery, which has 54 months of use), you can expect that the battery may fail to start the tractor. Many times that failure will come at the onset of cold weather when greater current demands are placed on the battery.

Batteries can also lose a charge when not used for extended time periods.

Often the battery can be recharged to prolong its usefulness.

This task sheet discusses the correct procedure to charge a 6- or 12-volt battery. For other voltage situations, consult the battery manufacturer's recommendations or your tractor Operator's Manual.

Some chargers can also be used to jumpstart a battery.

Battery Charging Procedures

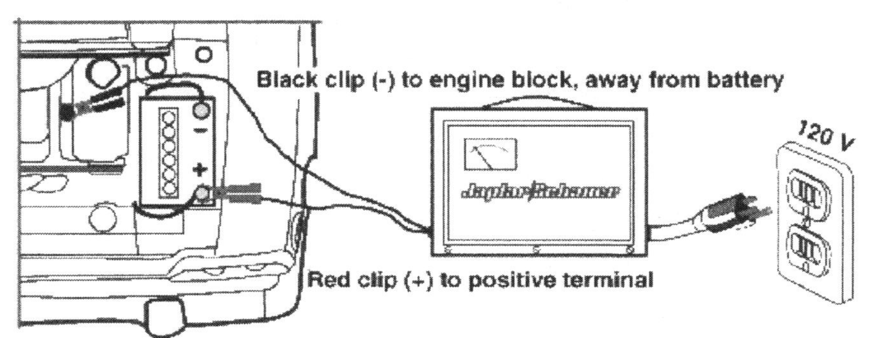

Figure 4.6.3.a. **Battery Polarity**: A battery has two poles or posts. The positive battery post is usually marked POS, P, or (+) and is larger than the negative post which is usually marked NEG, N, or (-).

Tools You Will Need:

- Safety glasses
- Approved battery charger
- Wrenches to remove battery cables
- Battery terminal cleaner
- Rubber gloves

Battery Charging:
The red cable goes to the POSITIVE (+) battery terminal, and the black cable goes to the NEGATIVE (-) battery terminal.

Learning Goals

- To safely use a battery charger to charge a weak battery
- To use all safety procedures to prevent chemical burn, explosion or fire, and electrical shock

Related Task Sheets:

Lead Acid Batteries	**4.6.2**
Using Jumper Cables	**4.6.4**

Cooperation provided by The Ohio State University and National Safety Council

Steps in Charging a Battery

If the tractor has a negative ground (most tractors do, but if you are not sure have it checked).

STEP 1. CONNECTING THE CHARGER TO BATTERY:

- If the charger has a switch with an OFF position, it MUST be set to OFF.
- The AC power cord to the charger MUST be unplugged.
- Connect the POSITIVE (RED) charger clip to the POSITIVE post of the battery.
- Next connect the NEGATIVE (BLACK) charger clip to the frame or engine block away from the battery.

CAUTION: Do not connect clip to carburetor, fuel lines, or sheet metal body parts. Connect to a heavy gauge metal part of the frame or engine block. This prevents sparks at the battery terminals, which can ignite hydrogen gas produced by the battery during a rapid charging situation.

STEP 2. TURNING THE CHARGER ON:

- If equipped with a voltage switch, set the switch to the voltage of the battery (normally 6 to 12 volts).
- If equipped with a rate switch, set the switch for the desired charge rate: normally 2, 6, 12, 30 amps.
- If equipped with a timer, set the timer to the charge time desired.
- Plug the AC cord into a grounded outlet. Stand away from the battery.
- Do not touch the charger clips when the charger is on.
- The charger should now be on and the ammeter showing the rate at which the battery is charging.
- The initial rate may be somewhat higher or lower than the charger's nameplate rating depending on battery condition and AC voltage at the outlet.

STEP 3. TURNING THE CHARGER OFF:

- Set the selector switch to OFF.
- Unplug the AC power cord from the outlet.
- Remove black charger clip connected to frame. If charging a battery outside of a vehicle, remove clip connected away from battery.
- Remove clip connected to positive battery post.

Connect the battery charger using the correct procedure to avoid sparks and possible explosions.

Safety Activities

1. With the help of an adult mentor, use a battery charger to charge a weak battery as described in this task sheet.
2. Use the Internet site www.autoeducation.com to ask questions about charging a battery.
3. Identify all the ways a battery's posts may be labeled to identify the positive and negative battery poles.

References

1. www.ask.com
2. www.autoeducation.com
3. www.battery-chargers.com
4. www.autobatteries.com

Contact Information

National Safe Tractor and Machinery Operation Program
The Pennsylvania State University
Agricultural and Biological Engineering Department
246 Agricultural Engineering Building
University Park, PA 16802
Phone: 814-865-7685
Fax: 814-863-1031
Email: NSTMOP@psu.edu

Credits

Developed by WC Harshman, AM Yoder, JW Hilton and D J Murphy, The Pennsylvania State University. Reviewed by TL Bean and D Jepsen, The Ohio State University and S Steel, National Safety Council. Revised 3/2013.

This material is based upon work supported by the National Institute of Food and Agriculture, U.S. Department of Agriculture, under Agreement Nos. 2001-41521-01263 and 2010-41521-20839. Any opinions, findings, conclusions, or recommendations expressed in this publication are those of the author(s) and do not necessarily reflect the view of the U.S. Department of Agriculture.

Cooperation provided by The Ohio State University and National Safety Council.

USING JUMPER CABLES

HOSTA Task Sheet 4.6.4

NATIONAL SAFE TRACTOR AND MACHINERY OPERATION PROGRAM

Introduction

If batteries are not cared for properly (see Task Sheet 4.6.2) or if they are nearing the end of their useful life expectancy, (e.g., a 60-month guaranteed battery which has 54 months of use), the battery may fail to start the tractor. Many times battery failure will come at the onset of cold weather.

Batteries can also lose a charge when not used for extended time periods. Using a booster battery and jumper (booster) cables to start the tractor, truck, or car may be necessary.

This task sheet discusses the correct procedures to boost or jumpstart a 6- or 12-volt battery to start an engine. For other voltage ratings, consult the tractor's or battery's Owner's Manual or manufacturer recommendations.

Jumpstarting a tractor: The red cable goes to the POSITIVE (+) battery terminal, and the black cable goes to the NEGATIVE (-) battery terminal on the good (charged) battery.

Battery Jumping Diagram

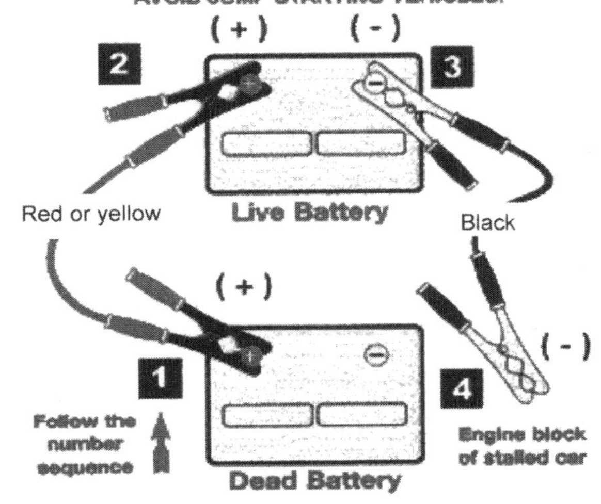

Figure 4.6.4.a. Battery Polarity: A battery has two poles or posts. The positive battery post is usually marked POS, P, or (+ or red) and is larger than the negative post, which is usually marked NEG, N, or (- or black). Connect positive to positive and negative to negative terminals to jumpstart the battery.

Tools you will need:

- Safety glasses
- Approved booster cables of 4-, 6-, or 8-gauge wire. Lighter wire (higher wire gauge number) will not carry enough current to jumpstart the battery.
- Wrenches to remove battery cables
- Battery terminal cleaner
- Booster battery usually from another tractor or vehicle.
- Rubber gloves

Learning Goals

- To safely use booster cables to jump-start a weak battery
- To use all safety procedures to prevent chemical burn, explosion or fire, and electrical shock

Related Task Sheets:

Cooperation provided by The Ohio State University and National Safety Council.

Steps to Jumping a Battery

Jumpstarting an engine with a drained battery is the same whether the drained battery is in a tractor, truck or car. Normally, you will use another tractor, truck or car battery to try and start the tractor with the drained battery.

IMPORTANT: Most vehicles have negative ground batteries. Be sure both the drained battery and the booster battery have negative grounds.

Follow these steps for jump-starting a tractor with a drained battery:

1. Pull the tractors next to each so they are not touching, and turn off both ignitions.
2. Connect the positive (+, yellow or red) clamp of the jumper cable to the drained battery's positive terminal.
3. Connect the other positive (+, yellow or red) clamp of the cable to the positive terminal of the booster battery.
4. Connect the negative (- or black) clamp of the cable to the negative terminal of the booster battery.
5. Connect the other negative (- or black) clamp of the cable to the vehicle's engine block or other metal surface of the tractor to be started away from the drained battery. This serves as your ground or connection point.

CAUTION: Do not connect clamp to carburetor, fuel lines, or sheet metal body parts. Connect to a heavy gauge metal part of the frame or engine block.

6. Make certain all cables are clear of fan blades, belts and other moving parts of both engines and that everyone is standing away from the vehicles.
7. Start the tractor with the booster battery.
8. Allow 1-5 minutes for the drained battery to accept a charge.
9. Try to start the tractor with the drained battery.

IF VEHICLE STARTS:

Allow the engine to return to idle speed. Remove the cables in the reverse order that you put them on.

1. Remove the negative (- or black) clamp from the frame of the vehicle with the drained battery.
2. Remove the negative (- or black) clamp from the booster battery.
3. Remove the positive (+, yellow or red) clamp from the booster battery.
4. Remove the positive (+, yellow or red) clamp from the formerly drained battery.

IF ENGINE DOES NOT START:

Wait a few moments and try again. If it still doesn't start, check for other problems.

Be sure to connect the jumper cables using the correct procedure to avoid sparks and damage to the battery or yourself.

Safety Activities

1. With the help of an adult mentor, use booster cables to boost a weak battery.
2. Use the Internet site www.autoeducation.com to ask questions about boosting a battery.

References

1. www.ask.com
2. www.autoeducation.com
3. www.battery-chargers.com
4. www.autobatteries.com

Contact Information

National Safe Tractor and Machinery Operation Program
The Pennsylvania State University
Agricultural and Biological Engineering Department
246 Agricultural Engineering Building
University Park, PA 16802
Phone: 814-865-7685
Fax: 814-863-1031
Email: NSTMOP@psu.edu

Credits

Developed by WC Harshman, AM Yoder, JW Hilton and D J Murphy, The Pennsylvania State University. Reviewed by TL Bean and D Jepsen, The Ohio State University and S Steel, National Safety Council. Revised 3/2013.

This material is based upon work supported by the National Institute of Food ad Agriculture, U.S. Department of Agriculture, under Agreement Nos. 2001-41521-01263 and 2010-41521-20839. Any opinions, findings, conclusions, or recommendations expressed in this publication are those of the author(s) and do not necessarily reflect the view of the U.S. Department of Agriculture

TIRE AND WHEEL CONDITION

HOSTA Task Sheet 4.6.5

NATIONAL SAFE TRACTOR AND MACHINERY OPERATION PROGRAM

Introduction

Tractors are traction machines! Better traction comes from good tires.

Tractor tires can cost several hundred dollars each. Estimates show that tractor tire repair and replacement comprise nearly 30% of the total repair costs during a tractor's lifetime.

You are responsible for protecting this valuable traction component.

This task sheet discusses tractor tire and wheel conditions for safe tractor operation.

Tire

Calcium fill line is 80% if calcium or similar solution is used.

Valve Stem

Rim

Figure 4.6.5.a. Tractor tire components include the tire, the rim or wheel, an inner tube with valve, and, many times, a calcium solution filling about 80% of the inner tube.

Tire Basics

These simple activities can extend the life of tractor tires:

- Check tire pressure regularly.
- Use wheel weights to reduce excess slippage, which can damage the tire.
- Drive carefully to avoid damaging objects.
- Make tire repairs promptly.

Tire and Wheel Hazards

Tractors are not built for high speed. *High speeds* on paved roads reduce tire life. Unpaved roads can do the same and also increase the chance for large stones to damage the tire as well.

Foreign objects can puncture tires. All farms have their share of sharp rocks, hidden field objects, and construction debris. Fields near rural roads may have glass bottles and metal cans which can cut tires. Be alert for those objects which can damage tires.

Improper use can ruin tires. Turning too tight and gouging the tire into towed equipment leads to cut tires. Most tractors have no shock absorbers; so the tire must absorb all ground shocks. Tire sidewall breaks can occur when objects are impacted.

Some rear tractor tires are filled with a calcium solution to add weight to the tractor to improve traction.

Learning Goals

- To identify faulty tire and wheel situations and take corrective action to remedy the problem

Related Task Sheets:

Preventative Maintenance and Pre-operation Checks	**4.6**

Cooperation provided by The Ohio State University and National Safety Council.

Tire and Wheel Defects

Fig. 4.6.5.b. Worn treads and dry rot make for poor traction and risk for downtime due to a blowout.

Fig. 4.6.5.c. Damaged rims from careless use may cause damaged tire beads and flat tires.

Tractor tires are expensive. They may cost hundreds of dollars to repair or replace.

Fig. 4.6.5.d. A leaking valve stem released calcium solution which rusted the rim. A major expense will be incurred, as well as a severe safety hazard in using this tractor.

Safety Activities

1. Call a local tire dealer who specializes in tractor tires, and ask for the price of a tractor tire that fits your tractor. For comparison purposes, call several dealers.

2. Have an adult mentor, leader, or teacher show you how to check air pressure in a calcium-filled tractor tire.

3. Find out how much a rear tractor tire weighs when it is filled with a calcium solution. You can use the *Yellow Pages* of the phone book to find a tractor tire repair service or tire dealer.

4. Ask a local tractor tire dealer what the recommendations are for filling tractor tires with liquid ballast (calcium solution, or similar solution).
5. Learn about the purpose of tractor ballast.

References

1. Farm and Ranch Safety Management, John Deere Publishing, 2009.

2. Safety Management for Landscapers, Grounds-Care Businesses and Golf Courses, 2001, 1st Edition, John Deere Publishing, Moline, Illinois.

Contact Information

National Safe Tractor and Machinery Operation Program
The Pennsylvania State University
Agricultural and Biological Engineering Department
246 Agricultural Engineering Building
University Park, PA 16802
Phone: 814-865-7685
Fax: 814-863-1031
Email: NSTMOP@psu.edu

Credits

Developed by WC Harshman, AM Yoder, JW Hilton and D J Murphy, The Pennsylvania State University. Reviewed by TL Bean and D Jepsen, The Ohio State University and S Steel, National Safety Council. Revised 3/2013.

This material is based upon work supported by the National Institute of Food and Agriculture, U.S. Department of Agriculture, under Agreement Nos. 2001-41521-01263 and 2010-41521-20839. Any opinions, findings, conclusions, or recommendations expressed in this publication are those of the author(s) and do not necessarily reflect the view of the U.S. Department of Agriculture.

THE OPERATOR PLATFORM

HOSTA Task Sheet 4.6.6

NATIONAL SAFE TRACTOR AND MACHINERY OPERATION PROGRAM

Introduction

If you compare the tractor operator platform to the cockpit of a jet fighter plane, both the tractor and jet fighter have:

- Steps to climb on board
- Adjustable operator seat with seat belt
- Multiple controls at hand and foot positions
- High visibility from the operator's seat

Keep these similar work areas free of obstructions for safe operation.

Could the pilot of the jet plane be able to fly to our defense in a moment's notice if:

- The steps were covered with mud and manure?
- The cockpit was filled with chains, grease guns, tools, and hitch pins?
- The windows were covered with pesticide spray drift or other materials?
- The pilot could not reach the controls because of a poorly adjusted seat?

This task sheet discusses the need for a clear tractor operator platform and an adjustable seat to safely reach the operating controls.

Figure 4.6.6.a. The operator's platform is not a tool box. You must have room to operate hand and foot controls. PTO levers, differential locks, foot throttles, and brake locks have to be engaged from the floor position. Soda cans and tobacco snuff containers can roll under control pedals and prevent correct, timely operation.

Operator Platform Workplace

Figure 4.6.6.b. Falls account for many farm injuries. Keep the steps and platform clean of mud, manure, and tools.

The tractor platform serves as the cockpit of this farm tool.

Figure 4.6.6.c. Tractors with ROPS come equipped with seat belts. Use them.

Figure 4.6.6.d. Keep windows and mirrors clean for good visibility.

Learning Goals

- To understand the need to keep steps and platform clear of tools and debris at all times
- To adjust the tractor seat and seat belt to safely reach all controls while your seat belt is buckled

Related Task Sheets:

Preventative Maintenance and Pre-Operation Checks	4.6

Cooperation provided by The Ohio State University and National Safety Council.

Seat Adjustment

Each person who operates the tractor will be a different size and weight. Check and adjust the seat adjustment so that you can comfortably reach all controls.

Seat controls may be levers or knobs and will be black in color. They may:

1. Release the seat to tilt it away from rain if the tractor is sitting outside.
2. Position the seat higher, lower, closer, farther, or to a different tilt position from the steering wheel and foot pedals.
3. Adjust the seat for the weight of the operator.
4. Be sure the seat belt is also adjusted for the seat.

Figure 4.6.6.e. Locate seat adjustments and know how they work. You may need the Operator's Manual.

Farm and Ranch Safety Management, John Deere Publishing, 2009. Illustrations reproduced by permission. All rights reserved.

Figure 4.6.6.f. The steering wheel should be adjusted as soon as you are seated. In the correct position, your arms are bent at a 90-degree angle as you hold the steering wheel. Your legs should remain slightly angled while the foot pedals are fully depressed.

Seat belts keep tractor drivers from being thrown out of the cab or off the seat during roll-overs. Wear your seat belt!

Safety Activities

1. Select any tractor at the farm where you work, and clean the tractor steps and platform. List how many different objects you can find there.
2. Use the NIOSH website to locate data on injuries due to falls in agricultural work. Are falls from getting on or off tractors considered a problem? If so, describe how serious it is.
3. Conduct a farm survey in the area with the help of your club or class members to determine how many tractors have seats or seat belts that can be easily adjusted.

References

1. www.cdc.gov/niosh/injury/trauma
2. Owners' Manuals for Specific Tractors.
3. Farm and Ranch Safety Management, John Deere Publishing, 2009. Illustrations reproduced by permission. All rights reserved.

Contact Information

National Safe Tractor and Machinery Operation Program
The Pennsylvania State University
Agricultural and Biological Engineering Department
246 Agricultural Engineering Building
University Park, PA 16802
Phone: 814-865-7685
Fax: 814-863-1031
Email: NSTMOP@psu.edu

Credits

Developed by WC Harshman, AM Yoder, JW Hilton and D J Murphy, The Pennsylvania State University. Reviewed by TL Bean and D Jepsen, The Ohio State University and S Steel, National Safety Council. Revised 3/2013.

This material is based upon work supported by the National Institute of Food and Agriculture, U.S. Department of Agriculture, under Agreement Nos. 2001-41521-01263 and 2010-41521-20839. Any opinions, findings, conclusions, or recommendations expressed in this publication are those of the author(s) and do not necessarily reflect the view of the U.S. Department of Agriculture.

STARTING AND STOPPING DIESEL AND GASOLINE ENGINES

HOSTA Task Sheet 4.7 **Core**

NATIONAL SAFE TRACTOR AND MACHINERY OPERATION PROGRAM

Introduction

Starting an engine is more than turning the ignition key. The safe operator is prepared to think clearly and to react to all the conditions surrounding the tractor being operated. Tractors may vary in design and layout of the instrument panel and ignition system, but starting and stopping gasoline and diesel engines involves only slightly different procedures. This task sheet discusses how to start and stop both diesel and gasoline tractor engines.

Before You Start the Engine

Review gasoline engine operation:

- Starter motor spins the engine
- Fuel and air mix enters combustion chamber; spark plug ignites mix
- Engine starts

OR

Review diesel engine operation:

- Starter motor spins the engine and activates the fuel pump
- Fuel droplets are sprayed into super hot combustion chamber
- Engine starts

Figure 4.7.a. Whether it is an older tractor or right out of the showroom, it may not be easy to see how to start the tractor without a demonstration or reading the owner's manual. If you are not sure, have someone show you first. That's the smart thing for a beginning operator to do.

Starting the engine is more than turning the ignition key.

Do not start any engine inside a building—gasses may kill you.

For both engines make a pre-operation check:

1. Check oil, fuel, and coolant level (cold engine only)
2. Check the tires
3. Check the controls for neutral positions

For all engines, avoid bypass starting. Many tractors have had their safe start systems bypassed. This is an unsafe practice. If the tractor is in gear, the tractor will move forward and crush you. Start the tractor from the seat only. The bypass starting hazard is discussed in Task Sheet 4.8.

Learning Goals

- To safely start and stop the engine of a gasoline tractor
- To safely start and stop the engine of a diesel tractor
- To explain the differences between gasoline and diesel engines

Related Task Sheets:

Tractor Instrument Panel	**4.4**
Engine Stop Controls	**4.5.1**
Mounting and Starting the Tractor	**4.8**

Cooperation provided by The Ohio State University and National Safety Council.

Figure 4.7.b. Choke the engine on cold days to allow more fuel than air to enter the combustion chamber. Release the choke after the engine has started as the fuel rich mixture will foul the spark plugs and valves of the engine.

Graphic courtesy of Hobart Publications

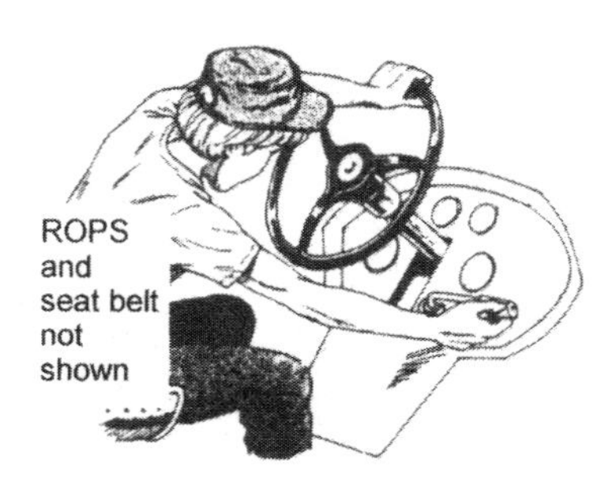

Figure 4.7.c. Set the throttle to 1/3 of the working range.

Graphic courtesy of Hobart Publications

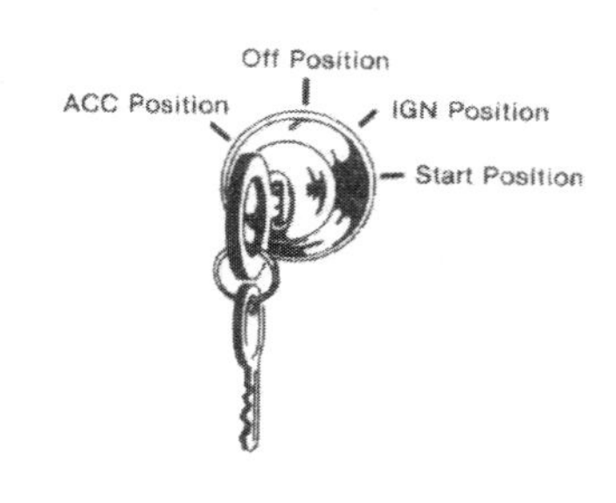

Figure 4.7.d. Turn the ignition key to the start position. Do not hold the key there for extended periods of time. This can burn up the starter motor or drain the battery of its charge.

Graphic courtesy of Hobart Publications

A cold engine must be choked to start easily. Choking increases the fuel to air ratio during cold weather.

Starting and Stopping Gasoline Engines

Follow these steps after you have fastened your seat belt.

1. Push the clutch in, and check that the tractor is in a neutral gear.
2. Adjust throttle to 1/3 open.
3. Choke the engine on cool days.
4. Turn starter key to "on."
5. Check indicator lights/gauges for oil pressure, temperature, and electrical charge.
6. Turn key to "start" position, but do not crank the engine for more than 10-30 seconds to avoid damage to the starter or running down the battery.
7. Re-check gauges—especially the oil gauge.
8. Warm up the engine at 800-1000 RPMs for a few minutes.

To stop the gasoline engine:

1. Throttle back to idle speed.
2. Place tractor in PARK or neutral and set brakes.
3. Turn off ignition key, and remove the key to prevent accidental starting by an untrained person.
4. If parking on a hill, place the transmission in a low gear with brakes set.

Figure 4.7.e. Be sure fuel pump shut-off knob is in the "on" position. If the conditions are cold, turn the ignition key to the glow plug setting and hold until the indicator shows that the combustion chamber is hot enough to ignite the diesel fuel droplets. If you don't know how to do this, ask someone to show you.

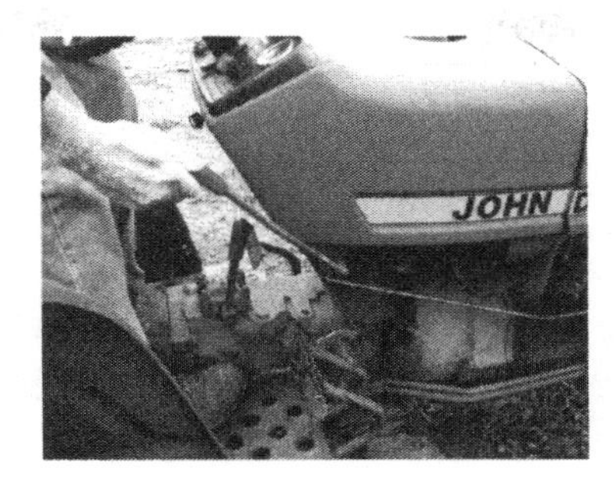

Figure 4.7.f. Set the throttle to 1/3 of its working range.

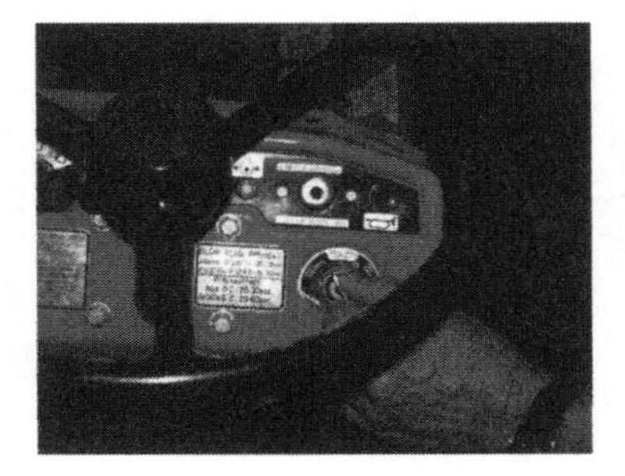

Figure 4.7.g. Turn the ignition key clockwise to the start position. Do not hold it there for extended periods of time. This can damage the starter motor or drain the battery of its charge.

Starting and Stopping Diesel Engines

Follow these steps after you have fastened your seat belt.

1. Push the clutch in, and check that the tractor is in a start or neutral gear.
2. Adjust throttle to 1/3 of the working range.
3. On cold days, turn ignition key to warm the glow plug (glow plugs pre-heat the combustion chamber air). ***Do not use an ether starter fluid.***
4. Check indicator lights/gauges for oil pressure, temperature, and electrical charge.
5. Turn key to "start" position, but do not crank the engine for more than 10-30 seconds to avoid damage to the starter or running down the battery.
6. Re-check gauges—especially oil gauge.
7. Warm up the engine at 800-1000 RPMs for a few minutes.

To stop the diesel engine:

1. Throttle back to idle speed.
2. Place tractor in PARK or neutral and set brakes.
3. Turn off ignition key, and remove it to prevent accidental starting by some untrained person.
4. Pull the "red" fuel pump shut-off control rod.
5. If parking on a hill, place the transmission in a low gear with brakes set.

Turning the key to the "off" position usually does not stop a diesel engine. You must shut off the fuel pump also.

Safety Activities

1. Using the procedures listed earlier, practice starting and stopping gasoline and diesel tractor engines.
2. Trace the linkage of the choke lever on the gasoline engine from the carburetor to the instrument panel. Draw a sketch of that linkage path.
3. Trace the linkage of the diesel fuel flow from the fuel tank to the fuel pump to the injectors. Draw a sketch of the linkage which leads from the fuel pump to the "red" fuel shut-off switch located on the operator's platform or instrument panel.
4. Answer these questions:

   A. True or False? Gasoline engines do not give off dangerous fumes.

   B. Choking an engine to start it on a cold morning means:

      1. Holding the key in the start position for as long as it takes.
      2. Providing more fuel than air for better ignition.
      3. Gassing the engine by pumping the throttle.
      4. Pouring extra fuel into the air cleaner to start the engine.

   C. Diesel engines do not have spark plugs. How is diesel fuel ignited in the cylinder?

   D. Why should a cold engine be allowed to warm up before pulling a heavy load?

   E. What can happen to the tractor's parts if you crank the starter motor too long?

   F. True or False? Diesel engines do not give off carbon monoxide.

   G. True or False? Diesel engines give off carbon dioxide gasses.

   H. What are the lethal gasses given off by a gasoline engine called?

      1. Carbon dioxide 2. Carbon trioxide 3. Carbon monoxide

   I. Where are glow plugs found, and what do they do?

References

1. Farm and Ranch Safety Management, John Deere Publishing, 2009.
2. Safety Management for Landscapers, Grounds-Care Businesses and Golf Courses, 2001, First Edition, John Deere Publishing, Moline, Illinois

Contact Information

National Safe Tractor and Machinery Operation Program
The Pennsylvania State University
Agricultural and Biological Engineering Department
246 Agricultural Engineering Building
University Park, PA 16802
Phone: 814-865-7685
Fax: 814-863-1031
Email: NSTMOP@psu.edu

Credits

Developed by WC Harshman, AM Yoder, JW Hilton and D J Murphy, The Pennsylvania State University. Reviewed by TL Bean and D Jepsen, The Ohio State University and S Steel, National Safety Council. Revised 3/2013.

This material is based upon work supported by the National Institute of Food and Agriculture, U.S. Department of Agriculture, under Agreement Nos. 2001-41521-01263 and 2010-41521-20839. Any opinions, findings, conclusions, or recommendations expressed in this publication are those of the author(s) and do not necessarily reflect the view of the U.S. Department of Agriculture.

MOUNTING AND STARTING THE TRACTOR

HOSTA Task Sheet 4.8 **Core**

NATIONAL SAFE TRACTOR AND MACHINERY OPERATION PROGRAM

Introduction

Safe tractor operation includes climbing onto the tractor in a safe way. Many operators have bruised shins and broken bones from slipping and falling while recklessly climbing or jumping onto tractors. Specific tractor pre-operation checks have been discussed, but there are other items to consider to safely start the tractor. This task sheet identifies the safe way to mount a tractor and the starting procedures to use.

Figure 4.8.a. Mount a tractor safely by using the handholds and steps. Scraped shins, along with worse injuries, have occurred when handholds and footsteps are not used.

Safe Tractor Mounting

Establish yourself as a good tractor operator by using these procedures each time you climb onto and sit down on a tractor seat.

- Keep the operator platform free of tools, equipment, mud or other debris.
- Use handholds and steps as you mount the tractor. Try to keep three points (two hands and one foot or two feet and one hand) on the tractor at all times.
- Adjust the seat and steering wheel (if necessary).
- Adjust and buckle the seat belt (if the tractor has ROPS).
- Check the major controls (PTO, hydraulics, gearshift stick) for the neutral (or PARK) position.

Before You Start the Engine

The safe operator will then think about and check many things before turning the key.

1. Is the area immediately around the tractor clear of persons and animals?
2. Is the tractor inside a building? If yes, is the building as open as possible to avoid a carbon monoxide fume buildup?
3. Do you understand the tractor's instrument panel?
4. Have pre-operation checks been made?

Now you are ready to start the tractor.

Always take your time and mount the tractor safely.

Learning Goals

- To mount a tractor safely
- To understand the hazards of bypass starting
- To know safe tractor startup procedures

Related Task Sheets:

Starting and Stopping Diesel and Gasoline Engines	4.7
Stopping and Dismounting the Tractor	4.9

Bypass Starting Dangers

Safety start systems have been in tractors for many years. The most common example of the safety start system is when the gearshift must be in neutral and the clutch must be depressed for the tractor to start. Some newer tractors may also have a switch in the seat that prevents the tractor from starting if the operator is not sitting in the seat. Safety start systems encourage operators to start their tractors while in the tractor seat—the safe place to be.

There are ways to bypass safe start systems. Unfortunately, the same operator who makes this mistake in judgment is also the operator who misjudges the location of the gearshift and has the tractor in gear while attempting to bypass start the tractor. The result is a tractor that lurches forward with the rear wheel running over and crushing the operator. Every year, experienced and inexperienced tractor operators die from bypass starting. Do not be one of them!

Figure 4.8.b. Newer tractors have covers on the starters to prevent bypass starting. Do not attempt to bypass start any tractor. *Safety Management for Landscapers, Grounds-Care Businesses, and Golf Courses, John Deere Publishing, 2001. Illustrations reproduced by permission. All rights reserved.*

Start the tractor engine from the seat only.

Safety Activities

1. Practice your safe mounting technique in front of a parent, instructor or classmate. Explain each step as you complete it.
2. Visit area farms and equipment dealers, and record how many tractors have some type of safety start system. See how many different systems you can find.
3. What are the dangers of bypass starting?

References

1. Owners' Manuals for Specific Tractors.
2. Safety Management for Landscapers, Grounds-Care Businesses, and Golf Courses, John Deere Publishing, 2001. Illustrations reproduced by permission. All rights reserved.

Contact Information

National Safe Tractor and Machinery Operation Program
The Pennsylvania State University
Agricultural and Biological Engineering Department
246 Agricultural Engineering Building
University Park, PA 16802
Phone: 814-865-7685
Fax: 814-863-1031
Email: NSTMOP@psu.edu

Credits

Developed by WC Harshman, AM Yoder, JW Hilton and D J Murphy, The Pennsylvania State University. Reviewed by TL Bean and D Jepsen, The Ohio State University and S Steel, National Safety Council. Revised 3/2013.

This material is based upon work supported by the National Institute of Food an Agriculture, U.S. Department of Agriculture, under Agreement Nos. 2001-41521-01263 and 2010-41521-20839. Any opinions, findings, conclusions, or recommendations expressed in this publication are those of the author(s) and do not necessarily reflect the view of the U.S. Department of Agriculture

STOPPING AND DISMOUNTING THE TRACTOR

HOSTA Task Sheet 4.9 **Core**

NATIONAL SAFE TRACTOR AND MACHINERY OPERATION PROGRAM

Introduction

Stopping and shutting off a tractor at the end of a day or for an extended period of time involves some specific procedures. Safe tractor operation includes climbing down from the tractor in a safe way. Many operators have ended up with twisted or broken bones from slipping and falling while recklessly jumping off tractors. This task sheet identifies safe tractor shutdown procedures and the safe way to dismount from a tractor.

Figure 4.9.a. To prevent falls, use the handholds and footsteps provided to dismount from the tractor. Falls while dismounting account for many farm injuries each year.

Shutting Down the Tractor

At the end of a day, there are many things to think about as you prepare to park and shut the tractor off for a period of time or for the night.

- Engine cool down—Manufacturers suggest cooling the engine for several minutes at a fast idle (800-1200 RPM) to prevent internal damage to hot engine parts. While letting the engine idle to cool, check all systems on the tractor. Then stop the engine.
- Hydraulic system—Even if you did not use the hydraulic system recently, static pressure keeps hydraulic lines pressurized. Work the hydraulic controls to relieve that pressure. It will be easier to attach the hydraulic lines later.
- Stop and park on the most level ground possible. Set the brakes (both brakes should be locked together) or place the gearshift in PARK.
- Lower all attached equipment to the ground.
- Place all controls and switches in an off, neutral, or locked position.
- Chock wheels if a heavy load is attached to the tractor to prevent runaways.

Stopping the tractor is more than turning the ignition key to the "off" position.

Learning Goals

- To know safe tractor shutdown procedures
- To learn to dismount a tractor safely

Related Task Sheets:

Starting and Stopping Diesel and Gasoline Engines	**4.7**
Mounting and Starting the Tractor	**4.8**

Cooperation provided by The Ohio State University and National Safety Council.

Safe Tractor Dismounting

The keys to safely dismounting are:

- Keep the operator platform free of tools, equipment, mud or other debris.
- Face the tractor, and use handholds and steps that are provided. Try to keep three points (two hands and one foot or two feet and one hand) on the tractor at all times.
- Take the key with you. Untrained operators, children, and visitors cannot accidentally start the engine if the keys are removed.

Figure 4.9.b. The key has been removed from the ignition. Untrained operators and little children cannot start the tractor if the key is removed from the ignition.

Remove ignition key to prevent untrained persons from starting the tractor.

Safety Activities

1. Practice your safe tractor shutdown procedure in front of a parent, instructor or classmate. Explain each step as you complete the procedure.
2. Ask the tractor owner(s) what policy they have for removing the keys from tractor ignition switches when the tractor is not in use.
3. If chock blocks are not available for wagons and implements at home, manufacture chock blocks in your school shop or home shop area.

References

1. Owners' Manuals for Specific Tractors.
2. Farm and Ranch Safety Management, John Deere Publishing, 2009

Contact Information

National Safe Tractor and Machinery Operation Program
The Pennsylvania State University
Agricultural and Biological Engineering Department
246 Agricultural Engineering Building
University Park, PA 16802
Phone: 814-865-7685
Fax: 814-863-1031
Email: NSTMOP@psu.edu

Credits

Developed by WC Harshman, AM Yoder, JW Hilton and D J Murphy, The Pennsylvania State University. Reviewed by TL Bean and D Jepsen, The Ohio State University and S Steel, National Safety Council. Revised 3/2013

This material is based upon work supported by the National Institute of Food and Agriculture, U.S. Department of Agriculture, under Agreement Nos. 2001-41521-01263 and 2010-41521-20839. Any opinions, findings, conclusions, or recommendations expressed in this publication are those of the author(s) and do not necessarily reflect the view of the U.S. Department of Agriculture.

MOVING AND STEERING THE TRACTOR

HOSTA Task Sheet 4.10 **Core**

NATIONAL SAFE TRACTOR AND MACHINERY OPERATION PROGRAM

Introduction

A safe and effective tractor operator can move the tractor in the proper direction and maneuver around field obstacles without damage. A well- trained operator can:

- Start the tractor moving without stalling, jerking, or lunging
- Steer the tractor with attached implements in and around buildings, fences, and crops without damage to the tractor, equipment, or property

Important: Tractors are traction machines. They are not made for speed or for fun. "Popping the clutch" or doing "wheelies" to show off can result in damage, injury or death. You must be able to move the tractor without rearing up the front end of the tractor.

Important: Tractors and implements are wider and longer than cars. You must judge how much room you need to turn or to drive between objects.

This task sheet discusses moving and steering the tractor by smoothly engaging the drive train and paying attention to the space occupied by the equipment.

Figure 4.10.a. Start the tractor moving smoothly. Excessive engine speed can the cause the tractor to start with a jerk.
Farm and Ranch Safety Management. John Deere Publishing, 2009. Illustrations reproduced by permission. All rights reserved.

Before the Tractor Moves

Do you know what makes the tractor move forward or backward?

The power train provides a means of transmitting power from the engine to the point of use (drive wheels). The mechanism that functions as a switch to disconnect the rotating crankshaft of the engine from the transmission gears. may be a clutch, a hydraulic device, or an electro-hydraulic mechanism. These serve three purposes:

- Allow for a smooth start
- Interrupt power while changing gears
- Interrupt power when stopping

There may be a foot control pedal or hand control lever(s)/joysticks to control tractor movement. Remember that these are orange color-coded controls. Ask for help if you do not understand the task.

An expert tractor operator moves the tractor without stalling or jerking.

Learning Goals

- To move a tractor without stalling or jerking through proper use of the clutch control pedal or lever
- To steer a tractor without damaging the tractor or towed or attached machine

Related Task Sheets:

Ground Motion Controls	**4.5.2**
Operating a Manual Shift Transmission	4.10.1
Tractor Transmissions	4.10.2
Operating the Tractor on Public Roads	**4.14**
Connecting Implements to the Tractor	5.1

Cooperation provided by The Ohio State University and National Safety Council.

Transmission and clutch types you may use

Transmissions can be divided into three general categories. They include:

- Manual shift transmissions, where the operator uses one or more shift levers to change gears and power range.
- Hydrostatic transmissions, where the operator pushes a control lever or pedal which engages a hydraulic pump to a hydraulic motor which turns the drive wheels.
- A combination of gear driven and hydraulically-assisted transmissions, where the transmission speeds can be altered by lever or button control and the direction of travel changed by way of a shuttle shift lever (reverser). These units may have a clutch pedal for stopping movement.

Tractor manufacturers may use combination of clutches and transmission controls, therefore you must be willing to ask your supervisor how to operate the specific model of tractor.

Skills for moving the tractor

Before attempting this skill, examine the Operator's Manual and have a qualified operator demonstrate what you must do.

To start moving the tractor:

1. Check the controls as you have learned in Task Sheet 4.5.2, adjust the seat, and fasten the seat belt.

Figure 4.10.b. Remember that ORANGE colored control levers indicate ground motion controls (Task Sheet 4.5.2). Study the gear shift pattern on the tractor you are operating. Use a lower gear to start the tractor moving. Use a higher gear for operation. Higher gear use with heavy loads may stall the engine or can cause rearward overturns. If the shift pattern is hard to locate, ask someone who is familiar with the tractor to show you the shift pattern.

2. Start the engine with the brake and clutch fully depressed You may need to be in PARK or a neutral-start position on many tractors.

3. Select a low starting gear to begin moving the tractor with or without a load.

4. After checking the area around the tractor, increase engine speed slightly; slowly engage the transmission until you feel the tractor begin to move.

5. Release the clutch and brakes fully when you are moving. Partial engagement (riding the clutch) can heat and place wear on the clutch parts

6. Increase speed and change gears as the task requires.

7. To stop movement activate the clutch control pedal or lever and apply the brakes.

8. When stopped place the shift lever in the Park position. Lock the brakes.

Use speeds appropriate to the task. Excessive ground speed can affect the operation of towed equipment.

Steering Involves Many Concepts

Steering involves several concepts each dealing with spacing. You must have knowledge of the:

- Width and length of the tractor.
- Width and length of the tractor and an attached implement.
- Space needed to corner the tractor and equipment around a building or object.
- Differences in the turning radius of narrow front-end versus wide front-end tractor steering.
- Individual wheel brakes on the tractor that can also be used to steer or control slippage on steeper slopes.

Brakes can help make steering corrections in tight places. Since the tractor's brakes may be used to brake each wheel separately, they can be used to make slight steering adjustments. Do not overdo this practice, as brakes can wear out quickly.

Caution: Lock brakes together for highway travel. Pushing one brake at high speeds can cause the tractor to be thrown sideways resulting in a side overturn.

Figure 4.10.c. Use both hands to steer the tractor. If you hit a hole in the field, the wheel can be jerked from your hand.

Cornering

Before attempting this skill, have a qualified operator demonstrate what you must do before you attempt cornering. Each tractor and implement will occupy a different space and corner differently as well. Know the relationship between the tractor and any towed implement. Too tight of a turn can cause the implement to pinch and possibly tear the tractor tire.

To turn a corner with the tractor and towed implement:

- Move as far away from a building and object as the roadway will permit.
- Drive in a long arc around the corner to prevent jack-knifing the tractor and machine.
- Observe the inside turning radius of the tractor and implement. Too tight of a turn can cause damage to the tractor, the tires, or to the towed equipment.
- As you complete the turn, observe the outside or opposite side of the tractor to be sure it has clearance from any other objects.

Wide turns on public roads will place the tractor and equipment into the opposite lane of traffic. This creates a hazard.

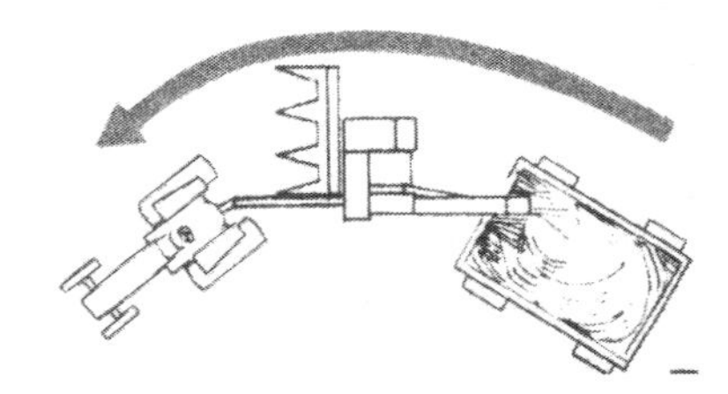

Figure 4.10.d. Too tight of a turning radius can damage tractor, tires, or implement.

Safety Activities

For Moving and Steering the Tractor:

1. On the tractor to which you are assigned, learn where the ground motion controls are found. This includes:
- Clutch control pedal or lever
- Gear shift pattern
- Shuttle shift and/or shift lever

2. In a large open area, practice starting a tractor, moving it forward, and slowly steering it in a figure 8 pattern. Then place the tractor in reverse gear and slowly back through the figure 8 pattern. Use a low range gear and a low-speed throttle adjustment.

3 Ask an experienced operator to show you how to move a tractor and implement uphill and downhill from a standing start.

For Steering the Tractor:

1. Use a 4-H or FFA Tractor Driving Course layout to practice driving a tractor through the course. You can also use the Driving Course Exam Layout from this program or develop your own challenging course..

2. Complete the obstacle course by using the reverse gear and backing through the course using the tractor alone.

3. Make the obstacle course a little larger (Use the course lay-out guide to determine the size of the driving course as described in the NSTMOP Program); repeat the practice with a tractor towing a two-wheeled implement.
   As you develop skill, reduce the size of the opening and practice further. You may make the course smaller as you achieve greater skill.

References

1. Farm and Ranch Safety Management, John Deere Publishing, 2009. Illustrations reproduced by permission. All rights reserved.
2. Operator's Manual for the model tractor you will operate

Contact Information

National Safe Tractor and Machinery Operation Program
The Pennsylvania State University
Agricultural and Biological Engineering Department
246 Agricultural Engineering Building
University Park, PA 16802
Phone: 814-865-7685
Fax: 814-863-1031
Email: NSTMOP@psu.edu

Credits

Developed by WC Harshman, AM Yoder, JW Hilton and D J Murphy, The Pennsylvania State University. Reviewed by TL Bean and D Jepsen, The Ohio State University and S Steel, National Safety Council.

Revised 3/2013

This material is based upon work supported by the National Institute of Food and Agriculture, U.S. Department of Agriculture, under Agreement Nos. 2001-41521-01263 and 2010-41521-20839. Any opinions, findings, conclusions, or recommendations expressed in this publication are those of the author(s)

OPERATING A MANUAL SHIFT TRANSMISSION

HOSTA Task Sheet 4.10.1

NATIONAL SAFE TRACTOR AND MACHINERY OPERATION PROGRAM

Introduction

To many people this is called a "stick shift". A gear shift lever on the tractor operator's platform allows the operator a choice of gears and speeds. The tractor operator must manually shift the gears to change speeds of the tractor. There is a skill involved in doing this task. If done properly a smooth change is noticed; if done improperly, gears will clash and transmission wear will increase.

This task sheet will assist the operator in being able to move a manual shift equipped transmission tractor smoothly with minimal transmission gear wear or damage.

How it works

When you push in on the clutch pedal you are disengaging the friction disk (clutch plate) that turns between the transmission and the engine. This allows the engine to continue to run, but the gears will stop turning.

As the gears slow down, the gear shift lever can be used to slide other gears into place to change direction (reverse) or change ground speeds. There may be only a few choices such as: reverse, 1st gear, 2nd gear, and 3rd gear, or twice as many gear choices if the tractor is equipped with a high and low range (using a second shift lever) transmission.

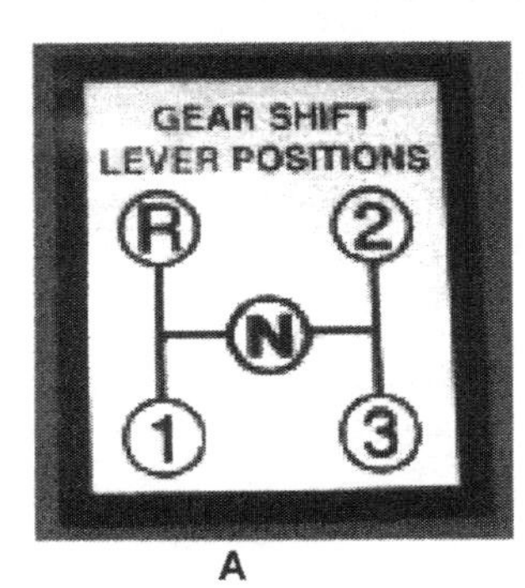

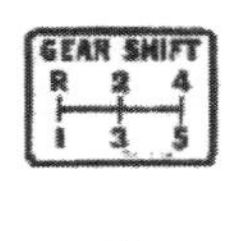

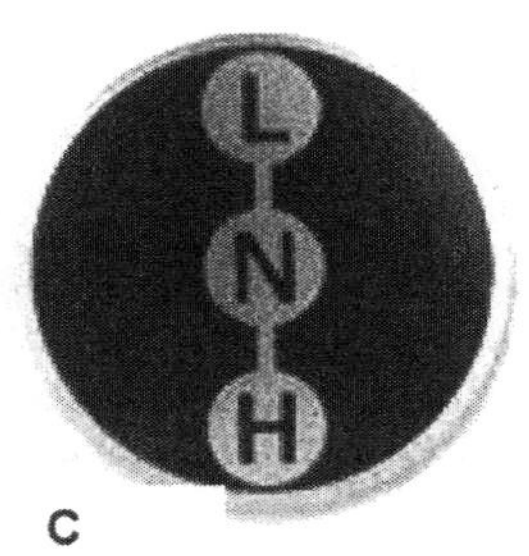

Figure 4.10.1.a. A). The typical "H" pattern gear shift choice. B). There may be 5 or 6 gear choices as well. C). A high-low range shift lever found near the "stick" shift provides a faster or slower speed for a particular gear. The middle or N position is neutral. These symbols may be missing or worn causing the operator to have to ask for assistance, or to locate the gear pattern themselves.

Gear shift pattern

As you sit in the operator's seat, find the gear shift lever. Notice that it operates within a notched device that holds the gear shift in that position until you wish to change gears. You may find that this notched part has a lettered decal or numbers embossed into the steel case or in the shift lever knob. Sometimes these are worn to the point you cannot see them. Usually the gear shift pattern resembles the letter H. (Figure 1 A,B, and C)

If you cannot see a pattern, ask your supervisor to show you. If no one is available to show you the gear shift pattern follow these steps:

1. Hold the brake tight.
2. Place the tractor in a gear while releasing the clutch slowly to see what direction or speed begins to occur. Try each gear.

If you don't know/ can't find the gear shift pattern, ask your supervisor to show you.

Learning Goals

- To be able to select and use the various gears of the standard shift transmission effectively without damage to the tractor.

Related Task Sheets:

Moving and Steering the Tractor **4.10**

Cooperation provided by The Ohio State University and National Safety Council.

What gear should I use?

The tractor is a traction machine. This means that we are using the tractor to pull a load. Driving fast is not the proper choice for a heavy load.

Tractors work more efficiently in a higher gear with a reduced throttle setting. When pulling a light load use a higher gear and reduce the throttle to maintain the desired ground speed. This is difficult to do when using the PTO (power take-off) since most PTO driven machines must be operated at a rated speed to perform properly. Engine speed/PTO speed is often displayed on the tractor instrument panel. Ask for help to understand the gear to select.

Moving the tractor

With the clutch and brake pedals pushed down to disengage the transmission and engage the brakes and with the tractor engine running, increase engine speed slightly and slowly let out on the clutch pedal until you feel movement; then slowly let off the brake pedal at the same time. When you are moving take your feet off the pedals because partial pressure on the clutch pedal (riding the clutch) will cause wear on the clutch plate and bearings in the transmission case.

To change gears, reduce engine speed, press in on the clutch pedal, move the gear shift lever to the next higher or lower gear and slowly release the clutch pedal. Gear up or down as needed. Be prepared to use the brakes if needed.

Clashing gears means you are not timing engine speed with gear shifting movements.

Figure 4.10.1.d . Depress the clutch pedal (A) and the brake pedals (B) at the same time. Start the engine. Select the proper gear to move the tractor. Then let out on the clutch pedal smoothly. As you feel movement let off the brake pedal smoothly as well. *Source: Manuals.deere.com*

Safety Activities

1. Locate the gear shift pattern decal or marking on the tractor you will operate. Was it easy to find? Was it worn off?
2. Sit in the operator's seat; push in on the clutch pedal and practice shifting the gears in the pattern shown.
3. If the gear shift pattern markings are gone; ask the supervisor to show you the shift pattern.

References

1. Google search, how to drive a manual transmission vehicle
2. Internet site, www.wikihow.com/Drive-Manual.

Contact Information

National Safe Tractor and Machinery Operation Program
The Pennsylvania State University
Agricultural and Biological Engineering Department
246 Agricultural Engineering Building
University Park, PA 16802
Phone: 814-865-7685
Fax: 814-863-1031
Email: NSTMOP@psu.edu

Credits

Developed by WC Harshman, AM Yoder, JW Hilton and D J Murphy, The Pennsylvania State University.

Version 3/2013

This material is based upon work supported by the National Institute of Food and Agriculture, U.S. Department of Agriculture, under Agreement No.2010-41521-20839. Any opinions, findings, conclusions, or recommendations expressed in this publication are those of the author(s) and do not necessarily reflect the view of the U.S. Department of Agriculture.

Cooperation provided by The Ohio State University and National Safety Council.

TRACTOR TRANSMISSIONS

HOSTA Task Sheet 4.10.2

NATIONAL SAFE TRACTOR AND MACHINERY OPERATION PROGRAM

Introduction

"Jump on that tractor and move it to the next farm down the road" said the supervisor. That should be a simple task for anyone who has driven a tractor. Maybe not! Tractor transmissions have come a long way since the manual stick shift. This task sheet briefly discusses tractor transmissions. There are many transmission types and combinations on the market and newer ones are introduced regularly. The operator should be trained on each tractor with a different transmission that he or she is expected to operate.

Figure 4.10.a. A tractor with a High-Low Range transmission has two shift levers (A). The transmission may have 3-5 gears with a high and low range for each. To start the tractor both shift levers may have to be in neutral. The shift pattern decal may be worn or missing requiring the operator to ask the employer for this information.

Know the gear shift pattern and how to smoothly shift gears without clashing.

Advice for beginners

With the many variations in transmission shift patterns, ask the tractor dealer, or employer to show you how the tractor is effectively shifted for various tasks.

Use the Operator's Manual for the specific tractor to learn more about the transmission use.

Use an Internet search tool such as "Google" to help you understand a specific type of transmission.

To prevent injury to yourself or others, and to prevent damage to the tractor or other property, ask for training before using the tractor.

Manual Transmissions

There are many "stick shift" tractors still in use and entry level employees may be assigned to these tractors until experience is gained.

Review Figure 4.10.s.a. for a reminder of what a stick shift gear pattern may look like, as well as where the gear shift lever is located.

The gear shift lever and high-low range shift is normally found on the transmission housing and between the operator's legs.

See Task Sheet 4.10.1, Operating a Manual Shift Transmission for more information.

Learning Goals

- To emphasize the need for on-going training in the many types of transmissions found on farm tractors.

Related Task Sheets:

Moving and Steering a Tractor	**4.10**
Operating a Manual Shift Transmission	**4.10.1**

Cooperation provided by The Ohio State University and National Safety Council.

Transmissions other than standard or stick shift

Changing gears with a stick shift is inefficient. Tractor manufacturer's developed a means of "shifting on the go" to improve tractor efficiency and increase operator comfort. The clutch pedal may still be used to disengage the clutch between the transmission and the engine, but chances are the newer model tractors will not use a pedal for shifting gears at all. Once on the move the simple movement of the power (gear) shift lever gives a greater range of power and speed.

There are many variations shown here. You must become familiar with the transmission you will be using. Study the Operator's Manual and ask for assistance from a knowledgeable person.

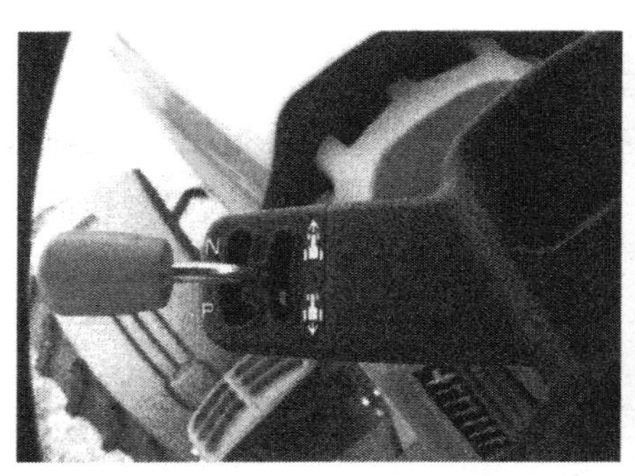

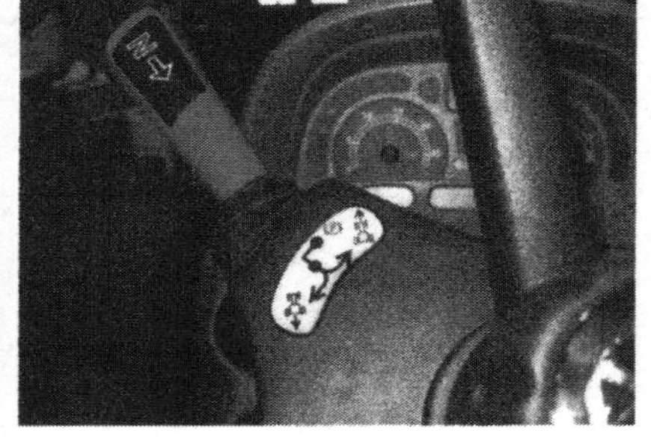

Figure 4.10.2.c. Look on the left side of the steering column. The shuttle shift lever, sometimes called the reverser, allows for ease of changing direction without using the clutch. This lever may serve as the positive park to be used with the parking brake. Some may have to be in a neutral position for the tractor to start. This may be the only time that the clutch pedal is used.

Figure 4.10.2.d. The clutch pedal on the left side of the floor may be used only to start the tractor and may not be used again until stopping.

Figure 4.10.2.f. The tractor should always be left in the positive park position with the parking brake locked (circled). The shift lever is moved from "P", park, to forward or reverse position. A range of numbers and/or letters (1-2-3-4, A-B-C-D) provides the power and speed ranges for the task.

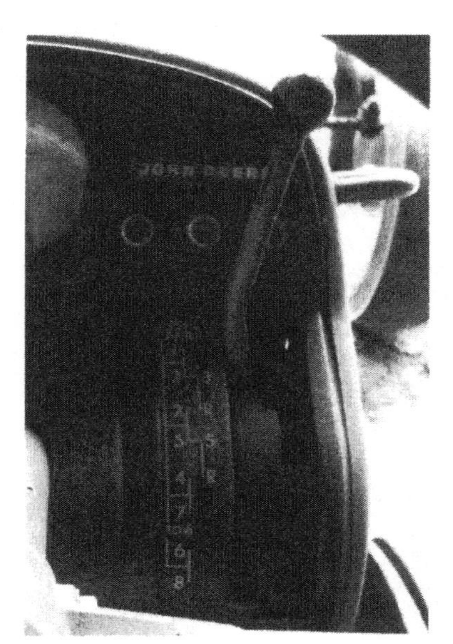

Figure 4.10.2.b. A synchronized shift transmission permits the operator to shift on the go within a range of "synchronized" gears.

Figure 4.10.2.e. A single orange colored proportional travel lever (shift lever) may be used to move the tractor forward or in reverse. There may be an adjustment knob or button built into the shift lever to further set the range of power and speed. The foot and hand accelerator are orange colored also.

Always look for the "orange" colored controls. Some may be faded or nearly invisible.

Cooperation provided by The Ohio State University and National Safety Council.

Figure 4.10.2.g. Visual monitoring of speed and power ranges are made to assist the operator in efficient operation. Someone must explain the system to you so that you can properly operate the tractor.

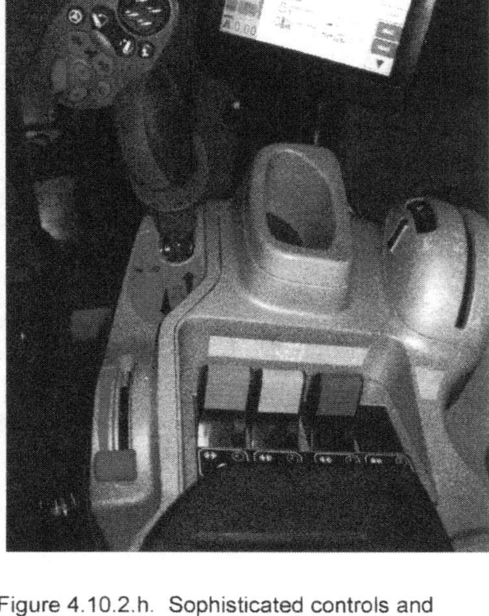

Figure 4.10.2.h. Sophisticated controls and monitoring systems are found in the contemporary tractor's operator's station. Use the Operator's Manual plus ask a reliable person to explain the tractor's operation.

Continued transmission evolution expected

The newer model tractor transmissions not only combine hydraulic power with a synchronized gear train, but have added an electronic monitoring system to further make for efficient, fuel-saving usage. Here are a few of the possibilities:

- Electronic monitoring of 100 checks per second to modify tractor engine speed in relation to power requirements of the job.
- A de-clutch button to enable gear shifting on the go without depressing the clutch pedal. It conveniently allows the operator to increase transport performance and comfort, helping to save additional time.
- Electro-hydraulic Hi-Lo range selection with the push of a button.
- Standard creeper range for work where slower speeds are required.

It is wise to ask for a demonstration of how the transmission of the tractor works. Tractors are a large investment which can be damaged if not used properly.

Figure 4.10.2.i. Tractor's imported into the USA may not use the same color coding standards for controls as tractors manufactured for the US market. Request training before using imported tractors. You won't be dismissed from your work by asking questions.

Safety Activities

1. Locate the speed and power shift control lever on the tractor which you will operate. Was it easy to find? What color is that control lever? Was the color worn to the point that you could not tell what color it was?

2. Sit in the operator's seat. Is there a clutch pedal? Can you move the speed and power control lever(s) while the engine is not running? Can you locate the shuttle shift, or reverser lever? Where is the shuttle shift, or reverser lever?

3. Try starting the tractor. Did it start when you followed the directions in the Operator's Manual? If it did not start, can you explain why it did not start?

4. On late model tractors, use the Operator's Manual to locate information that would tell you what speed and power setting to use for various field operations. Answer these:

Field work to be done	Transmission setting	Engine RPM
A. Light load	________________	_________
B. Heavy load	________________	_________

References

1. Operator's Manuals from various tractor manufacturers, specifically the tractor which you will operate.
2. Internet sources for various tractor manufacturers.
3. Knowledgeable, trained individuals who can demonstrate how and why the transmission is used for a specific tractor.

Contact Information

National Safe Tractor and Machinery Operation Program
The Pennsylvania State University
Agricultural and Biological Engineering Department
246 Agricultural Engineering Building
University Park, PA 16802
Phone: 814-865-7685
Fax: 814-863-1031
Email: NSTMOP@psu.edu

Credits

Developed by WC Harshman, AM Yoder, JW Hilton and D J Murphy, The Pennsylvania State University.
Version 3/2013

This material is based upon work supported by the National Institute of Food and Agriculture, U.S. Department of Agriculture, under Agreement No. 2010-41521-20839. Any opinions, findings, conclusions, or recommendations expressed in this publication are those of the author(s) and do not necessarily reflect the view of the U.S. Department of Agriculture.

OPERATING THE TRACTOR IN REVERSE

HOSTA Task Sheet 4.11 **Core**

NATIONAL SAFE TRACTOR AND MACHINERY OPERATION PROGRAM

Introduction

Being able to steer the tractor in the proper direction of operation without damage to the tractor and machinery is important. The safe and effective tractor operator can make the tractor move where it is supposed to go.

Not all operations will be in a forward direction. You must be able to operate the tractor and equipment in the reverse direction. Hitching to equipment, backing equipment, unloading crops, and even storing machinery is done in reverse.

Reverse travel and steering is generally done without looking at the direction of the steered wheels. These wheels will usually be out of your line of sight. You must master the concept that the steered wheels are pointing the rear of the tractor in the direction you want it to go.

To back safely, check your line of travel, back slowly, and have someone help direct you if needed.

This task sheet instructs how to correctly steer a tractor in reverse. Master this task without any equipment hooked to the tractor before beginning to back a two-wheel or four-wheel machine or implement.

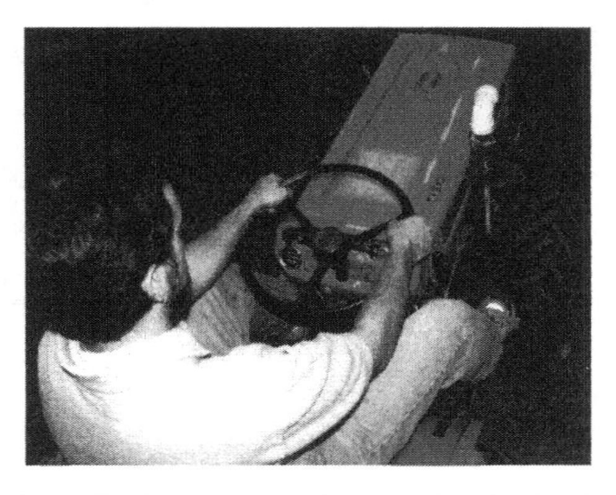

Figure 4.11.a. Which way should I turn the steering wheel to make the tractor turn to the left when using the reverse gear?

Reverse Direction Hazards

People, animals, or other objects may be in the line of travel. As you look from the front of the machine to the rear, you may lose sight of such obstacles. Skid loaders and industrial equipment have reverse gear alarms to warn others, but tractors usually do not.

Workers who help you by hitching the implement to the tractor can be crushed if your foot slips from the clutch pedal, if you are driving too fast in reverse, or if you fail to steer in the correct direction in reverse. Do not permit the helper to go between the tractor and implement to be hitched before you stop the tractor and turn off the engine.

There is a tendency to shift slightly from the operator's seat when looking to the rear. You must stay in good contact with foot and hand controls to be safe.

Imagine which way the rear end of the tractor will go as you turn the steering wheel when moving in reverse.

Learning Goals

- To safely drive a tractor in reverse gear to a specific location with few directional corrections
- To spot the tractor drawbar to the hitch of the machine with no more than three changes of direction

Related Task Sheets:

Cooperation provided by The Ohio State University and National Safety Council.

Tips for Backing a Tractor With an Implement

Follow these safety tips when driving a tractor in reverse without an implement:

1. Be sure seat and controls are adjusted for you.
2. Be sure all persons, animals, and objects are clear of the tractor.
3. Engage the clutch slowly, use a low engine speed, and maintain foot contact with the clutch and brake.
4. Turn the top of the steering wheel in the direction you want the rear of the tractor to move. If you wish to move the rear of the tractor to the left, move the steering wheel to the left. If you wish to move the rear of tractor to the right, move the steering wheel to the right.
5. To back with a two-wheeled implement, you must use the rear of the tractor to force the implement to go where you want it. To move the implement to the left, steer the tractor to the right. To move the implement to the right, steer the tractor to the left. This must be done slowly.

Figure 4.11.b Which way would you turn the steering wheel of this tractor to back the manure spreader to the right?

Remember, when backing an implement, turn the steering wheel in the opposite direction.

Safety Activities

1. With a clutch-type tractor, practice pushing the clutch in all the way and then releasing the clutch slowly to get the feel of the clutch pressure. Do this without starting the tractor; then practice this task in a clear area with the engine running.
2. Practice starting a tractor, moving it forward, slowly steering in a figure 8. Place tractor in reverse gear and slowly back in a figure 8 pattern. Use a low-range, low-throttle adjustment.
3. Ask an experienced operator to demonstrate how to back a two-wheeled implement.
4. Practice turning and backing with a two-wheeled implement attached to the tractor.

References

1. Farm and Ranch Safety Management, John Deere Publishing, 2009.

Contact Information

National Safe Tractor and Machinery Operation Program
The Pennsylvania State University
Agricultural and Biological Engineering Department
246 Agricultural Engineering Building
University Park, PA 16802
Phone: 814-865-7685
Fax: 814-863-1031
Email: NSTMOP@psu.edu

Credits

Developed by WC Harshman, AM Yoder, JW Hilton and D J Murphy, The Pennsylvania State University. Reviewed by TL Bean and D Jepsen, The Ohio State University and S Steel, National Safety Council. Revised 3/2013.

This material is based upon work supported by the National Institute of Food and Agriculture, U.S. Department of Agriculture, under Agreement Nos. 2001-41521-01263 and 2010-41521-20839. Any opinions, findings, conclusions, or recommendations expressed in this publication are those of the author(s) and do not necessarily reflect the view of the U.S. Department of Agriculture.

TRACTOR STABILITY

HOSTA Task Sheet 4.12 **Core**

NATIONAL SAFE TRACTOR AND MACHINERY OPERATION PROGRAM

Introduction

No other machine is more identified with the hazards of farming as the tractor. Nearly 50% of tractor fatalities come from tractor overturns. Tractors are used for many different tasks. Because the tractor is a versatile machine, operators sometimes stretch the use of the tractor beyond what the machine can safely do. For example, an operator may turn a corner too quickly for the tractor to stay upright. The use of a rollover protective structure (ROPS) and a seat belt can save your life if a tractor overturns while you are driving.

This task sheet explains the four major reasons and forces that allow tractors to overturn, gives rules for how to prevent tractors from overturning, and discusses the use of tractor ROPS with a seat belt.

How Tractors Overturn

Center of gravity (CG). A center of gravity is the point where all parts of a physical object balance one another. When you balance a pencil on your finger, you have found the pencil's CG. This is the part of the pencil that is resting on your finger. On a two-wheel drive tractor, CG is about 10 inches above and 12 inches in front of the rear axle. Figure 4.12.a shows the normal position of a tractor's CG.

Figure 4.12.a. Expected position of a tractor's center of gravity. *Safety Management for Landscapers, Grounds-Care Businesses, and Golf Courses. John Deere Publishing, 2001. Illustrations reproduced by permission. All rights reserved.*

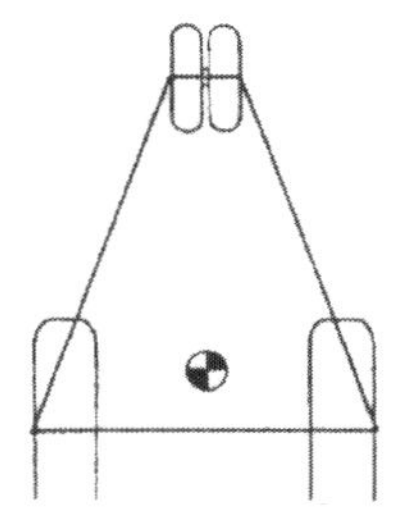

Figure 4.12.b. The tractor's center of gravity is inside the stability baseline.

Look at Figure 4.12.b. This shows that the CG is inside a tractor's stability baseline. Drawing a line to connect all the wheels of the tractor as the wheels set on level ground forms a tractor stability baseline. The line connecting the rear tire ground contact points is the rear stability baseline. The lines connecting the rear and front tire on the same side are the right and left side stability baselines. Front stability baselines exist but have limited use in tractor overturn discussions.

There are two very important points to remember about tractor CG and stability baselines:

- The tractor will not overturn if the CG stays inside the stability baseline.
- The CG moves around inside the baseline area as you operate the tractor.

As you can see in figure 4.12.b, a wide front-end tractor provides more space for the CG to move around without going outside the stability baseline.

CG

Center of Gravity

Learning Goals

- To explain the role that center of gravity plays in tractor overturns
- To list reasons the center of gravity moves within a stability baseline
- To explain how to be protected during a tractor overturn

Related Task Sheets:

Agricultural Tractors	4.1
Tractor Hazards	4.2
Moving and Steering the Tractor	4.10
Using the Tractor Safely	4.13
Operating the Tractor on Public Roads	4.14

Cooperation provided by The Ohio State University and National Safety Council.

Reasons the CG Moves Around

There are five main reasons why a tractor's CG moves outside the stability baseline.

1. The tractor is operated on a steep slope.
2. The tractor's CG is raised higher from its natural location 10 inches above the rear axle.
3. The tractor is going too fast for the sharpness of the turn.
4. Power is applied to the tractor's rear wheels too quickly.
5. The tractor is trying to pull a load that is not hitched to the drawbar.

How Center of Gravity and Centrifugal Force Result in an Overturn

When a tractor is on a slope, the distance between the tractor's CG and stability baseline is reduced. Figure 4.12.c shows how this occurs. On steep slopes, the tractor is already close to an overturn. A small bump on the high side, or a groundhog hole on the low side, may be all that is needed for the tractor to overturn.

A front-end loader or other attachment mounted on a tractor can raise the tractor's CG. When the bucket is raised high, the balance point for the whole tractor is also raised. Figure 4.12.d shows how a raised CG makes it easier for a tractor to turn over sideways.

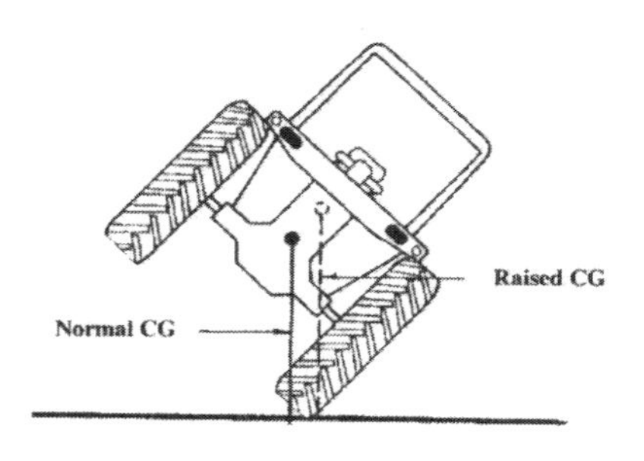

Figure 4.12.c. When a tractor is on a slope, the distance between the tractor's CG and stability baseline is reduced.

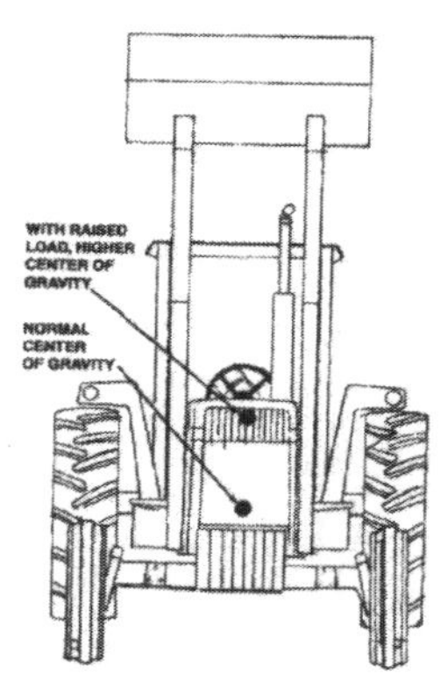

Figure 4.12.d. A raised CG makes it easier for a tractor to turn over sideways. *Safety Management for Landscapers, Grounds-Care Businesses, and Golf Courses, John Deere Publishing, 2001. Illustrations reproduced by permission. All rights reserved.*

Centrifugal force (CF) is the outward force nature exerts on objects moving in a circular fashion. During tractor overturns, CF is that force trying to roll the tractor over whenever the tractor is turning. Centrifugal force increases both as the turning angle of the tractor becomes sharper (decreases), and as the speed of the tractor increases during a turn. For every degree the tractor is turned tighter, there is an equal amount of increased CF.

The relationship between CF and tractor speed, however, is different. Centrifugal force varies in proportion to the square of the tractor's speed. For example, doubling tractor speed from 3 mph to 6 mph increases the strength of CF four times ($2^2 = 2 \times 2 = 4$). Tripling tractor speed from 3 mph to 9 mph increases CF nine times ($3^2 = 3 \times 3 = 9$).

Centrifugal force is what usually pushes a tractor over when the tractor is driven too fast during a turn or during road travel. During road travel, rough roads may result in the tractor's front tires bouncing and landing in a turned position. If the tractor starts to veer off the road, over correction of steering can result in side overturns. Centrifugal force is often a factor in tractor side overturns. When the distance between the tractor's CG and side stability baseline is already reduced from being on a hillside, only a little CF may be needed to push the tractor over.

Engaging the clutch of a tractor results in a twisting force, called torque, to the rear axle. Under normal circumstances, the rear axle (and tires) should rotate and the tractor will move ahead. If this occurs, the rear axle is said to be rotating about the tractor chassis. If the rear axle cannot rotate, then the tractor chassis rotates about the axle. This reverse action results in the front end of the tractor lifting off the ground until the tractor's CG passes the rear stability baseline. At this point, the tractor will continue rearward from its own weight until the tractor crashes into the ground or other obstacle. See Figure 4.12.e.

Cooperation provided by The Ohio State University and National Safety Council.

The CG of a tractor is found closer to the rear axle than the front axle. A tractor may only have to rear to about 75 degrees from a level surface before its CG passes the rear stability baseline and the tractor continues flipping over. This position is commonly called the "point of no return." As Figure 4.12.e shows, this point can be reached more quickly than an operator can recognize the problem.

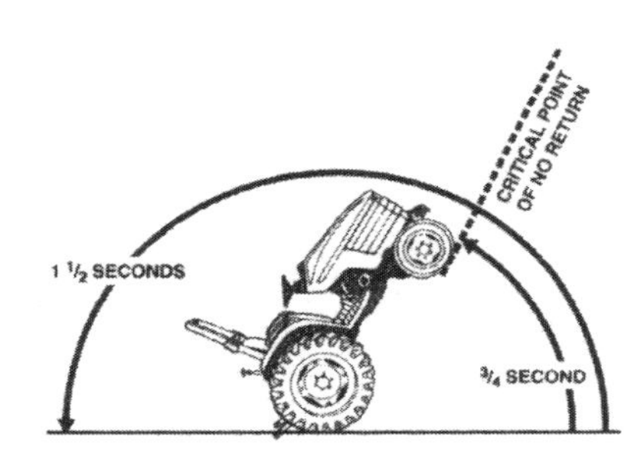

Figure 4.12.e. The point of no return is reached in 3/4 of a second. *Safety Management for Landscapers, Grounds-Care Businesses, and Golf Courses, John Deere Publishing, 2001. Illustrations reproduced by permission. All rights reserved.*

Common examples of this type of tractor overturn are: the rear tires are frozen to the ground; tires stuck in a mud hole; or tires blocked from rotating by the operator. Rear overturns can also happen on a slope if an operator applies too much power too quickly to the rear axle. When a tractor is pointed up a slope, there is less rise needed to reach the point of no return because the CG has already moved closer to the stability baseline. Figure 4.12.f shows how this occurs.

When a two-wheel drive tractor is pulling a load, the rear tires push against the ground. At the same time, the load attached to the tractor is pulling back and down against the forward movement of the tractor. The load is described as pulling down because the load is resting on the earth's surface. This backward and downward pull results in the rear tires becoming a pivot point, with the load acting as a force trying to tip the tractor rearward. An "angle of pull" is created between the ground's surface and the point of attachment on the tractor.

A tractor, including the drawbar, is designed to safely counteract the rearward tipping action of pulled loads. When loads are attached to a tractor at any point other than the drawbar, the safety design of the tractor for pulling loads is defeated.

The heavier the load and the higher the "angle of pull," the more leverage the load has to tip the tractor rearward. Figures 4.12.g, 4.12.h, and 4.12i. show important information about safe hitching points.

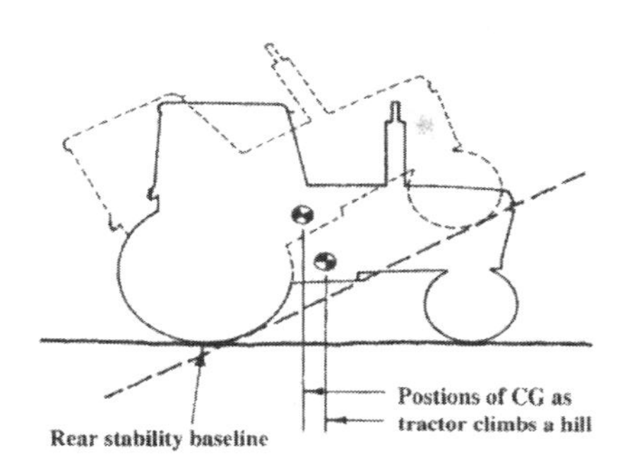

Figure 4.12.f. When a tractor is pointed up a slope, the CG is closer to the rear stability baseline.

Figure 4.12.g. Only hitch to the drawbar. *Safety Management for Landscapers, Grounds-Care Businesses, and Golf Courses, John Deere Publishing, 2001. Illustrations reproduced by permission. All rights reserved.*

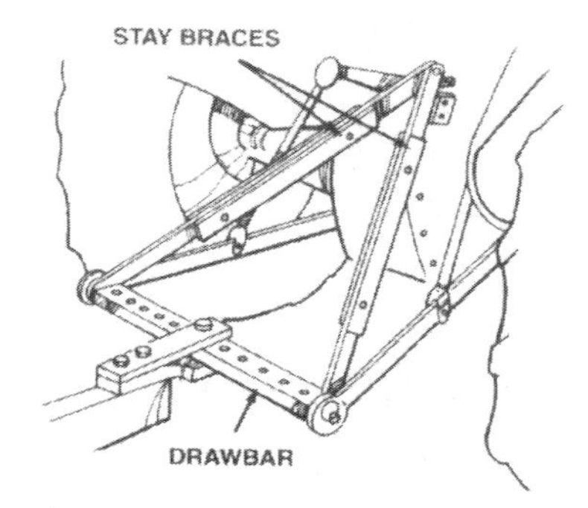

Figure 4.12.h. Never hitch to the top link of a 3-point hitch. *Safety Management for Landscapers, Grounds-Care Businesses, and Golf Courses, John Deere Publishing, 2001. Illustrations reproduced by permission. All rights reserved.*

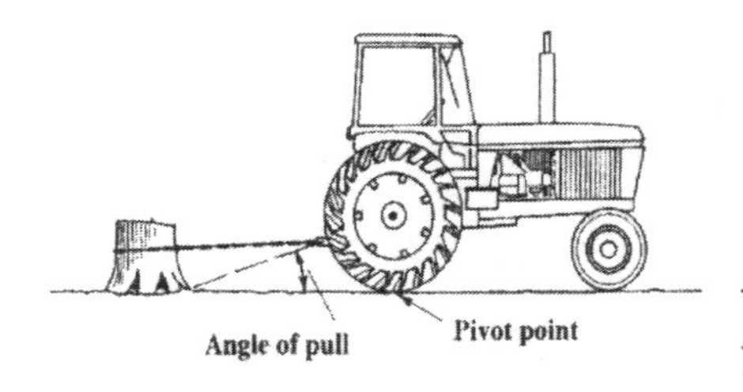

Figure 4.12.i. The angle of pull should be kept to a minimum.

Protect Yourself in a Tractor Overturn

The rollover protective structure (ROPS) and seat belt, when worn, are the two most important safety devices to protect operators from death during tractor overturns. Remember the ROPS does not prevent tractor overturns, but can prevent the operator from being crushed during an overturn. The operator must stay within the protective frame of the ROPS (Zone of Protection) in order for the ROPS to work as designed. This means the operator must wear the seat belt. Not wearing the seat belt may defeat the primary purpose of the ROPS.

A ROPS often limits the degree of rollover, which may reduce the probability of injury to the operator. A ROPS with an enclosed cab further reduces the likelihood of serious injury because the sides and windows of the cab protect the operator. This assumes that cab doors and windows are not removed.

To prevent tractors from overturning in the first place, follow the safety recommendations that are illustrated in Task Sheet 4.13.

Note: ROPS are available in folding and telescoping versions for special applications, such as orchards and vineyards and low-clearance buildings. Some ROPS may be a protective frame only and not an enclosed cab.

Figure 4.12.j. A rollover protective structure (ROPS) and a seat belt can protect you in the event of an overturn. If you are in the cab of a ROPS-equipped tractor, fasten the seat belt. *Safety Management for Landscapers, Grounds-Care Businesses, and Golf Courses, John Deere Publishing, 2001. Illustrations reproduced by permission. All rights reserved.*

Safety Activities

1. Use a toy scale model or a full-size tractor to illustrate the five main reasons tractors overturn.
2. Invite a farmer whom you know who has survived a tractor rollover to speak to the class about the experience.
3. Conduct a survey of area farm people to find out instances of tractor overturns in the last five years. How many overturns resulted in a fatality? How many survived an overturn? Did a ROPS play a role in their

References

1. Safety Management for Landscapers, Grounds-Care Businesses, and Golf Courses, John Deere Publishing, 2001. Illustrations reproduced by permission. All rights reserved.

2. www.cdc.gov/Type agriculture tractor overturn hazards in search box/Click on 1 0.67 Tractor Overturn Hazards, August 2002.

3. Farm and Ranch Management, John Deere Publishing, 2009.

Contact Information

National Safe Tractor and Machinery Operation Program
The Pennsylvania State University
Agricultural and Biological Engineering Department
246 Agricultural Engineering Building
University Park, PA 16802
Phone: 814-865-7685
Fax: 814-863-1031
Email: NSTMOP@psu.edu

Credits

Developed by WC Harshman, AM Yoder, JW Hilton and D J Murphy, The Pennsylvania State University. Reviewed by TL Bean and D Jepsen, The Ohio State University and S Steel, National Safety Council. Revised 3/2013.

This material is based upon work supported by the National Institute of Food and Agriculture, U.S. Department of Agriculture, under Agreement Nos. 2001-41521-01263 and 2010-41521-20839. Any opinions, findings, conclusions, or recommendations expressed in this publication are those of the author(s) and do not necessarily reflect the view of the U.S. Department of Agriculture.

USING THE TRACTOR SAFELY

HOSTA Task Sheet 4.13 **Core**

NATIONAL SAFE TRACTOR AND MACHINERY OPERATION PROGRAM

Introduction

Tractors can be operated safely if they are used as designed and operated following recommended practices.

There are an estimated 300 farm tractor fatalities each year. Read these short examples.

- Teenager killed using tractor to spotlight deer in the woods.
- Man killed when tractor rolled onto him while dragging logs in the woods.
- Grandfather killed, but passenger grandson lives when tractor goes over an embankment while going for a fun ride.
- Tractor overturns while towing stalled pickup full of firewood.
- Tractor upsets sideways while high lift bucket is in a raised position while traveling across a rough slope.

This task sheet will identify several proper and improper uses of tractors.

Figure 4.13.a. When tractors are operated for their intended purpose, the American farm worker produces more food than in any other country in the world.

A tractor is designed to do work.

Proper Use Defined

Tractors are made to work, not to be treated as ATVs, four-wheelers, dune buggies, or as other recreational vehicles.

Tractors serve four purposes:

1. They are a remote power source.
2. They carry/pull machines.
3. They move loads.
4. They transport materials.

If you are not sure of a specific use for your tractor, consult the Owner's Manual.

Learning Goals

- To recognize proper uses of the tractor
- To recognize improper uses of the tractor

Related Task Sheets:

Agricultural Tractors **4.1**

Proper Use Means Avoiding Improper Use

Figure 4.13.b. Tractors are designed for the operator only. No passengers allowed!

Figure 4.13.c. Tractors provide remote power to machinery. This turning shaft, the PTO, must be guarded to prevent entanglement hazards such as this.

Recognize when a driver is operating the tractor in an unsafe manner.

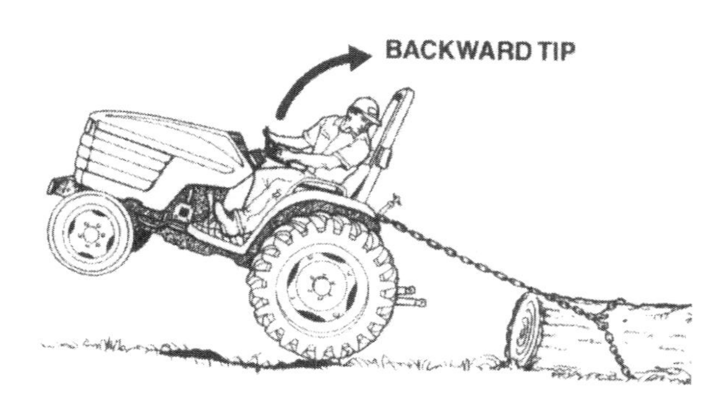

Figure 4.13.d. Hitch loads only to the drawbar. The drawbar has been engineered to pull heavy loads without risking a rear overturn hazard. *Safety Management for Landscapers, Grounds-Care Businesses, and Golf Courses, John Deere Publishing, 2001. Illustrations reproduced by permission. All rights reserved.*

Figure 4.13.e. If you are stuck or need to be towed, you will need help from a second tractor. Use the strongest and best tow strap, cable, or chain that is available. Hitch only to the drawbar. The best advice for a young operator is to get adult help to pull the disabled or stuck tractor. *Farm and Ranch Safety Management, John Deere Publishing, 1994. Illustrations reproduced by permission. All rights reserved.*

Cooperation provided by The Ohio State University and National Safety Council.

Figure 4.13.f. Avoid ditch embankments. Tractors are heavy and embankments can give way. For example, if the ditch is 6 feet deep, stay back at least 6 feet. *Safety Management for Landscapers, Grounds-Care Businesses, and Golf Courses, John Deere Publishing, 2001. Illustrations reproduced by permission. All rights reserved.*

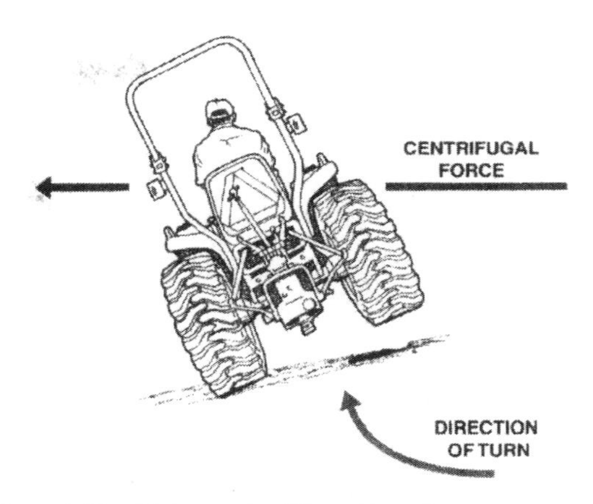

Figure 4.13.g. High speeds while making a turn can cause a sideways overturn. Make sure brakes are locked together. Reduce speed before entering the turn. *Safety Management for Landscapers, Grounds-Care Businesses, and Golf Courses, John Deere Publishing, 2001. Illustrations reproduced by permission. All rights reserved.*

Figure 4.13.h. Avoid obstacles as you operate the tractor. Some tractor operators will check the field before beginning the operation. Stumps, rocks, animal dens, etc, can upset a tractor.

Figure 4.13.i. Tractors are powerful, but each one has a limit to its pulling power. Overloading a tractor could stall the engine, but rearward overturns can occur as well. *Farm and Ranch Safety Management, John Deere Publishing, 1994. Illustrations reproduced by permission. All rights reserved.*

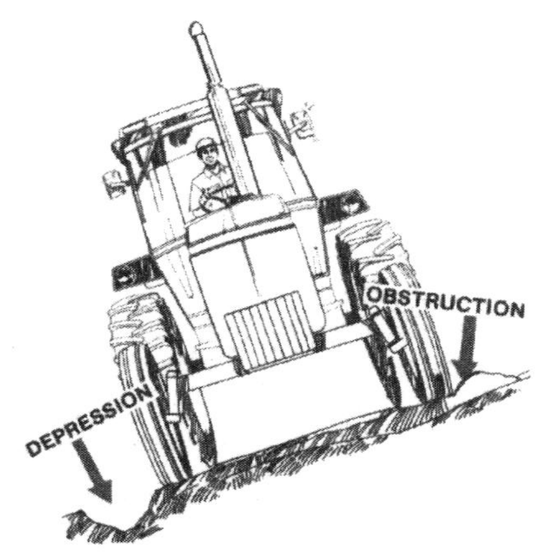

Figure 4.13.j. Field conditions pose special hazards to tractor operation. The operator must know where these obstructions and depressions are located. *Farm and Ranch Safety Management, John Deere Publishing, 1994. Illustrations reproduced by permission. All rights reserved.*

Figure 4.13.k. When operating a high-lift bucket with a load or without a load, keep the bucket as low to the ground as possible while in transport. Sideway overturns are possible if you try to travel with the bucket in the up position. *Farm and Ranch Safety Management, John Deere Publishing, 1994. Illustrations reproduced by permission. All rights reserved.*

Figure 4.13.l. A tractor stuck in mud is immovable without help. Adult supervision is necessary.

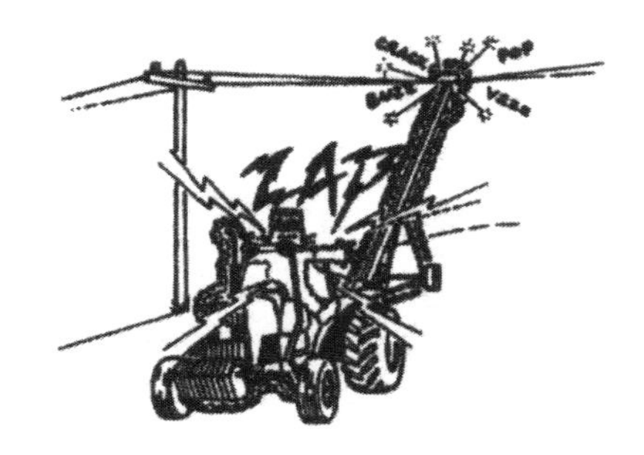

Figure 4.13.m. Avoid overhead power lines while transporting equipment to avoid risk of electrocution.

Safety Activities

1. Start a collection of farm accident reports from magazines, newspapers, and farm newsletters.

2. Using a camera or video recorder, take photos or video film of unsafe tractor use situations. Make a display for your club, classroom, employee room or farm shop.

3. Try this Word Search Game to find words related to proper tractor use. Words or phrases may be spelled forward, backward, up, down, or diagonally.

S A D Z C D E F G H
J A T T I T U D E I
K N F D L M G N O P
Y O V E R T U R N X
T R T E T S A Y R Q
U I V P R Y R O W X
D D C S E B D A Z Y
L E F W L G S H I J
K R L O A M N O P Q
T S V L V W X Y Z R

Use this word list: attitude, safety, guards. no riders, overturn, alert, low speed, pto

References

1. The Ten Commandments of Tractor Safety, 1999, Kubota Tractor Corp., Compton, California.
2. www.cdc.gov/niosh/injury/traumagric.html, September 2002.
3. Safety Management for Landscapers, Grounds-Care Businesses, and Golf Courses, John Deere Publishing, 2001. Illustrations reproduced by permission. All rights reserved.
4. www.cdc.gov/niosh/nasd/At search box, type tow ropes, cables and chains.
5. Farm and Ranch Safety Management, John Deere Publishing, 2009. Illustrations reproduced by permission. All rights reserved.

Contact Information

National Safe Tractor and Machinery Operation Program
The Pennsylvania State University
Agricultural and Biological Engineering Department
246 Agricultural Engineering Building
University Park, PA 16802
Phone: 814-865-7685
Fax: 814-863-1031
Email: NSTMOP@psu.edu

Credits

Developed by WC Harshman, AM Yoder, JW Hilton and D J Murphy, The Pennsylvania State University. Reviewed by TL Bean and D Jepsen, The Ohio State University and S Steel, National Safety Council. Revised 3/2013.

This material is based upon work supported by the National Institute of Food and Agriculture, U.S. Department of Agriculture, under Agreement Nos. 2001-41521-01263 and 2010-41521-20839. Any opinions, findings, conclusions, or recommendations expressed in this publication are those of the author(s) and do not necessarily reflect the view of the U.S. Department of Agriculture

OPERATING THE TRACTOR ON PUBLIC ROADS

HOSTA Task Sheet 4.14 **Core**

NATIONAL SAFE TRACTOR AND MACHINERY OPERATION PROGRAM

Introduction

Today's farmers are traveling more miles than ever before on public roads to plant, grow, and harvest crops. Slow-moving tractors and implements are no match for the general public's high-speed travels. Most crashes between farm equipment and motor vehicles occur during daylight and in good weather. You can never let your guard down when traveling on a public road with farm equipment.

This Task Sheet discusses operation of the tractor and equipment on public roads.

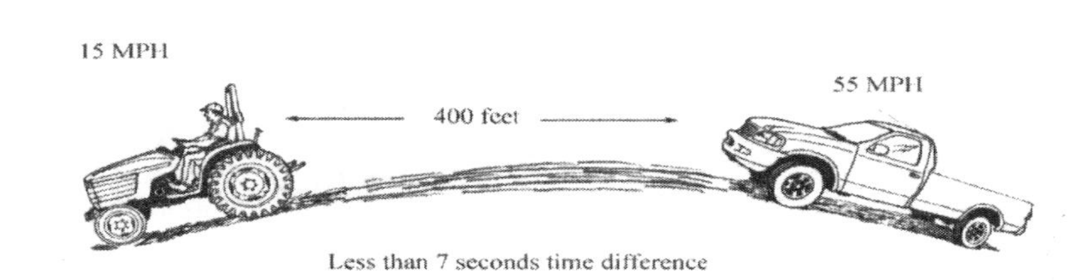

Figure 4.14.a. An automobile traveling 55 mph will cover the 400 ft. distance between the car and tractor in less than 7 seconds. Will that be enough time or space to get slowed or stopped to prevent a rear-end crash?

Movement Hazards

These traffic situations are created by operating tractors on public roadways.

- Pulling slowly onto roads with long and heavy loads
- Slow tractor travel speeds
- Left turns across traffic into narrow field lanes
- Swinging into the left lane to make a right turn into a field
- Wide machinery being transported
- Potential for spilled loads

All rules of vehicle safety, as well as all rules of courteous driving, must be followed to prevent traffic problems.

Obeying the Law

Each state varies in their highway regulations regarding the ages and places where one may operate a farm tractor. States seldom require a driver's license for a tractor, but many do limit 14- and 15-year-old drivers to crossing over public roadways only or to operating equipment on roads that bisect or adjoin their farm.

Check with your local state police to learn more about the laws in your area.

You must also obey all traffic laws and signs as well.

Tractor operators are at a great disadvantage when traveling on busy highways.

Learning Goals

- To understand the difference between farm equipment road use and normal highway vehicle road use
- To use all safe and courteous traffic driving practices to prevent farm equipment and motor vehicle crashes

Related Task Sheets:

Safety and Health Regulations	1.2
Reaction Time	2.3
Hand Signals	2.9

Cooperation provided by The Ohio State University and National Safety Council.

Figure 4.14.b. Lighting and marking standards may or may not be the standards for your state. Check your state laws.

Using the proper lighting and marking standards gives motorists ample warning that farm equipment is using the public roadway.

Lighting and Marking

American Society of Agricultural Engineers (ASAE) Standards for lighting and marking are summarized in Table 4.14.a. Most farm equipment delivered from the factory today will have used these standards. Does the equipment that you will use measure up to these standards? If not, can the equipment be improved with retrofit kits of lights and reflectors?

Although not included in the ASAE standard, rotary beacons and back-up alarms are optional accessories which may be add-ons depending upon your needs. If accessories have been added, they should be in working order.

Figure 4.14.c. Be sure that a work light that points to the rear is off during road travel at night. Single white lights may not be recognized as slow-moving or as a tractor light. Also if SMV emblems are worn or obsolete, replace them with newer more reflective SMV emblems.

Table 4.14.a. Recommendations from ASAE for lighting and marking.

Item	Recommendation
Headlights	Two white lights mounted at the same level
Taillights	Two red lights mounted at the rear
Hazard Flashers	Two or more lamps with amber color to the front and red color to the rear
Turn Indicators	Two amber to the front and two red-colored lights to the rear mounted with flashers
SMV Emblem	One visible at 1000 ft. mounted to the rear and 2-10 ft. above the ground
Reflectors	Two red reflectors (on rear outside corners) and 2 yellow reflectors (on the front outside corners) of the machine
Conspicuity Material	Red retro-reflective and red-orange fluorescent color visible to mark the rear. Yellow retro-reflective material to mark the front.

Towing Safety

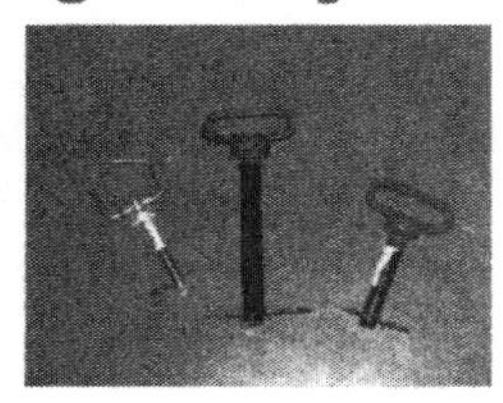

Figure 4.14.d. Secure hitch pins with locking clips as shown.

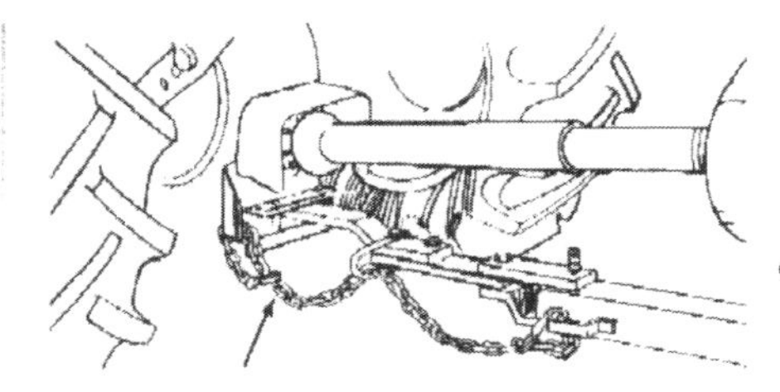

Figure 4.14.e. Use safety chains to insure load hitching safety when possible. Trucks pulling farm loads should have safety chains also. *Safety Management for Landscapers, Grounds-Care Businesses, and Golf Courses, John Deere Publishing, 2001. Illustrations reproduced by permission. All rights reserved.*

Figure 4.14.f. SMV emblems are required on vehicles designed to travel less than 25 mph while occupying public roadways. SMV emblems should be visible from no less than 1000 feet to the rear of the tractor or towed implement. Therefore, mounting height may vary from 2 to 10 feet above the road surface. Replace faded, damaged SMV emblems.

General Practices for Tractors on Highways

Think about the following when traveling on the highway with farm machinery.

- Time of day–Is it possible to avoid the busy times of the day to move equipment? Hauling large loads during early morning or late afternoon while people hurry to and from work creates traffic problems for both of you. Moving loads after nightfall may be better timing, but lighting becomes a necessary consideration.
- Courtesy– Try to be as watchful of others as possible. Let the high-speed traffic go first. Your best manners on the highway will be the first safe practice to follow.
- Blind spots– Are there locations which pose problems with visibility? Avoid them if possible.
- Shifting loads– If you upset a load of hay, spill a load of manure or a tank of pesticide mixture, or coat the road with mud from the field, you are responsible for getting help for cleanup and alerting traffic to be cautious. If manure or chemical spills endanger waterways, notify your employer who may have reporting requirements with state environmental officials.
- Safe Equipment– Your walk-around inspection should have shown you if you have damaged equipment. Be sure damaged equipment does not create a road hazard. For example, a loose wheel on a hay rake could cause a disaster.

Figure 4.14.g. A best practice for transporting wide loads on a public roadway is to use an escort vehicle to assist in alerting other motorists. Be a courteous tractor operator to bring good public relations to the farm community.

Pull completely off the road to let traffic flow past if possible. DO NOT SIGNAL THEM TO PASS YOU. Signaling to motorists to pass makes you responsible for them.

LEFT TURN

Figure 4.14.h. Use accepted hand signals to inform other drivers of your intentions. *Safety Management for Landscapers, Grounds-Care Businesses, and Golf Courses, John Deere Publishing, 2001. Illustrations reproduced by permission. All rights reserved.*

Safety Activities

1. Measure the length of the longest tractor and implement combination with which you will work. Then have someone time how long it takes you to move the front end of that tractor to the rear end of the towed implement past a point or across the highway in front of the farm. How many seconds did it take to cross the road? ___________seconds.

2. A car approaching the farm driveway is traveling at 60 mph. How many feet will that car travel in 1 second? _________seconds.

   Hint: 60 mph =1 mile/minute Calculate what distance in feet will be covered in 1 second. Remember that 5,280 feet equals 1 mile.

   1 mile / minute = ___________ feet / second.

3. Multiply the answer (feet/second) in question number 2 by the time you recorded in question number 1.

   This is the distance the car going 60 mph will travel in the time it takes you to cross the road. Record the answer here. _________ Can you see that far as you pull out to cross the roadway?

4. Conduct a survey of the lighting, marking, and hitching of the tractors on your farm or farm of employment. Does it meet the safety requirements of your state?

5. Practice the hand signals for right, left, and stop that you will use while operating a tractor not equipped with turn signals.

References

1. Safety Management for Landscapers, Grounds-Care Businesses, and Golf Courses, John Deere Publishing, 2001. Illustrations reproduced by permission. All rights reserved.
2. Farm and Ranch Safety Management, John Deere Publications, 2009.
3. State Laws and Regulations.

Contact Information

National Safe Tractor and Machinery Operation Program
The Pennsylvania State University
Agricultural and Biological Engineering Department
246 Agricultural Engineering Building
University Park, PA 16802
Phone: 814-865-7685
Fax: 814-863-1031
Email: NSTMOP@psu.edu

Credits

Developed by WC Harshman, AM Yoder, JW Hilton and D J Murphy, The Pennsylvania State University. Reviewed by TL Bean and D Jepsen, The Ohio State University and S Steel, National Safety Council. Revised 3/2013.

This material is based upon work supported by the National Institute of Food and Agriculture, U.S. Department of Agriculture, under Agreement Nos. 2001-41521-01263 and 2010-41521-20839. Any opinions, findings, conclusions, or recommendations expressed in this publication are those of the author(s) and do not necessarily reflect the view of the U.S. Department of Agriculture.

LIGHTING AND MARKING

HOSTA Task Sheet 4.14.1 **Core**

NATIONAL SAFE TRACTOR AND MACHINERY OPERATION PROGRAM

Introduction

Today's farmers are traveling more miles on public roads than ever before to tend to livestock and plant, grow, and harvest crops. Slow-moving tractors and implements are no match for the public's high-speed travels.

Most crashes between farm equipment and motor vehicles occur during daylight hours and in good weather. You must be careful when traveling on public roads with farm equipment. Tractors and equipment must be clearly identified as slow-moving vehicles using recognizable lighting and marking.

This task sheet discusses lighting and marking as it relates to moving tractors and equipment on public roadways.

Lighting and Marking

The American Society of Agricultural Engineers (ASAE) Standard for Lighting and Marking are summarized in Table 4.14.1.a. See the reference section to access the asae.org information.

Most farm equipment manufactured today will use this standard. Exceptions to the standard may occur with equipment manufactured outside the United States. Many states use a similar standard in their Motor Vehicle Codes to specify lighting and marking of slow-moving vehicles and farm equipment.

Lighting and marking on older equipment can be improved to meet this standard with add-on lights and reflectors kits.

Figure 4.14.1.a. Use of public roadways dictates that farm tractors and equipment be visible and identified as slower moving and wider than the usual vehicle traveling the road.

Proper equipment marking and lighting is a must for roadway travel.

Table 4.14.1.a Recommendations from ASAE for lighting and marking

Item	Recommendations
Headlights	Two white lights mounted at the same level
Taillights	Two red lights mounted at the rear
Hazard Flashers	Two or more lamps with amber color to the front and red color to the rear
Turn Indicators	Two amber lamps to the front and two red-colored lights to the rear mounted with flashers
SMV Emblem	One visible at 1000 ft. mounted to the rear and 2-10 ft. above the ground
Reflectors	Two red reflectors (on rear outside corners) and two yellow reflectors (on the front out-side corners) of the machine
Conspicuity Material	Red retro-reflective and red-orange fluores-cent color visible to mark the rear. Yellow retro-reflective material to mark the front.

Learning Goals

- To understand the recommendations for lighting and marking of farm tractors and machinery

Related Task Sheets:

Safety and Health Regulations	**1.2**
OSHA Act	**1.2.2**
State Vehicle Codes	**1.2.5**
Hand Signals	**2.9**
Operating the Tractor on Public Roads	**4.14**

Cooperation provided by The Ohio State University and National Safety Council.

Is Your Lighting and Marking Adequate?

Highway transport of farm equipment at night requires lighting and marking. Older equipment must meet these requirements as well. The requirements are:

- Slow-moving speed shown by SMV emblem
- Extremities of width defined by side marker lights or decals
- Ability to warn of turns by recognizable signals

If the tractor and equipment or self-propelled equipment does not meet these requirements, the operator increases the risk of injury to him or herself and the public.

Figure 4.14.1.b. Can you identify the lighting and marking features used on this tractor and grain drill? Number 1 is__________. Number 2 are ___________. Number 3 represents_____________. Is the equipment safely marked for roadway travel?

Safety Activities

1. Clean all the lights, SMV emblems, and reflective markers daily on the farm equipment you will operate.
2. Conduct an inspection of all tractors and equipment on a local farm. Make a list of lighting and marking deficiencies you find.
3. Use the website www.asae.org to learn more about machinery and equipment lighting and marking standards.
4. Using the Internet, search your favorite brand of tractor or machinery to access pictures that show the lighting and marking methods used. Do the methods meet the ASAE standards?

References

1. American Society of Agricultural and Biological Engineers, ANSI/ASABE, S279.12 Lighting and Marking of Agricultural Equipment on Highways, St. Joseph, MI.
2. Farm and Ranch Safety Management, John Deere Publishing, 2009.

Contact Information

National Safe Tractor and Machinery Operation Program
The Pennsylvania State University
Agricultural and Biological Engineering Department
246 Agricultural Engineering Building
University Park, PA 16802
Phone: 814-865-7685
Fax: 814-863-1031
Email: NSTMOP@psu.edu

Credits

Developed by WC Harshman, AM Yoder, JW Hilton and D J Murphy, The Pennsylvania State University. Reviewed by TL Bean and D Jepsen, The Ohio State University and S Steel, National Safety Council. Revised 3/2013.

This material is based upon work supported by the National Institute of Food and Agriculture, U.S. Department of Agriculture, under Agreement Nos. 2001-41521-01263 and 2010-41521-20839. Any opinions, findings, conclusions, or recommendations expressed in this publication are those of the author(s) and do not necessarily reflect the view of the U.S. Department of Agriculture.

CONNECTING IMPLEMENTS TO THE TRACTOR

HOSTA Task Sheet 5.1 **Core**

NATIONAL SAFE TRACTOR AND MACHINERY OPERATION PROGRAM

Introduction

"The owner says that I should be able to connect (hitch) the rake to the tractor and be in the nearby field within 5 minutes. It has been 10 minutes, and I still can't seem to get the drawbar of the tractor lined up with the hitch on the rake."

Can you steer in reverse? Can you use the clutch and brakes smoothly? If not, review the lessons on steering in reverse and moving and steering the tractor.

Do you understand where to hitch to the load to insure tractor stability? If not, review the lessons on tractor stability.

This task sheet provides an overview of safe and efficient hitching of implements to the tractor. See Task Sheet 5.2 or 5.3 for additional details.

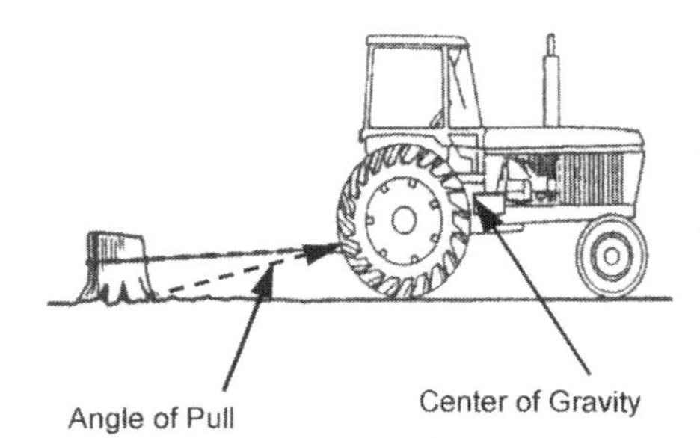

Figure 5.1.a. An example of safe hitching. The drawbar will lower if the front end lifts off the ground. This reduces the "angle of pull" and the risk of a rear overturn.

Pulling a load with the downward and rearward force above the tractor's center of gravity will result in a rear overturn. You must hitch only to the drawbar to prevent the tractor from rearing up and turning over. Even small lawn and garden-size tractors can flip rearward if not properly hitched to a load.

Figure 5.1.c. The tractor drawbar is the only safe place to connect a load. Do not hitch higher than the drawbar so all pulling forces stay below the tractor's center of gravity. For most operations, the drawbar should be placed midpoint between the rear tires to maximize pulling power. Hillside operations may require a drawbar adjustment to one side to balance the pulling forces.

Hitch to the drawbar only! Hitching anywhere else can result in rear turnover and death.

Hitching and the Center of Gravity

In Task Sheet 4.12, *Tractor Stability*, you learned about the tractor's center of gravity and stability baseline. Tractor hitches are designed so the downward and rearward force during a pull are below the center of gravity (Figure 5.2.a.). To maintain tractor stability, the "angle of pull" should be kept as low as possible by hitching to the drawbar only.

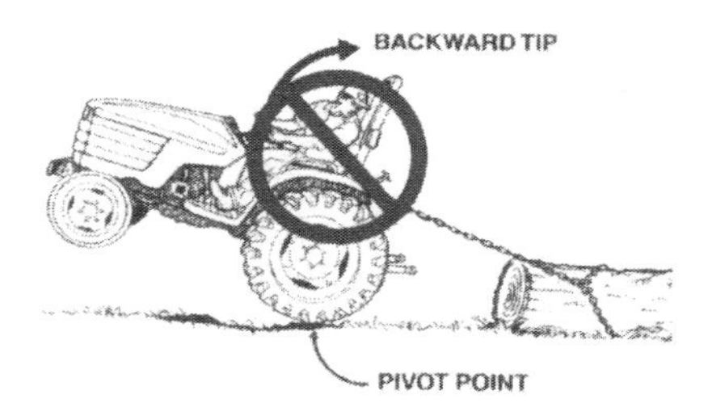

Figure 5.1.b. The log is fairly immovable. A chain hooked above the center of gravity of the tractor (e.g., top of 3-point hitch bracket), allows a rearward tip of the tractor. Improper hitching has overridden safe tractor engineering design. Many people have lost their lives as a result. *Safety Management for Landscapers, Grounds-Care Businesses, and Golf Courses. John Deere Publishing, 2001. Illustrations reproduced by permission. All rights reserved.*

Learning Goals

- To safely connect an implement to the tractor's drawbar
- To safely connect an implement to the tractor's 3-point hitch

Related Task Sheets:

Tractor Stability	4.12
Using the Tractor Safely	4.13
Operating the Tractor on Public Roads	4.14
Using Drawbar Implements	5.2
Using 3-Point Hitch Implements	5.3

Cooperation provided by The Ohio State University and National Safety Council.

Figure 5.1.d. Tractor drawbars are designed at the correct height from the ground to keep the pull forces below the center of gravity. See Table 5.1.a. Only use the drawbar to tow a load. A swinging or floating drawbar permits adjustment of the center line of pull to be maintained even on a hillside.

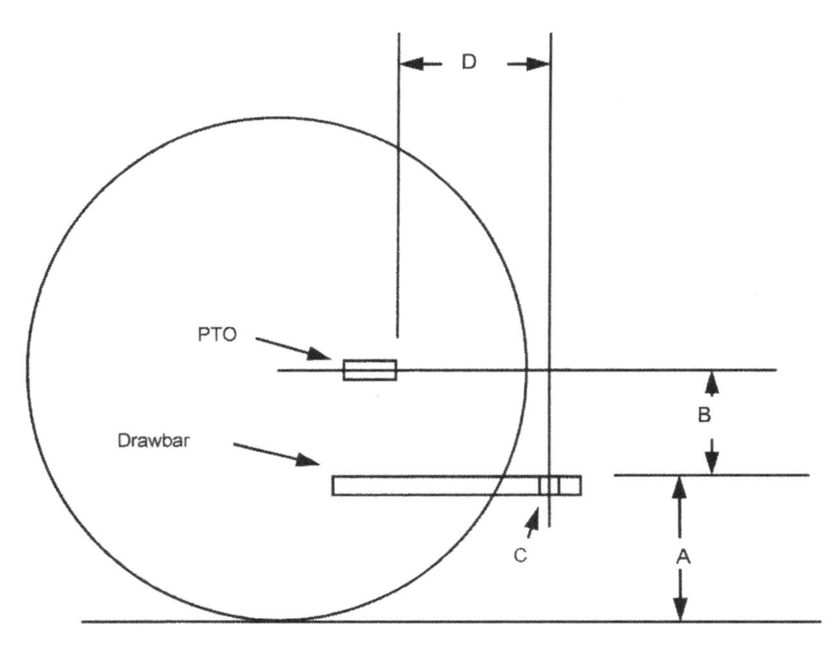

Figure 5.1.e. The tractor power take-off and drawbar position are designed with specific measurements for the size and horsepower rating of the tractor. The operator should not make changes to these design standards by changing the hitch point. Table 5.1.a lists the measurements and relationships at points A, B, C, and D above for each range of tractor size.

A bolt laying around the farm shop is not a substitute hitch pin! Hitch pins are designed for specific drawbar loads and power ratings and must fit the drawbar hole.

Drawbar Hitch Category

Item	I	II	III	IV
Tractor HP	20-45	40-100	80-275	180-400
Drawbar Height above ground (A)	15"+/-2"	15"+/-2"	19"+/-2"	19"+/-2"
Drawbar to PTO (B)	8"-12"	8" - 12.5"	8.5" -14"	10" -14"
Hitch-Pin Hole Size(C)*	1.1"	1.3"	1.7"	2.1"
Nominal Hitch Pin Size*	1.0"	1.2"	1.6"	2.0"
Drawbar Dimensions (Thickness x width)	1-3/16"x2.0"	1-9/16"x2.5"	2"x 3-3/16"	2-3/8"x 4-7/8"
Regular Size PTO Stub Shaft to Drawbar Hitch Hole (D)	14-20"	14-20"	14-20"	14-20"

* The measurement has been rounded to the nearest 1/10 (0.1) inch. Hitch pins must fit the hitch-pin hole without excessive movement.

Table 5.1.a. Drawbar Sizing and Positioning Standards (ASAE S482)

 Cooperation provided by The Ohio State University and National Safety Council.

The 3-Point Hitch

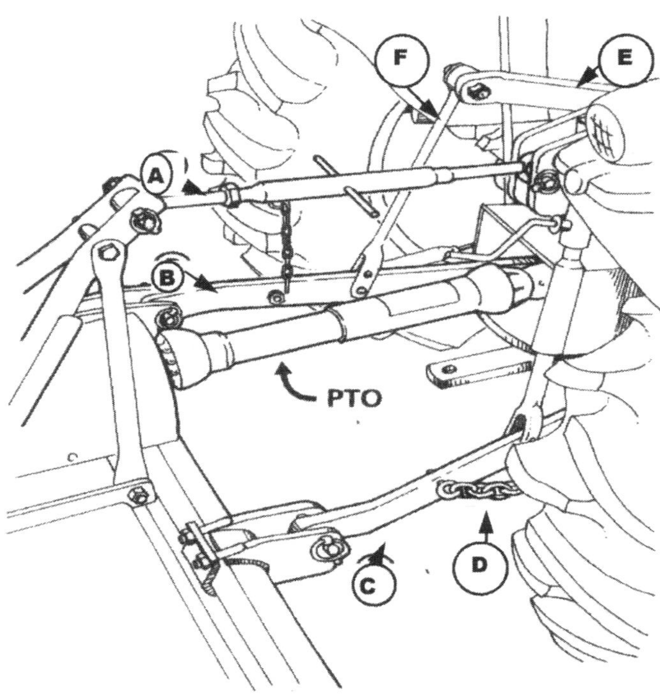

Figure 5.1.f. Parts of the 3-Point Hitch

A. Top Link B,C. Draft Arms D. Anti-sway bar or chain

E. Lift Arm F. Lift Rod

Safety Management for Landscapers, Grounds-Care Businesses, and Golf Courses, John Deere Publishing, 2001. Illustrations reproduced by permission. All rights reserved.

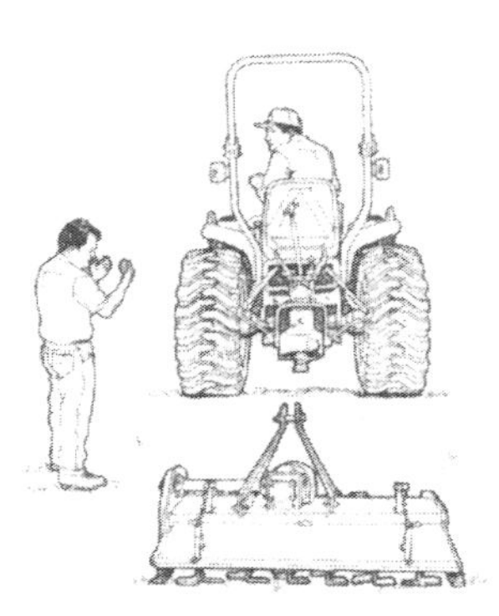

Figure 5.1.g. Never let another person stand between the tractor and the implement during hitching. Too fast of an approach or the operator's foot slipping from the clutch can lead to injury or fatality to the person standing nearby. *Safety Management for Landscapers, Grounds-Care Businesses, and Golf Courses, John Deere Publishing, 2001. Illustrations reproduced by permission. All rights reserved.*

Implement Hitching

Follow these steps for hitching to a drawbar: Also see Task Sheet 5.2.

1. Position the tractor to align the hole in the drawbar with the hole in the implement hitch. This is called spotting. You may need to practice this skill.
2. Stop the engine, put the tractor in park, or set the brakes.
3. Attach the implement using the proper-sized hitch pin and security clip.
4. Raise the implement jack stand and remove chock blocks from the wheels.
5. Connect the PTO shaft, hydraulic hoses, and/or electrical connections as required. Refer to the appropriate task sheets on these subjects.

Follow these steps for hitching to a 3-point hitch attachment: Also see Task Sheet 5.3.

1. Move the stationary tractor drawbar forward for clearance.
2. Position the tractor so the pin holes of the draft arms are closely aligned with the implement hitch points.
3. Raise or lower the draft arms to match the implement hitch points.
4. Stop the engine, securely park the tractor, set the brakes.
5. First attach left draft arm to the implement hitch point using the proper size hitch pin and security clip. Right arm is adjustable and is connected next.
6. Remount and start the tractor to use the hydraulic system to raise the lift arms if needed.
7. Match the top link of the 3-point hitch to the implement's upper hitch point. Raise the lift arms to lengthen, or drive ahead with implement down to shorten to adjust if needed. The implement may not be level if the upper link has been adjusted too many times. If it is out of level, the machine may not work properly *If you cannot level the machine, ask for help.*
8. Securely attach the upper hitch pin with the proper size hitch pin and security clip.

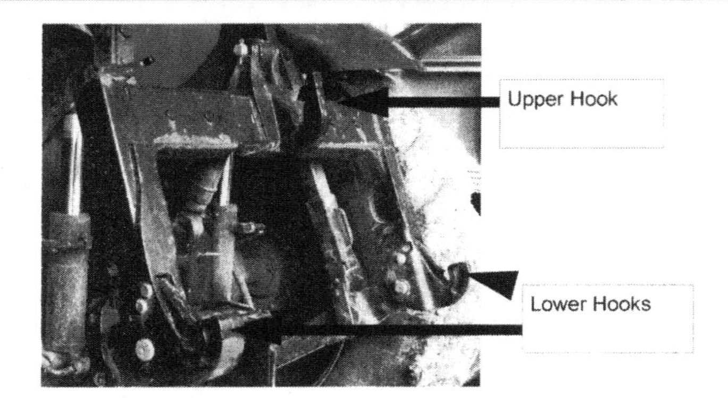

Figure 5.1.h. Heavy-duty quick-attach couplers are mounted on to the tractor's 3-point hitch and can safely handle large 3-point hitch implements without a person moving between the tractor and the implement. See the circled areas which show these hook and latch points. Refer to the Owner's Manual for additional instructions on their use or have a qualified operator demonstrate to you the correct procedure to use a quick-attach coupler.

Safety Activities

1. Practice backing a tractor with a drawbar to an implement to "spot" the hole in the drawbar to the hole in the implement tongue. You should be able to perform this skill with a minimum number of changes of direction to be a proficient tractor operator.

2. Practice backing a tractor with a 3-point hitch to an implement to adjust the pin hole in the draft arms to the lower hitch pins on the implement's 3-point hitch attachment. As you become more able to align these points, securely park the tractor. Attach the draft arm hitch pins, restart the tractor, adjust the draft arms to align, and connect the upper link point. You should be able to perform this skill with a minimum change of direction to be a proficient tractor operator.
3. On a tractor you can easily measure, take measurements and record the following:
   a. distance from ground to drawbar _________inches
   b. dimensions of drawbar (width and thickness) _____x_____inches
   c. hitch-pin hole opening in drawbar ______inches
   d. vertical distance from drawbar to center of PTO stub shaft _____inches
   How do these measurements compare with the standards shown on Table 5.2a?

4. Using a battery-operated toy truck or tractor, devise a place to hitch a load at a point above the toy's axle. Make a sled from sheet metal or cardboard, and attempt to pull a load of small objects such as nuts, bolts, etc. What happens as the toy attempts to pull the load? Change the height and length of the angle of pull, and record the reaction of the toy truck or tractor to the changes made.

References

1. American Society of Agricultural and Biological Engineers, ANSI/ASABE, S482 Drawbars, St. Joseph, MI.
2. American Society of Agricultural and Biological Engineers, ANSI/ASABE, S217 Three-Point Free Link Attachment, St. Joseph, MI.
3. Safety Management for Landscapers, Grounds-Care Businesses, and Golf Courses, John Deere Publishing, 2001. Illustrations reproduced by permission. All rights reserved.

Contact Information

National Safe Tractor and Machinery Operation Program
The Pennsylvania State University
Agricultural and Biological Engineering Department
246 Agricultural Engineering Building
University Park, PA 16802
Phone: 814-865-7685
Fax: 814-863-1031
Email: NSTMOP@psu.edu

Credits

Developed by WC Harshman, AM Yoder, JW Hilton and D J Murphy, The Pennsylvania State University. Reviewed by TL Bean and D Jepsen, The Ohio State University and S Steel, National Safety Council. Revised 3/2013

This material is based upon work supported by the National Institute of Food and Agriculture, U.S. Department of Agriculture, under Agreement Nos. 2001-41521-01263 and 2010-41521-20839. Any opinions, findings, conclusions, or recommendations expressed in this publication are those of the author(s) and do not necessarily reflect the view of the U.S. Department of Agriculture.

USING DRAWBAR IMPLEMENTS

HOSTA Task Sheet 5.2 **Core**

NATIONAL SAFE TRACTOR AND MACHINERY OPERATION PROGRAM

Introduction

Several agricultural implements are ground-driven (the power comes from the wheels turning on the ground). Use of the PTO is unnecessary. If you stop moving forward with these implements, the machine stops operating. The beginning tractor operator is often assigned to hitch to and use these types of drawbar implements.

A qualified operator should demonstrate how to safely use equipment before expecting you to operate the equipment successfully.

This task sheet will focus on towed equipment which is ground-driven. Other task sheets will provide information regarding PTO and hydraulic and electrical connections between the tractor and the implement.

Hitching Review

Follow these steps for drawbar hitching to an implement equipped with a height positioning jack:

1. Back to the correct position to align, or "spot," the hole in the drawbar with the hole in the tongue. (Figure 5.2.a.)
2. Stop the engine, securely park the tractor, and set the brakes.
3. Dismount from the tractor to adjust the implement tongue height using the support jack.
4. Remount and start the tractor to make final adjustment to the "spot." If necessary stop the engine, securely park the tractor, and set the brakes.
5. Attach the implement using the proper hitch pin and security clip, and move the jack to the transport position.

Hitching Safely

Backing a tractor in reverse to connect an implement can be an easy and safe task. Figure 5.2.a. shows how to spot the hitch to the drawbar. The caption explains how "spotting" to the drawbar can create a hazard.

Practice backing the tractor to align the drawbar with the implement hitch or tongue. You should not need more than three changes of direction to do this job.

Figure 5.2.a. If the tongue of the implement is not adjusted for the height of the drawbar, the drawbar can push the implement to the rear, knock the jack stand over, or harm someone standing behind the implement. Take a preliminary measurement of the drawbar height and the implement tongue before hitching makes aligning the two points much easier. A stable jack stand should be used. *Safety Management for Landscapers, Grounds-Care Businesses, and Golf Courses, John Deere Publishing, 2001. Illustrations reproduced by permission. All rights reserved.*

Figure 5.2.b. If you need help to hitch to a machine, insist that the helper stand off to the side of the operation. Many people have been crushed between tractors and machines trying to help connect the machine to the tractor. This picture shows an unsafe act. Can you explain what is unsafe?

If your foot slips from the clutch while hitching to a machine, a helper could be crushed.

Learning Goals

- To safely attach implements to a tractor's drawbar
- To safely use drawbar implements during transport, field use, turns, and backing operations

Related Task Sheets:

Moving and Steering the Tractor	**4.10**
Using the Tractor Safely	**4.13**
Operating the Tractor on Public Roads	**4.14**
Connecting Implements to the Tractor	**5.1**

Cooperation provided by The Ohio State University and National Safety Council.

Using Ground-Driven Machinery

Disks, harrows, hay rakes, windrow inverters, and older manure spreaders are a few of the ground-driven implements assigned to beginning tractor operators. Use them safely by remembering these points.

1. Make sure you know how wide the machine is compared to the tractor.
2. Be sure the machine width is reduced to the "transport" position for travel on public roadways.
3. Shift the machine to the wider "field" position when ready to use it.
4. Stop the engine, securely park the tractor, and set the brakes before dismounting to engage the machine operation mechanism (levers, pins, etc.) allowing the wheels to turn the machine.
5. Pay attention to field boundary fences and obstacles before you begin field operations.
6. Allow plenty of space at ends of rows or fields to turn the equipment without "jack-knifing."
7. Be sure to return the implement to the transport position before using public roads or passing through narrow farm gates.

Figure 5.2.c. Implements can be wider than the tractor in transport or road position. Passing through farm gates or using public roadways can create a hazardous situation.

If you have never used a particular implement, ask for a demonstration before you try the job.

Safety Activities

1. Practice spotting the tractor drawbar to the tongue of the implement so that you can hitch to a machine quickly and safely.
2. Demonstrate the safety procedures to use when backing a tractor to hitch a machine by showing a helper where to stand to safely help you spot the drawbar and implement tongue.
3. Inspect the ground-driven machines you may use to learn:
   a. how they are moved from transport to field position and vice versa.
   b. what mechanism is used to engage the ground wheels with the turning parts of the machine.
4. Check the machinery and tractors you may use for the hitch pins that will be used. Are they available, of the proper size, and have a securing clip? Where are the hitch pins stored on the farm?
5. Practice raising and lowering the various jack stands you find on agricultural equipment.

References

1. Safety Management for Landscapers, Grounds-Care Businesses, and Golf Courses, John Deere Publishing, 2001. Illustrations reproduced by permission. All rights reserved.
2. Operators' Manuals for Tractor and Machinery.
3. American Society of Agricultural and Biological Engineers, ANSI/ASABE, S485 Jacks, St. Joseph, MI.

Contact Information

National Safe Tractor and Machinery Operation Program
The Pennsylvania State University
Agricultural and Biological Engineering Department
246 Agricultural Engineering Building
University Park, PA 16802
Phone: 814-865-7685
Fax: 814-863-1031
Email: NSTMOP@psu.edu

Credits

Developed by WC Harshman, AM Yoder, JW Hilton and D J Murphy, The Pennsylvania State University. Reviewed by TL Bean and D Jepsen, The Ohio State University and S Steel, National Safety Council. Revised 3/2013

This material is based upon work supported by the National Institute of Food and Agriculture, U.S Department of Agriculture, under Agreement Nos. 2001-41521-01263 and 2010-41521-20839. Any opinions, findings, conclusions, or recommendations expressed in this publication are those of the author(s) and do not necessarily reflect the view of the U.S. Department of Agriculture.

USING 3-POINT HITCH IMPLEMENTS

HOSTA Task Sheet 5.3 **Core**

NATIONAL SAFE TRACTOR AND MACHINERY OPERATION PROGRAM

Introduction

Once you can successfully connect an implement to a tractor's 3–point hitch, you are ready to start using the machine. Some machines are powered by the PTO, while others are ground-driven (the power comes from the wheels turning on the ground). A qualified operator should demonstrate how to safely use equipment before expecting you to use the machinery.

This task sheet discusses 3-point hitch equipment which is both ground- and PTO-driven. Later task sheets will provide information regarding hydraulic connections and electrical connections between the tractor and the implement.

Hitching Review

Follow these steps for connecting implements to a 3-point hitch.

1. Remove the drawbar, or move the drawbar forward or to the side for clearance.
2. Back the tractor so the pin holes of the tractor's draft arms are nearly aligned with the implement's lower hitch pins. See Figure 5.3.a.
3. From the tractor seat and using the hydraulic lift controls, raise or lower the draft arms to match the implements lower hitch pins. See Figure 5.3.a.

Figure 5.3.a. Back slowly to the implement to be attached. Using the hydraulic controls, raise or lower the draft arms to nearly match the implement's lower hitch pins.

4. Stop the engine, securely park the tractor, and set the brakes.
5. Attach each draft arm to the implement, and secure with the hitching pins and security clips. See page 2, Figure 5.3.c.
6. Remount and restart the tractor, and slowly raise the tractor's draft arms with the hydraulic lift controls to closely align the upper hitch points.
7. Stop the engine, securely park the tractor, and set the brakes.
8. Attach the tractor's upper hitching point of the 3-point hitch to the top hitch point of the implement with the proper size pin and securing clip. See page 2, Figure 5.3.d. The upper link may need to be lengthened or shortened to fit. Ask for help if there is a problem you cannot solve.

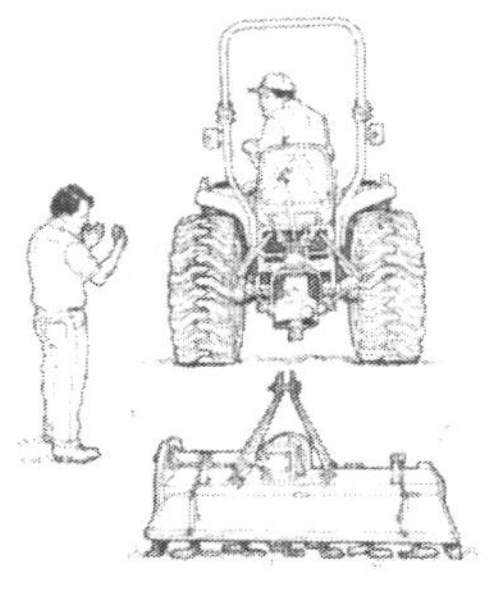

Figure 5.3.b. When connecting drawbar implements, never let a helper stand between the tractor and 3-point hitch implement. Crushing injuries and death can result. *Safety Management for Landscapers, Grounds-Care Businesses, and Golf Courses, John Deere Publishing, 2001. Illustrations reproduced by permission. All rights reserved.*

Do as many hitching operations as you can with the engine shut off and the tractor securely parked.

Learning Goals

- To safely connect a 3-point hitch implement
- To safely use a 3-point hitch implement
- To safely disconnect a 3-point hitch implement

Related Task Sheets:

Stopping and Dismounting the Tractor	**4.9**
Moving and Steering the Tractor	**4.10**
Using the Tractor Safely	**4.13**
Operating the Tractor on Public Roads	**4.14**
Connecting Implements to the Tractor	**5.1**
Making PTO Connections	**5.4**

 Cooperation provided by The Ohio State University and National Safety Council.

Figure 5.3.c. On both the left and right sides of the implement, insert the draft arm attachment pin of the tractor into the pin holes of the implement's lower hitch assembly. Secure the hitch with the proper size hitch pins and security clip.

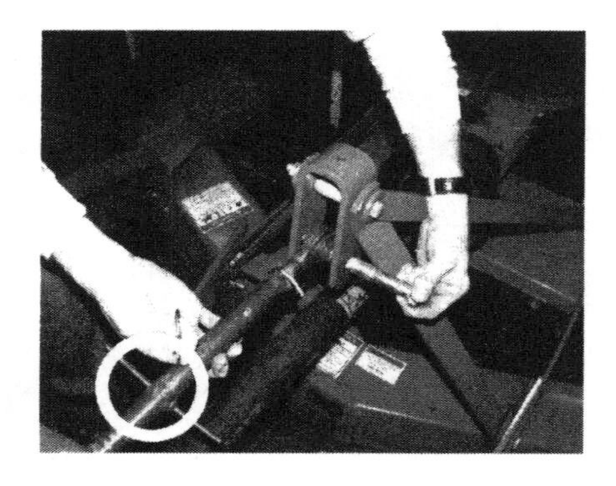

Figure 5.3.d. After adjusting the upper link of the tractor's 3-point hitch to align with the upper hitch point of the implement, secure the equipment with the proper size hitch pin and security clip. The circled area indicates where the upper link may be adjusted for fit. The implement must be in a level position after the connection is made. See Task Sheet 5.2.

The 3-point hitch works because the three hitch pins secure the implement to the tractor. Do not use chains or other temporary pins to attach the implement.

Figure 5.3.e. A firm grip will be needed to press in on the detent lock of the PTO shaft. This lock engages the groove in the stub shaft to secure the PTO driveline shaft to the stub shaft. Other forms of locking the PTO shaft will be found in Task Sheet 5.4.

3-Point Hitches and PTOs

After connecting the implement to the tractor, power is needed to operate the machine if it is not ground-driven. A PTO driveline, hydraulic motors, and electrical devices are used. The PTO is the most common source of remote power. Three examples of PTO-driven implements that a young agricultural worker may use or assist in using include: rotary mowers (bush hogs), fertilizer spreaders (spin spreader), and post hole diggers.

To attach the PTO shaft of a 3-point hitch implement, follow these steps.

1. Connect the 3-point hitch of the implement using the approved steps to align the hitch and to park the tractor securely.
2. Attach the implement driveline shaft to the PTO stub shaft of the tractor.

Here are some suggestions to make connecting the PTO easier.

A. Align the implement PTO shaft splines with the splines of the stub shaft of the tractor. See Task Sheet 5.4.

B. Press the detent lock (Figure 5.3.e) inward as you slide the implement shaft onto the tractor PTO stub shaft.

C. Slide the implement shaft forward far enough to make sure the detent lock has snapped into the lock position.

Figure 5.3.f. To attach the PTO shaft, you will be operating in a crowded space. Be sure the tractor engine is shut off and is securely parked.

Cooperation provided by The Ohio State University and National Safety Council.

Hitching Precautions for 3-Point Hitch Drawbars

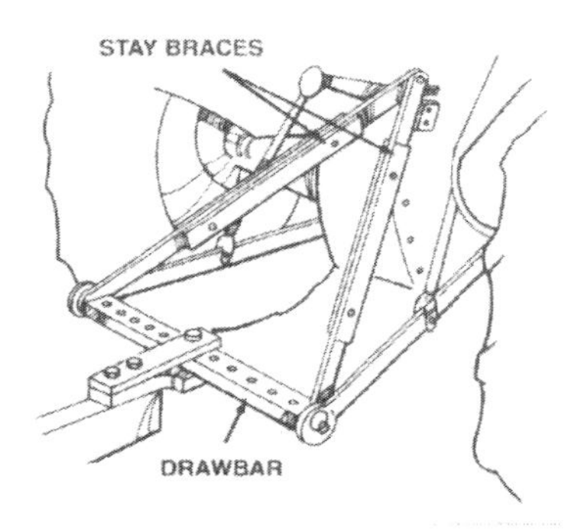

Figure 5.3.g. Stay braces prevent the 3-point hitch drawbar from being lifted too high. *Safety Management for Landscapers, Grounds-Care Businesses, and Golf Courses, John Deere Publishing, 2001. Illustrations reproduced by permission. All rights reserved.*

Never pull a load with the 3-point hitch drawbar more than 13-17 inches above the ground or the pulling forces will be higher than the tractor's center of gravity. A rear overturn hazard may develop as the tractor moves forward.

Using the 3-Point Hitch Implement

Ground-driven 3-point hitch implements are often assigned to the beginning tractor operator. A few ideas are presented here to help you safely operate these implements.

- Make sure you know how wide the machine is compared to the tractor.
- Be sure the machine is in "transport," or "up" position for travel on public roadways.
- Lower the machine to the "field" position when you are ready to use it. This keeps the load pulling below the center of gravity.
- Engage the machine operation mechanism (levers, pins, etc) for the wheels to power the machine if you are using a ground-driven machine. A qualified operator should demonstrate this procedure for each machine.
- Begin field operation of the machine by paying attention to field boundary fences and obstacles.
- Allow space at ends of rows or fields to lift the equipment with the 3-point hydraulic lift.
- Do not make turns with a 3-point hitch implement in or on the ground. This places undue force on the 3-point hitch draft and lift arms which can damage the machine.
- Backing a 3-point hitch implement, such as a small planter, while it is lowered onto the ground can plug the seed drops of the planter. Lift the implement before reversing the direction you are going to prevent possible damage to the implement or 3-point hitch draft and lift arms.
- Lift the implement to the transport position before using public roads or passing through narrow farm gates. Ground-driven implements operated on roadways can damage the road surface.

Figure 5.3.h. If the 3-point hitch is equipped with an extension to the lower draft arm, release the lock and pull or extend the draft arm extension to the rear before nearing the implement to be attached.

Lift the 3-point hitch implement from contact with the ground before turning, backing or transporting.

Figure 5.3.i. The telescopic extension to the draft arm is fully extended. In some cases, this must be done to align with the lower lift points of the 3-point hitch implement. Be sure the extension is pushed back into the draft arm until locked into place when you are finished attaching the implement.

Safety Activities

1. Practice spotting the tractor 3-point hitch draft arms to the 3-point hitch attachment points of the implement for quick and safe hitching.

2. Demonstrate the rules you will use when backing a tractor to connect to a 3-point hitch implement by showing your helpers where to stand to safely assist you in spotting the 3-point hitch to the implement.

3. Inspect the ground-driven machines you may use to learn:
   a. how are they moved from transport to field position and vice versa, if applicable?
   b. what mechanism is used to engage the ground wheels with the turning parts of the machine?

4. Inspect all hitch pins and security clips on 3-point hitch attachments. Did you find any problems or missing hitch pins?

5. Inspect a 3-point hitch quick attaching coupler for cracks or damage to upper and lower lift hooks. Report any problems to your employer, mentor, leader or instructor.

References

1. Safety Management for Landscapers, Grounds-Care Businesses and Golf Courses, 2001, 1st Edition, John Deere Publishing, Moline, Illinois.
2. www.nagcat.org/Click on Guidelines/Select item T4 from Tractor Fundamentals, 3-Point Implements (hitch/unhitch), July, 2012.
3. Operators' Manuals for specific tractors and equipment.

Contact Information

National Safe Tractor and Machinery Operation Program
The Pennsylvania State University
Agricultural and Biological Engineering Department
246 Agricultural Engineering Building
University Park, PA 16802
Phone: 814-865-7685
Fax: 814-863-1031
Email: NSTMOP@psu.edu

Credits

Developed by WC Harshman, AM Yoder, JW Hilton and D J Murphy, The Pennsylvania State University. Reviewed by TL Bean and D Jepsen, The Ohio State University and S Steel, National Safety Council. Revised 3/2013.

This material is based upon work supported by the National Institute of Food and Agriculture, U.S. Department of Agriculture, under Agreement Nos. 2001-41521-01263 and 2010-41521-20839. Any opinions, findings, conclusions, or recommendations expressed in this publication are those of the author(s) and do not necessarily reflect the view of the U.S. Department of Agriculture.

MAKING PTO CONNECTIONS

HOSTA Task Sheet 5.4 **Core**

NATIONAL SAFE TRACTOR AND MACHINERY OPERATION PROGRAM

Introduction

After spotting the hitch to connect the tractor to the implement, the operator must attach the PTO shaft of the tractor to the implement by way of the implement input driveline (IID). See Task Sheet 5.4.1. These connecting shafts can be heavy, greasy, and difficult to manipulate in the cramped space between the tractor and the equipment. The youthful operator must have a strong grip and will often have to work at an awkward angle. *Check the NAGCAT website to determine if you can handle the task of PTO connection.*

This task sheet discusses PTO design and how to make PTO connections through knowledge of that design.

PTO Stub Shaft Design

PTO Speeds: Tractor PTOs are designed to rotate at 540 rpm or 1000 rpm. Shiftable, dual-speed PTOs may reach a maximum design speed of 630 rpm or 1170 rpm.

PTO Splines: By counting the number of splines, or teeth on a PTO stub shaft, the beginning operator can identify the speed of the PTO shaft in rpms. A 540 rpm PTO shaft will have 6 splines or teeth. A 1000 rpm PTO shaft may have 20 or 21 splines or teeth.

The faster the PTO speed, the more teeth that are used to make the PTO connection between the tractor and the implement.

PTO Sizes: PTO stub shaft diameter for a 540 rpm shaft is 1 3/8 inch. The 1000 rpm stub shaft with 21 splines or teeth is 1 3/8 inch. The 1000 rpm stub shaft with 20 splines or teeth has a diameter of 1 3/4 inch.

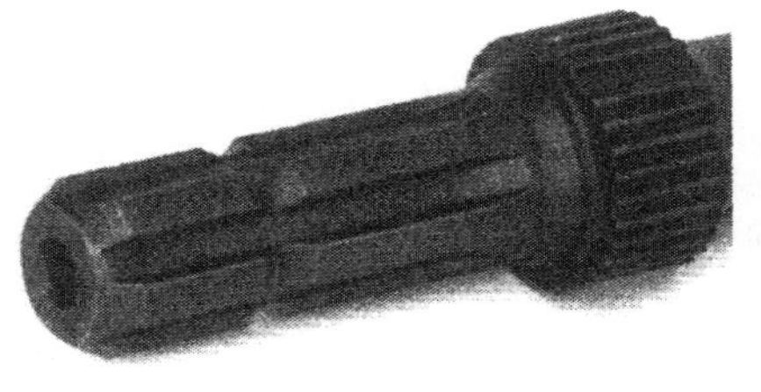

Figure 5.4.a. The 540 rpm PTO stub shaft has 6 splines or teeth and is 1 3/8 inch in diameter. *Farm and Ranch Safety Management, John Deere Publishing, 1994. Illustrations reproduced by permission. All rights reserved.*

Figure 5.4.b. The 1000 rpm PTO stub shaft has either 20 splines or teeth with a 1 3/4 inch diameter or may have 21 splines or teeth with a 1 3/8 inch diameter.

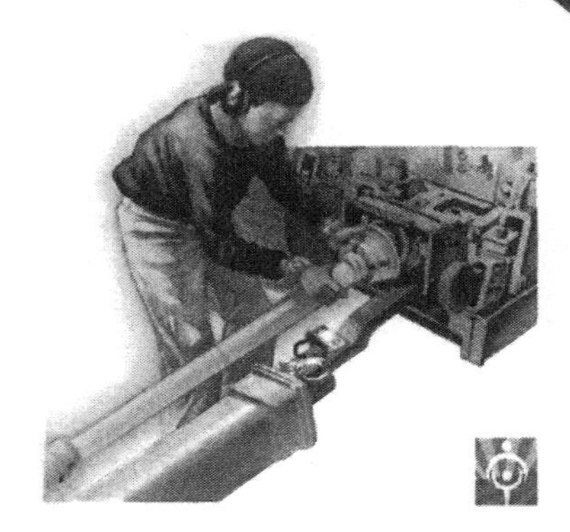

Figure 5.4.c. NAGCAT recommends that youthful farm workers wear snug-fitting clothes, non-skid shoes, and hearing protection while working around machinery. The youth's ability to lift and connect the PTO shaft must be evaluated by an adult who understands the physical development of children.

540 rpm PTOs have 6 splines or teeth. 1000 rpm PTOs have 20 or 21 splines or teeth.

Learning Goals

- To be able to attach the PTO driveline between the tractor and the implement

Related Task Sheets:

Reaction Time	2.3
Age-Appropriate Tasks	2.4
Mechanical Hazards	3.1
Using 3-Point Hitch Implements	5.3
Using Power Take-Off Implements	5.4.1

Cooperation provided by The Ohio State University and National Safety Council.

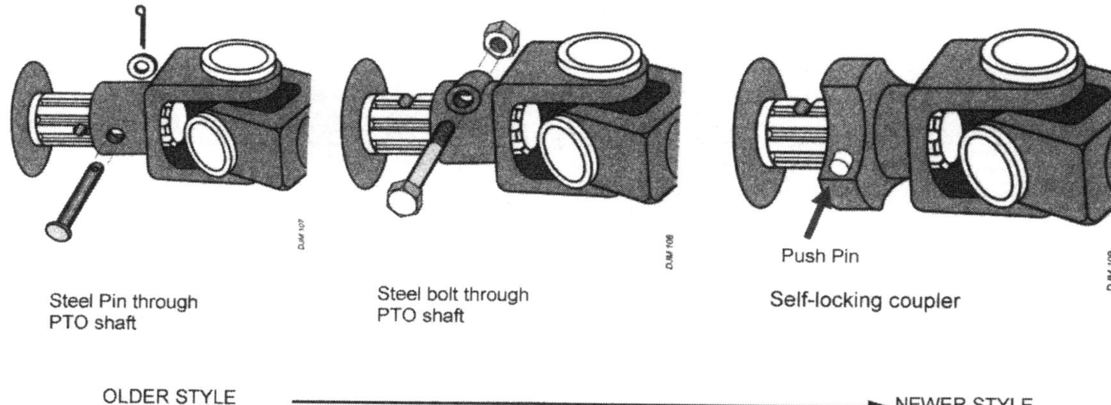

Figure 5.4.d. Various means to secure the PTO shaft to the stub shaft have been used over the years. Besides those connection methods shown above, another popular style is the push pin detent locking type shown in Figure 5.5.e. All types of locking device areas must be guarded as they are wrap points where the operator can become entangled in the PTO.

PTOs must be guarded to prevent an entanglement hazard.

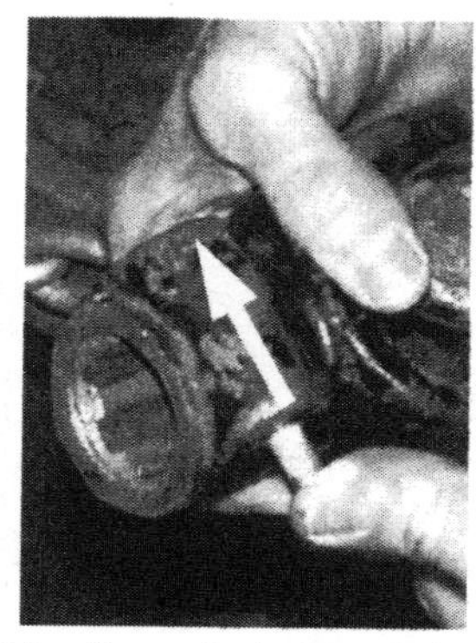

Figure 5.4.e. The push pin detent lock on the PTO driveline has a metal rod which fits in the PTO stub shaft groove to secure it. A firm grip is needed to press the pin. Do you have enough hand strength to push this pin in all the way?

Connecting the PTO

Follow these steps to attach the PTO shaft of a 3-point hitch implement.

1. Connect the tractor to the drawbar or to the 3-point hitch of the implement using the approved steps. See Task Sheets 5.1, 5.2, and 5.3.
2. Attach the PTO shaft of the implement to the PTO stub shaft of the tractor.

Here are some suggestions to make the PTO connection easier.

A. Align the driveline PTO shaft splines with the splines of the stub shaft of the tractor. If the splines will not align, try turning the tractor PTO stub shaft slightly, or use the implement flywheel to move the implement's PTO shaft. Have this procedure shown to you if necessary.

B. Press the detent lock push pin inward (Figure 5.4.e) as you slide the implement shaft onto the tractor stub shaft.

C. Slide the implement shaft forward far enough to make sure the detent pin has snapped into the lock position.

PTO Care and Use

Dirt and grease can make the PTO shaft difficult to grasp and connect. Keep the PTO shaft off the ground. Wipe the excess grease from the PTO shaft with a cloth.

Important: A new PTO shaft has paint inside the splines. This may prevent the shaft from fitting over the PTO stub. The paint must be removed.

Cooperation provided by The Ohio State University and National Safety Council.

PTO Phasing

Older PTO shafts can be separated or pulled apart. The two parts are made so that one part fits into the other. The PTO must be able to telescope in and out to permit machine operation over irregular terrain. If the parts become separated, they must be re-assembled "in phase" to avoid placing extra strain on the universal joints. Many shafts are designed to prevent this from happening.

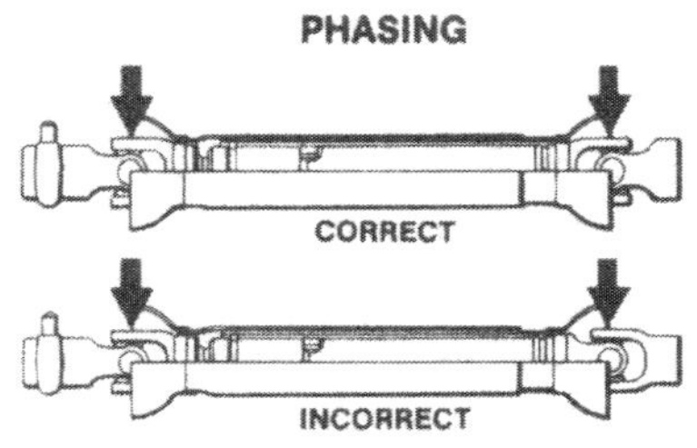

Figure 5.4.f. The upper portion of the drawing illustrates a correctly placed universal joint. You may wish to check the phasing on a PTO shaft.

NAGCAT Guidelines

NAGCAT recommendations for connecting and disconnecting a PTO shaft are shown in this section. These recommendations were developed by a knowledgeable group of safety experts as a means of helping parents to match youthful agricultural workers with the tasks that are appropriate to their development.

The PTO guidelines are presented here.

Adult Responsibilities:

- Be sure implement is in working order.
- Be sure that all safety features are in place.
- Be sure the work area has no hazards.
- Be sure the youth has long hair tied up out of the way, has non-skid shoes, and snug-fitting clothes. Hearing protection is recommended as well.

The adult in charge should also evaluate you using the following questions:

1. Can the youth drive the tractor skillfully?
2. Can the youth hitch and unhitch implements?
3. Does the PTO shaft weigh more than 10-15% of the youth's body weight? To avoid back injury, this should be the maximum weight you should be asked to lift.
4. Can the youth follow a 5-step process?
5. Has the youth been trained in proper lifting techniques?
6. Has an adult demonstrated connecting and disconnecting a PTO?
7. Can the youth do the job 4 or 5 times under direct supervision?
8. Can an adult provide the recommended supervision?

Your experience level may be acceptable to you, but proof of your expertise should be evaluated by a qualified tractor operator.

Figure 5.4.g. This is what the task of connecting a PTO looks like. You must lift a heavy object at an awkward angle while squeezing in the lock mechanism detent pin. Watch someone else connect a PTO several times before doing this job. Continue practicing connecting a PTO on your own with supervision.

Connecting a PTO shaft will be easier after practicing the job several times.

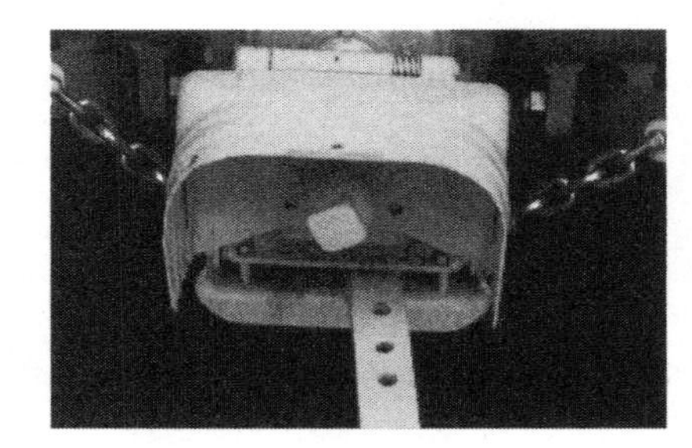

Figure 5.4.h. This PTO stub shaft is protected with a tractor master shield and stub shaft cover. To remove the stub shaft cover, grip the cover firmly and turn counterclockwise. Store the stub shaft cover where it will be available to replace when the job is done.

Safety Activities

1. Using an Internet search engine, type NAGCAT and view the many guidelines presented for the various farm jobs you may be assigned to do. Are you ready to accept these jobs based upon the guidelines presented?
2. Do a survey of the tractors on a farm to determine how many have 540 rpm PTO shafts and how many have 1000 rpm PTO shafts. Record the results.
3. Practice lifting a PTO shaft right handed while squeezing the locking mechanism of the PTO shaft connector. Practice lifting a PTO shaft left handed while squeezing the locking mechanism of the PTO shaft connector. From which side were you able to lift and squeeze best?
4. Check the phasing of three PTO shafts. Make a drawing of the universal joints on each end of the PTO shaft. Did you find any PTO shafts that were out of phase? If so, label this drawing to show what was wrong.

5. Fill in the blanks:

   A. A PTO shaft with 6 teeth on the shaft is designed for ________rpms of speed.
   B. A PTO shaft that has 20 teeth on the shaft is designed for ______rpms of speed.
   C. A PTO shaft that has 21 teeth on the shaft is designed for ______rpms of speed.
   D. What is the maximum weight that a 14– or 15-year-old worker should be expected to lift without straining the back muscles?________________ % of their body weight.

6. Word scramble. Unscramble the following words. Then fill in the blanks to form a safety message about PTOs.

   _________ all PTO _______.

   d a u g r = ___ ___ ___ ___ ___
   s s t a f h = ___ ___ ___ ___ ___ ___

7. From this phrase "implement input driveline," write a word list using as many letters as you can. The words must have at least four letters. No two-letter or three-letter words are permitted. Letters may only be used as many times as they appear in the phrase. Example: RIVET can be found in the phrase.

References

1. www.nagcat.org/Click on Guidelines/Scroll through the list to find a topic. July, 2012.
2. American Society of Agricultural and Biological Engineers, ANSI/ASABE, S203 Power Take Off, St. Joseph, MI.
3. Safety Management for Landscapers, Grounds-Care Businesses and Golf Courses, 2001, 1st Edition, John Deere Publishing, Moline, Illinois.
4. Farm and Ranch Safety Management, John Deere Publishing, 2009. Illustrations reproduced by permission. All rights reserved.

Contact Information

National Safe Tractor and Machinery Operation Program
The Pennsylvania State University
Agricultural and Biological Engineering Department
246 Agricultural Engineering Building
University Park, PA 16802
Phone: 814-865-7685
Fax: 814-863-1031
Email: NSTMOP@psu.edu

Credits

Developed, written and edited by WC Harshman, AM Yoder, JW Hilton and D J Murphy, The Pennsylvania State University. Reviewed by TL Bean and D Jepsen, The Ohio State University and S Steel, National Safety Council. Revised 3/2013

This material is based upon work supported by the National Institute of Food and Agriculture, U.S. Department of Agriculture, under Agreement Nos. 2001-41521-01263 and 2010-41521-20839. Any opinions, findings, conclusions, or recommendations expressed in this publication are those of the author(s) and do not necessarily reflect the view of the U.S. Department of Agriculture.

USING POWER TAKE-OFF (PTO) IMPLEMENTS

HOSTA Task Sheet 5.4.1 **Core**

NATIONAL SAFE TRACTOR AND MACHINERY OPERATON PROGRAM

Introduction

The power take-off (PTO) shaft, or Implement Input Driveline (IID), is an efficient means of transferring mechanical power between farm tractors and implements. This power transfer system helped to revolutionize North American agriculture during the 1930s. The PTO is also one of the oldest and most persistent hazards associated with farm machinery. This task sheet discusses several aspects of PTO safety.

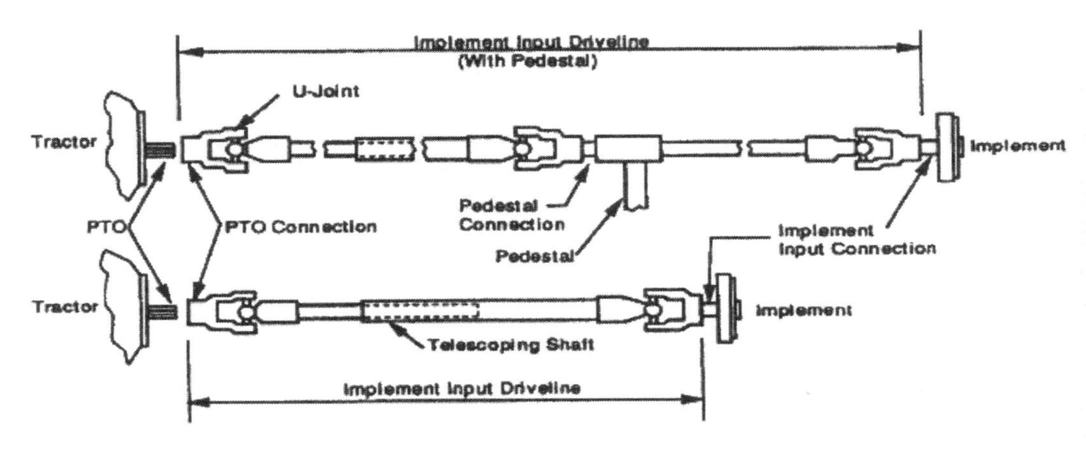

Figure 5.4.1.a. The major components of a PTO system.

PTO Components

Figure 5.4.1.a. is a diagram of the components of an implement PTO system. Two typical PTO system arrangements are shown. The top drawing is of a PTO system involving a pedestal connection, such as one found on many types of towed implements (hay balers, forage choppers, large rotary mowers, etc.). The lower drawing is of a PTO system where the implement's input driveline connects directly to the tractor PTO stub. Examples of this type of connection include three-point hitch-mounted equipment, such as post hole diggers, small rotary mowers, fertilizer spreaders, and augers.

Connections from the tractor to the implement are made through the flexible universal joints. The "U-joints" are connected by a square rigid shaft which turns inside another shaft. The PTO shaft can telescope in and out for use in turns or over uneven terrain.

The combination of universal joints and turning shafts provides the remote power source to a farm implement. Without proper guarding, a serious threat to the operator's safety is created. Study this task sheet carefully.

Learning Goals

- To identify the components of a PTO system
- To identify the hazards involved with PTO use
- To develop safe habits when using a PTO

Related Task Sheets:

Reaction Time	**2.3**
Mechanical Hazards	**3.1**
Making PTO Connections	**5.4**

Cooperation provided by The Ohio State University and National Safety Council

PTO Stub

- Transfers power from the tractor to the machine
- Rotates at 540 rpm (9 times/sec.) or at 1,000 rpm (16.6 times/sec.)

Master Shield

- Protects the operator from the PTO stub
- Is often damaged or removed and never replaced.

Figure 5.4.1.b. The major components of a PTO system found on the tractor.

The PTO is one of the oldest and most persistent hazards associated with farm machinery.

PTO Entanglement

This information is taken from the Purdue University source listed at the end of this fact sheet. This reference is the most comprehensive study of power take-off injury incidents to date. The data shown includes fatal and nonfatal injury incidents. Generally, PTO entanglements:

- involved the tractor or machinery operator 78 percent of the time
- occurred when shielding was absent or damaged in 70 percent of the cases
- were at the PTO coupling, either at the tractor or implement connection nearly 70 percent of the time
- involved a bare shaft, spring-loaded push pin, or through bolt component at the point of contact in nearly 63 percent of the cases
- occurred with stationary equipment, such as augers, elevators, post-hole diggers, and grain mixers in 50 percent of the cases
- involved semi-stationary equipment, such as self-unloading forage wagons and feed wagons in 28 percent of the cases
- happened mostly with incidents involving non-moving machinery, such as hay balers, manure spreaders, rotary mowers, etc., at the time of the incident (the PTO was left engaged).
- occurred 4% of the time when no equipment was attached to the tractor. This means the tractor PTO stub was the point of contact at the time of the entanglement.

PTO Guards

Implement Input Connection (IIC) Shield

- Protects the operator from the IIC, including the implement input stub and the connection to the IID

Safety Chain

- Keeps the integral journal shield from spinning
- Shows that the shield is not attached to the IID
- Should be replaced immediately if damaged or broken

Figure 5.4.1.c. The major guards of a PTO system.

Master Shield

- Protects the operator from the PTO stub and the connection of the IID to the PTO stub

Integral Journal Shield

- Completely encloses the IID
- May be made of plastic or metal
- Mounted on bearings to allow it to spin freely from the IID
- Always check before operation for free movement

PTO Safety Practices

There are several ways to reduce the risk of PTO injuries and fatalities. These safety practices offer protection from the most common types of PTO entanglements.

- Keep all components of PTO systems shielded and guarded.
- Regularly test driveline guards by spinning or rotating them to ensure they have not become stuck to the shaft.
- Disengage the PTO and shut off the tractor before dismounting to clean, repair, service, or adjust machinery.
- Walk around tractors and machinery rather than stepping over a rotating shaft.
- Always use the driveline recommended for your machine. Never switch drivelines among different machines.
- Position the tractor's drawbar properly for each implement used. This will help prevent driveline stress and separation on uneven terrain and in tight turns. See Task Sheet 5.1.
- Reduce PTO shaft abuse by observing the following: avoid tight turns that pinch rotating shafts between the tractor and machine; keep excessive telescoping to a minimum; engage power to the shaft gradually; and avoid over tightening of slip clutches on PTO-driven machines.

If PTO guards are removed or damaged, they should be replaced immediately.

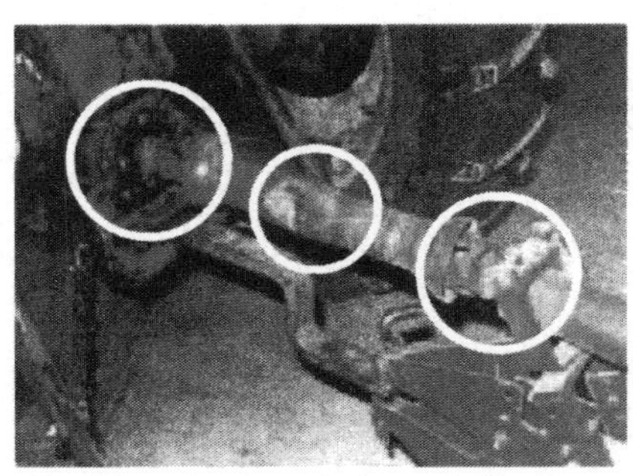

Figure 5.4.1.d. A bent shaft guard offers no protection from a spinning PTO shaft. Also notice the missing master shield and the inadequate guarding of the universal joint near the PTO pedestal.

PTO Safety Activities

1. Fill in the blanks in the following figure of the major components of a PTO system based on the information in this sheet.

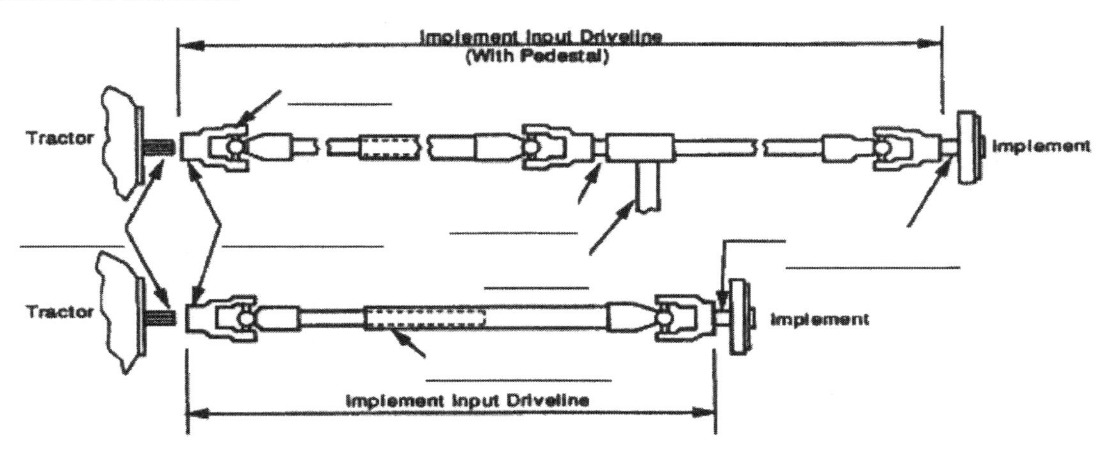

2. You are working with another tractor operator. He/she is sitting on the tractor seat and is able to reach the PTO control. If your shoelace is caught in the PTO shaft, how long does the shoelace need to be in order for the tractor operator to have enough time to shut off the PTO before your foot is pulled into the PTO shaft? The PTO shaft is spinning at 540 rpms, the shaft diameter is 3 inches (d), and the operator can react by shutting off the PTO in 3 seconds.

a. Find the circumference of the PTO shaft.
   Circumference = π d = 3.14 x 3 inches = ________ inches

b. How many times does the PTO shaft rotating 540 revolutions per minute rotate in one second?
   $\frac{540 \text{ revolutions}}{1 \text{ Min}} \times \frac{1 \text{ Min}}{60 \text{ sec}} = \frac{540 \text{ revolutions}}{60 \text{ sec}} =$ ________ $\frac{\text{revolutions}}{\text{sec}}$

c. How many times does the PTO shaft rotate in 3 seconds?
   Answer b x 3 sec = _________ revolutions

d. How much shoelace will become wrapped up in the PTO in 3 seconds?
   Answer a (in inches) x *Answer c* (in revolutions) = ______inches of shoelace.

PTO Safety Activity
Answers
1. See Figure 5.4.a
2a. 9.42 inches
2b. 9 revolutions per second
2c. 27 revolutions in 3 seconds
2d. 254.34 inches or 21.20 feet

References

1. Campbell,W.P.1987.*The Condition of Agricultural Driveline System Shielding and Its Impact in Injuries and Fatalities.*M.S.Thesis.Department ofAgricultural Engineering,Purdue University,WestLafayette,IN.
2. Farm and Ranch Safety Management, John Deere Publishing, 2009.
3. Murphy, D.J. 1992 *Power Take-Off (PTO) Safety.* Fact Sheet E-33. The Pennsylvania State University. University Park, PA.

Contact Information

National Safe Tractor and Machinery Operation Program
The Pennsylvania State University
Agricultural and Biological Engineering Department
246 Agricultural Engineering Building
University Park, PA 16802
Phone: 814-865-7685
Fax: 814-863-1031
Email: NSTMOP@psu.edu

Credits

Developed by WC Harshman, AM Yoder, JW Hilton and D J Murphy, The Pennsylvania State University. Reviewed by TL Bean and D Jepsen, The Ohio State University and S Steel, National Safety Council. Revised 3/2013

This material is based upon work supported by the National Institute of Food and Agriculture, U.S. Department of Agriculture, under Agreement Nos. 2001-41521-01263 and 20101-41521-20839. Any opinions, findings, conclusions, or recommendations expressed in this publication are those of the author(s) and do not necessarily reflect the view of the U.S. Department of Agriculture.

IMPLEMENTS WITH HYDRAULIC COMPONENTS

HOSTA Task Sheet 5.5 **Core**

NATIONAL SAFE TRACTOR AND MACHINERY OPERATION PROGRAM

Introduction

Hitching a machine to a tractor using a drawbar or 3-point hitch is the beginning skill in using the attached machinery. Many implements are powered by a PTO shaft (Task Sheets 5.4 and 5.4.1), while others are powered by hydraulics (or fluids), electrical connections, or some combination of these.

This task sheet will help you to understand and properly care for and use the hydraulic systems located on the tractor and used with the implement.

Hydraulic Power

The term "hydraulic" refers to fluids under pressure. Any liquid can be placed under pressure, but not all liquids are used for hydraulic work. An undrained garden hose left lying in the sun serves as an example. When we turn the nozzle on, solar-heated water erupts from the hose with great force. Water, however, becomes steam at 212 degrees Fahrenheit and could not be used as a working hydraulic fluid.

Oil is the common hydraulic fluid used with farm equipment. Hydraulic oil system components are briefly shown on page 2, Figure 5.5.b. Turn to Figure 5.5.b. before reading further.

Hydraulic fluids work through systems with very small openings and are under great pressure. There are several precautions which users must observe.

Wear safety glasses or a face shield and gloves when checking hydraulic systems.

Precautions When Using Hydraulics

To safely and correctly operate hydraulic systems, understand these three points:

- Clean oil needs
- Heat generated by use
- Oil leaks under pressure

Be sure you understand each point. If necessary, discuss these points with a knowledgeable farmer or mechanic.

Clean Oil Needs:

Hydraulic pumps and control valves operate with minute clearances (tolerances). Grit, grime, and dirt pushed through these openings can eventually wear the surfaces and damage the system. Clean hydraulic oil must be used. The fill area and connections must be kept clean as well. Dirt is the greatest source of hydraulic system damage.

Figure 5.5.a. Missing hydraulic connection covers signify a problem with this tractor. What will happen to the hydraulic system if the dust covers are not kept in place?

Hydraulic connector covers should be in place to keep out dust, dirt, grease, and moisture. Clean connections keep systems working longer.

Learning Goals

- To safely and correctly connect hydraulic components

Related Task Sheets:

Tractor Controls	**4.5**
Using 3-Point Hitch Implements	**5.3**
Using Power Take-Off (PTO) Implements	**5.41**

Cooperation provided by The Ohio State University and National Safety Council.

A Simple Hydraulic System

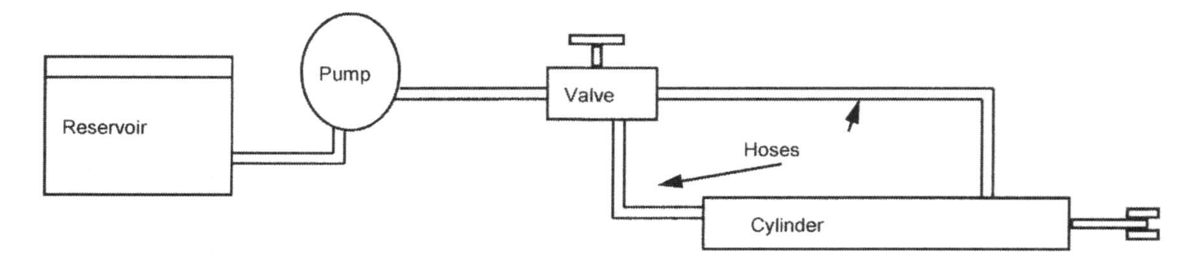

Figure 5.5.b. Hydraulic systems are closed systems which move and control fluid (hydraulic oil) for the purpose of operating cylinders and/or motors. This drawing gives a general look at hydraulic components. Much more detail is involved in these systems than this drawing shows. Filters, pressure relief valves, accumulators, etc. are included as well. Consult a machinery owner's manual to see drawings of more complex hydraulic systems.

Hydraulic system pressure may exceed 2000 psi. Pin hole leaks can develop.

Hydraulic Use Precautions (Continued)

Heat Generated by Use Hazards:

As hydraulic fluid moves through the closed system, the fluid meets resistance from the load to be lifted or moved. Pressure increases and heat from friction builds. Under extreme load conditions, the reinforced hoses can become hot, however, metal connections, fittings, and piping can become super-heated. *Place your hand near the connection to sense for heat before touching the connection.* If hot, allow the hydraulic system to cool down before touching the heated connections.

High Pressure Oil Leaks:

Pressure within the hydraulic system can exceed 2000 pounds per square inch (psi). Reinforced hoses develop pin hole leaks and hydraulic connections can vibrate loose.

Hydraulic leaks may be hard to see. Never check for these leaks with your hand. The high pressure can inject oil droplets under your skin. Oil injected under your skin is a medical emergency and will require immediate medical care. Gangrene can occur, and limb amputation may be necessary.

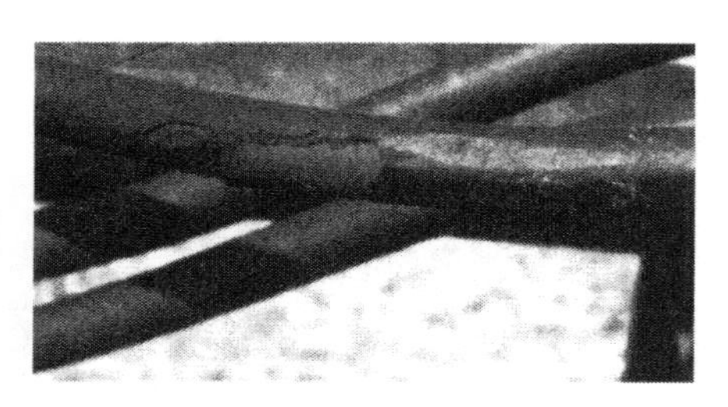

Figure 5.5.c. Hydraulic hoses may be reinforced, but damage to the outer covering—plus pin holes from high pressure—can cause serious injury (e.g. amputation) and machinery down time.

Figure 5.5.d. Hydraulic hoses and fittings can become hot during use. Place your hand near them to check for heating. Do not just grab them!

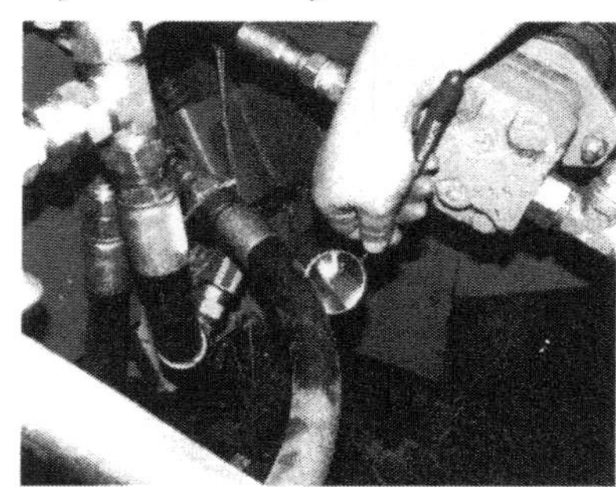

Figure 5.5.e. Use a mirror or piece of cardboard to check for high pressure hydraulic leaks. Do not use your skin! Pin hole leaks are often invisible.

Fittings and Connections

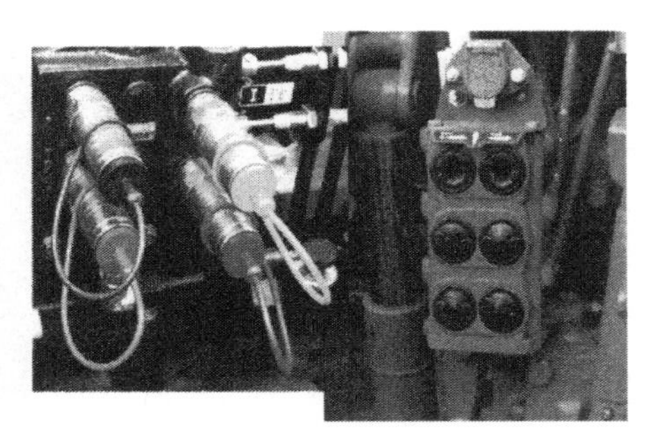

Figure 5.5.f. The female half of the hydraulic coupler is considered to be part of the tractor. Dust covers protect the quick release fitting which includes the lock ring.

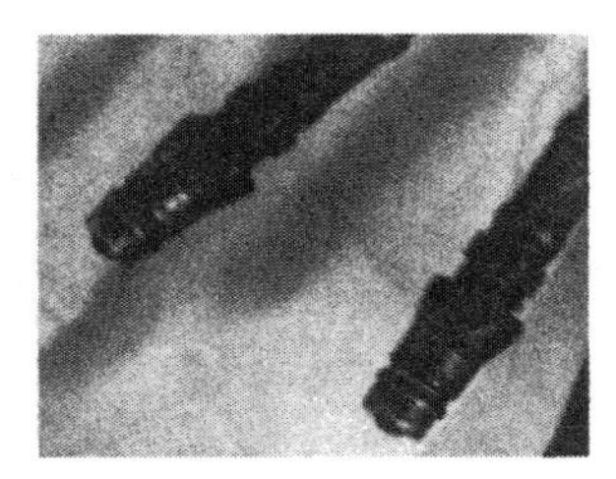

Figure 5.5.g. The male half of the coupler is part of the cylinder or hydraulic device system. Be sure to wipe dirt and grime from the hose end fitting.

Connecting Hydraulic Hoses to Couplers

Hydraulic couplers make the connections quick and simple to use. Follow these steps.

1. Use gloves or a wipe cloth to remove dirt and grit from the couplers.
2. Remove the dust covers from the couplers.
3. Push the couplers together until the lock ring snaps the two parts securely. See figure 5.5.h and 5.5.i. Older style lock levers and manual pull lock rings may be found also. Ask for a demonstration of these.

If you cannot easily make the connection, try the following:

a. While seated on the tractor where no hydraulic lift arms or other moving parts can crush you, move the hydraulic control levers back and forth to release the static pressure. The previous operator may have failed to do this.

b. Move the locking ring of the female coupler back and forth to be sure that dirt has not blocked its movement.

In some circumstances, the hoses leading to the hydraulic cylinders may have become reversed. The system will still operate. However, using the system with hoses reversed will result in the control valves/levers causing the opposite action of what is expected. This can lead to hazardous situations where operators must react quickly and adjust their knowledge and skills to the new condition.

To correct the reversal problem, disconnect the hydraulic hoses and switch them to the opposite female coupler.

If hydraulic repairs have changed the standard coupling set-up, you may find that you must ask for help in determining which hose goes with which coupler.

Hydraulic systems in operation can produce pressure in excess of 2000 psi. Oil trapped in a hydraulic component may still be under enough pressure to cause mechanical problems or hazardous situations to develop. Someone's faulty repairs may have created several problems that the beginning operator cannot solve.

Figure 5.5.h. The lock ring of the female end of the coupling will secure the two fittings together.

Disconnecting Hydraulic Hoses

To disconnect hydraulic hoses:

- relieve the static pressure
- push back on the lock ring
- remove the hydraulic hose
- replace the dust caps on each connector
- hang the hoses on the implement
- keep hoses off the ground

Figure 5.5.i. A firm grip will be needed to insert the hydraulic hoses from the implement (male end) into the coupling on the tractor (female end).

Safety Activities

1. On several different tractors, identify all the hydraulic system components that are external to the tractor. You may wish to name the parts and their purpose to a friend or mentor.
2. To supplement your knowledge of the hydraulic systems components, examine a log splitter and identify all the hydraulic components. You may wish to demonstrate the location, the name, and the function of each part to a friend or mentor.
3. Check the hydraulic fluid level of several tractors.
4. Practice connecting the hydraulic hoses to the tractor coupler.
5. Use the tractor hydraulic system for practice:
   a. raising and lowering the 3-point hitch arms
   b. raising and lowering a high-lift bucket
   c. tilting a high-lift bucket.

Note: NAGCAT recommends that 14- and 15-year-old youth operate front-end loaders on tractors of less than 20 horsepower only.

6. Answer these questions:

A. What is the greatest source of damage to a hydraulic system?

1. Water 2. Dirt 3. Air 4. None of these

B. The term hydraulic refers to:

1. Fluid under pressure 2. Air under pressure 3. Gas under pressure

C. Hydraulic pressures on farm equipment may exceed ______________psi

1. 2000 psi 2. 4000 psi 3. 10,000 psi

D. The safe way to check for pin hole leaks in the hydraulic system is to:

1. Rub your hand over the hose.
2. Hold a match near where you suspect the leak.
3. Hold a piece of metal or cardboard near where you suspect the leak.

References

1. American Society of Agricultural and Biological Engineers, ANSI/ASABE, S366 Hydraulic Couplers, St. Joseph, MI.
2. Safety Management for Landscapers, Grounds-Care Businesses and Golf Courses, 2001, 1st Edition, John Deere Publishing, Moline, Illinois.
3. www.nagcat.org/Select Guidelines topic, July, 2012.

Contact Information

National Safe Tractor and Machinery Operation Program
The Pennsylvania State University
Agricultural and Biological Engineering Department
246 Agricultural Engineering Building
University Park, PA 16802
Phone: 814-865-7685
Fax: 814-863-1031
Email: NSTMOP@psu.edu

Credits

Developed by WC Harshman, AM Yoder, JW Hilton and D J Murphy, The Pennsylvania State University. Reviewed by TL Bean and D Jepsen, The Ohio State University and S Steel, National Safety Council. Revised 3/2013

This material is based upon work supported by the National Institute of Food and Agriculture, U.S. Department of Agriculture, under Agreement Nos. 2001-41521-01263 and 2010-41521-20839. Any opinions, findings, conclusions, or recommendations expressed in this publication are those of the author(s) and do not necessarily reflect the view of the U.S. Department of Agriculture.

IMPLEMENTS WITH ELECTRICAL CONNECTIONS

HOSTA Task Sheet 5.6 **Core**

NATIONAL SAFE TRACTOR AND MACHINERY OPERATION PROGRAM

Introduction

Hitching an implement to a tractor using a drawbar or 3-point hitch is the beginning skill in using the attached machinery. Many implements are powered by a PTO shaft (Task Sheets 5.4 and 5.4.1), while others are powered by hydraulics (or fluids), electrical connections, or some combination of these.

This task sheet will help you to understand and properly care for and use the electrical systems located on the tractor and used with the implement.

Electrical Needs

Modern farm implements come equipped with many features which need electrical power.

Lights are added to implements to allow nighttime field operation, provide lighting for nighttime repair work, and to serve as warning signals during public roadway transport.

Electrical *sensors* are used to measure equipment operation functions and can stop the implement if problems exist.

Monitors signal the operator when a machine function is disrupted. For example, corn planter monitors can signal the tractor operator to discontinue the planting operation. The planter may not be dropping seeds due to a plugged seed drop tube.

Warning devices can be activated by using the reverse gear which sounds an alarm while backing the tractor and implement. Horns and flashing lights also serve to warn bystanders of your actions.

Convenience outlets, using a wiring harness, permit connection to a trailer or wagon for proper lighting for public road use.

Modern tractors and equipment rely not only on PTO and hydraulic systems, but on electrical accessories to complete the work package.

Figure 5.6.a. Connected implements need electrical power. Connectors may have prongs with screw in couplings to keep them from vibrating loose. Other connectors may be plug types that are simply pushed together. Recognize the difference to avoid damage when connecting or disconnecting them.

Figure 5.6.b. Electrical connectors are protected from dust, dirt, grease, and grime with dust caps. Keep these covers repaired and in place to insure that electrical connections can be made without delay.

Electric connector covers should be in place to keep out dust, dirt, grease, and moisture. Clean connections keep systems working without delays.

Learning Goals

- To safely and correctly connect electrical components

Related Task Sheets:

Tractor Controls	**4.5**
Using Drawbar Implements	**5.2**
Using 3-Point Hitch Implements	**5.3**

Cooperation provided by The Ohio State University and National Safety Council.

Using Electrical Connections

To properly care for and use electrical connections, follow these instructions:

- Turn the powered device to the "off" position before connecting or disconnecting the electrical apparatus. Power surges can damage electronic components.
- Wipe away moisture and dirt before making the connections.
- Carefully lift the protective cover to make the connections. Protective caps can be broken and the electrical contacts exposed to moisture, dust, and dirt.
- Slowly and carefully align the prongs or plugs of the connectors. Do not force connections together as you may damage them.
- Grip the connector body when disconnecting the circuit. *Do not pull on the wires.* Grasp the connectors firmly, and separate them using a straight line pull. Expect some connections to be tighter than others. A threaded connector must be unscrewed first! Others require a half-turn before disconnecting them.
- Consult the Operator's Manual for other precautions in using electrical components.

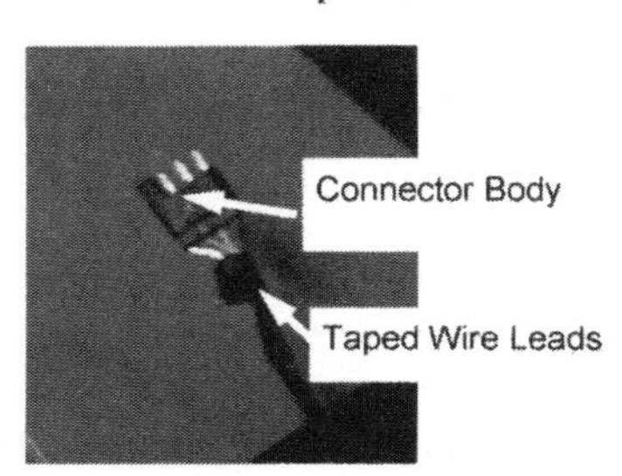

Figure 5.6.c. Grip the connector body to disconnect the wiring harness. Do not pull on the wires themselves. Become familiar with other connector styles before disconnecting them.

Figure 5.6.d. Align the pins of the electrical connection before attaching. Do not force the connector if the pins are not aligned. Threaded connectors do not pull loose; unscrew them first.

Be sure the electrical cables cannot be caught or pinched by hydraulic lift arms or moving parts of the equipment.

Safety Activities

1. Practice connecting electrical wiring harnesses together to get the feel of how easily the connection can be made.
2. Examine several tractor and implements to learn the positions of electrical connections and control switches or knobs that activate the circuits they connect.
3. Locate Operators' Manuals to learn more about machinery monitors, crop sensors, and remote lighting features of a machine.
4. Ask a qualified tractor operator to demonstrate a tractor's electrical components for you.

References

1. Electronic and Electrical Systems, Fundamentals of Service, John Deere Publishing, 2005.

Contact Information

National Safe Tractor and Machinery Operation Program
The Pennsylvania State University
Agricultural and Biological Engineering Department
246 Agricultural Engineering Building
University Park, PA 16802
Phone: 814-865-7685
Fax: 814-863-1031
Email: NSTMOP@psu.edu

Credits

Developed by WC Harshman, AM Yoder, JW Hilton and D J Murphy, The Pennsylvania State University. Reviewed by TL Bean and D Jepsen, The Ohio State University and S Steel, National Safety Council. Revised 3/2013

This material is based upon work supported by the National Institute of Food and Agriculture, U.S. Department of Agriculture, under Agreement Nos. 2001-41521-01263 and 2010-41521-20839. Any opinions, findings, conclusions, or recommendations expressed in this publication are those of the author(s) and do not necessarily reflect the view of the U.S. Department of Agriculture.

SKID STEERS

HOSTA Task Sheet 6.1

NATIONAL SAFE TRACTOR AND MACHINERY OPERATION PROGRAM

Introduction

Skid steer loaders are versatile machines. They fit into small spaces, can turn within a tight radius, and are easy to operate. Young farm workers can enjoy much work success with the skid steer loader.

This task sheet discusses the safe use of a skid steer loader. Skid steer loaders are safe to use if the operator works within the machine's limitations. As in all machinery use, the operator must know the machine's proper use, as well as its limitations.

Skid Steer Loader Basics

Hydraulic Power

A skid steer loader is a hydraulic workhorse. A hydrostatic transmission controls forward and reverse direction. Hydrostatic valves control the flow of hydraulic oil to steer the machine by "skidding" it sharply around corners. Hydraulic cylinders raise and lower lift arms and tilt the load bucket. Task Sheet 5.5 serves as a review of hydraulic power.

Hydraulic power is positive power. The machine moves the instant you move the hydraulic control levers or pedals. The skid steer will move forward, reverse, or "skid" steer. The load bucket will lift, roll or tilt. Bumping the control levers can cause the machine to move unintentionally.

Weight and Stability

A skid steer can move heavy loads. Operators of a skid steer may attempt to lift or move more weight than the skid steer is designed to handle. The skid steer's center of gravity is low and between the wheels. A load carried too high raises the center of gravity and increases the risk of a turnover. See Task Sheet 4.12, Tractor Stability, and Task Sheet 4.13, Using the Tractor Safely, as a review of center of gravity.

Machine Hazards

Skid steer loaders function to push, scrape, scoop, lift, and dump materials. Lift arms raise and lower a load bucket near the operator's cab. The load bucket is mounted in front of the operator and can be rolled forward or tilted back within inches of the operator.

Control levers, pedals, and a parking brake are arranged compactly within the operator's space. It is easy to bump these controls. Workers have been crushed between lift arms and the skid steer. Load buckets have dropped onto workers and killed them. Load buckets have rolled back and crushed a worker's legs.

Pinch points, shear points, and crush points exist within close reach of the operator's space. See Task Sheet 3.1, Mechanical Hazards, to review pinch point, shear point, and crush point hazards.

Figure 6.1.a. Skid steers are controlled by hand levers or joy sticks. Push the levers forward to travel forward; pull back to go in reverse. Let go to stop. The levers also steer the machine.

Over 50% of skid loader fatalities are due to crushing by lift arms and load buckets.

Learning Goals

- To safely use a skid steer loader

Related Task Sheets:

Hazard Warning Signs	2.8
Hand Signals	2.9
Mechanical Hazards	3.1
Noise Hazards and Hearing Protection	3.2
Tractor Hazards	4.2
Preventative Maintenance and Pre-Operation Checks	4.6
Starting and Stopping Diesel and Gasoline Engines	4.7
Tractor Stability	4.12
Using the Tractor Safely	4.13
Using Implements With Hydraulic Components	5.5

Cooperation provided by The Ohio State University and National Safety Council.

Use both hands and both feet to control the skid steer's work.

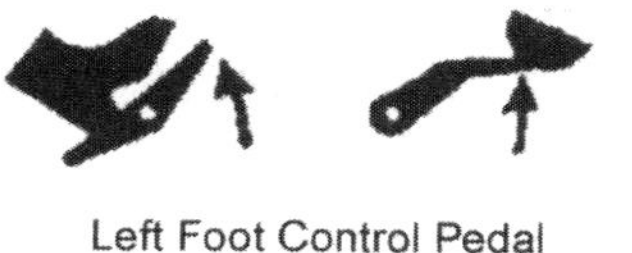

Left Foot Control Pedal

Right Foot Control Pedal

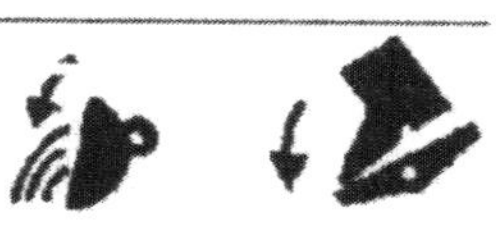

Figure 6.1.b. The hands and feet control the skid steer. Foot controls raise the lift arms (boom) and tilt or roll the bucket. The left heel raises the lift arms. The left toes lowers the lift arms. The right heel rolls the bucket back to load. The right toes tilt the bucket to dump the contents of the bucket. Practice these actions before proceeding to operate the skid steer.

Operating the Skid Steer Loader

Preventative maintenance

Before using the skid steer, complete a maintenance inspection of the machine. Check the oil level, tire pressure, coolant level, and fuel. See Task Sheet 4.6 to review similar items to check on a tractor.

Entering and exiting the skid steer

Before entering the machine, observe the following points.

- Lift arms and bucket should be completely lowered. Do not reach into the cab from the ground level to move hydraulic levers or pedals to position the lift arms and bucket. Crushing can result.
- The seat and floor should be clear of obstructions. Objects can roll beneath foot control pedals and interfere with the machine's operation.

To enter the skid steer, use the grab bars (hand holds) and the tread plates mounted on the load bucket. A three-point hold provides the safest footing. The load bucket and machine surfaces can be slippery when wet or muddy. Exit from the machine in the same manner. When seated, lower the restraint bar and/or fasten the seat belt immediately.

Controls

Before using the skid steer, become familiar with the controls. A qualified person should demonstrate how to start and stop the engine, how to move the machine forward and reverse, how to steer the skid steer, and how to raise, lower, and tilt the bucket attachment. It is a good idea to know how to safely change attachments. If an attachment to the skid steer uses hydraulic power, ask for a demonstration of how to engage the remote hydraulic unit.

Skid steer loaders are controlled by hand levers and foot pedals. The beginning operator should understand the following points:

- *Movement controls*: Grasp the right and left hand control levers; push both levers forward to move forward, or pull the levers rearward to move in reverse. Let go of the levers to stop the movement.
- *Steering controls:* To control the steering direction, push one hand lever forward while pulling the other lever back. Pushing the left lever forward while pulling the right lever back will make the skid steer travel to the right.
- *Lift controls*: Foot pedals control the lift arms and load bucket. The left pedal raises and lowers the lift arms, while the right foot pedal tilts the bucket to dump or rolls the bucket back. See Figure 7.1.b. and page 3 for more details.

Joystick controls are also used to control the functions of the skid steer. Get advice on there use.

Using the Lift Arm and Load Bucket Pedals

Foot pedals on the skid steer are used to control the high lift (boom) work of the skid steer. Toe and heel movements are needed to activate these controls. See Figure 7.1.b. *Note:* Some models use the hand controls to make these movements.

Raising the lift arms (left pedal):

The left pedal raises or lowers the lift arm (boom). Use the left heel to push on the back of the pedal to raise the lift arms and bucket. Use the left toes to push on the front of the pedal to lower the bucket. These movements must be done smoothly. Hard-soled shoes give better feel for the pressure needed on the pedal.

Tilting the bucket (right pedal):

The right pedal controls the load bucket. Use the right heel to push on the back of the pedal to roll the bucket back while loading. Use the right toes to push on the front of the pedal to dump the bucket while unloading.

Skid Steer Safety

Skid steer loaders can work in small areas, but they have similar limitations as does a tractor. Follow these skid steer safety recommendations:

- One seat and one seat belt means one operator. No passengers are permitted on the bucket.
- Lower the safety restraint bar and/or fasten the seat belt every time you enter the machine.
- Be sure area around skid steer is clear of children, bystanders, pets, and farm animals
- Do not work near overhead utility lines.
- Lower the load bucket for travel.
- Use slower speeds over rough ground
- Do not overload the bucket. Skid steers have a Rated Operating Capacity. Exceeding that capacity with a lifted load will result in forward or sideways tipping of the machine. See Figure 7.1.c.
- When moving up a slope, keep the heaviest weight up the hill. With an empty bucket, back up the hill. With a full bucket, drive forward up the hill. See Figure 7.1.d.
- Avoid crossing steep slopes.
- Avoid ditches and stream banks to prevent overturns.
- Lower the boom and bucket, stop the engine, and set the park brake before dismounting the machine. Do this every time.
- Never stand or lean where lift arms or load bucket movements could crush you.
- Use the lift arm locks (boom locks) to prevent lift arms from dropping downward if repairs must be made to the machine.
- Prevent load rollback by securing loads in the bucket and filling the bucket only to rated levels.
- Do not reach outside of the cab while the skid loader is running. All adjustments and connections of attachments should be made with the engine stopped.

Safe skid steer loader work requires attention to the machine, the surroundings, and the work being done.

Figure 6.1.c. Skid steer loaders can tip forward if overloaded. This is an important reason to wear the seat belt, as well as understand the skid loader's load limitations.

Some skid steer models use hand controls (joystick-type) to raise and lower the lift arms and to tilt the load bucket.

Figure 6.1.d. With no load in the bucket, the safest practice is to back up a steep slope. With a loaded bucket, drive up a steep slope with the bucket lowered.

Cooperation provided by The Ohio State University and National Safety Council.

Safety Activities

1. Use the Internet to visit manufacturers' websites (John Deere, New Holland, Bobcat, etc). Assemble a picture chart of as many skid steer loader attachments as you can find.

2. Set up a skid steer loader course to practice moving the skid steer around and through obstacles. Be sure that one part of the obstacle course involves using the load bucket.

3. With adult supervision and a blind fold (skid steer parked and brakes locked), raise and lower the lift (boom) arms and tilt and roll the bucket as the supervisor commands you. You must be able to use the proper controls to operate the skid steer without errors.

4. Matching. Match the skid steer control position with the resulting action to be expected.

Skid steer control position	Resulting action to be expected
______A. Left foot pedal pushed forward with toes	1. Skid steer spins in circles to the left
______B. Left foot pedal pushed downward with heel	2. Lift arm raises
______C. Right foot pedal pushed forward with toes	3. Bucket tilts forward to unload
______D. Right foot pedal pushed downward with heel	4. Bucket rolls back to load
______E. Right hand control lever pushed fully forward, left hand control lever pulled fully back	5. Lift arm lowers
______F. Right hand control lever pulled backward, left hand control lever pulled back	6. Skid steer moves forward
	7. Skid steer moves in reverse

5. Determine how the joystick controlled skid steer performs the functions in Question 4.

References

1. Safety Management for Landscapers, Grounds-Care Businesses, and Golf Courses, 2001, First Edition, John Deere Publishing, Moline, Illinois.
2. www.cdc.gov/niosh/nasd/Click on search by topic/Scroll to Skid Steer.
3. www.cdc.gov/niosh/At search box, type Preventing Injuries and Deaths from Skid Steer Loaders.

Contact Information

National Safe Tractor and Machinery Operation Program
The Pennsylvania State University
Agricultural and Biological Engineering Department
246 Agricultural Engineering Building
University Park, PA 16802
Phone: 814-865-7685
Fax: 814-863-1031
Email: NSTMOP@psu.edu

Credits

Developed, written and edited by WC Harshman, AM Yoder, JW Hilton and D J Murphy, The Pennsylvania State University. Reviewed by TL Bean and D Jepsen, The Ohio State University and S Steel, National Safety Council. Revised 3/2013

This material is based upon work supported by the National Institute of Food and Agriculture, U.S. Department of Agriculture, under Agreement Nos. 2001-41521-01263 and 2010-41521-20839. Any opinions, findings, conclusions, or recommendations expressed in this publication are those of the author(s) and do not necessarily reflect the view of the U.S. Department of Agriculture.

STARTING AND STOPPING A SKID STEER

HOSTA Task Sheet 6.1.1

NATIONAL SAFE TRACTOR AND MACHINERY OPERATION PROGRAM

Introduction

Don't be surprised if you can't just jump on the skid steer, turn the key and be gone. Skid steer manufacturer's safety decals warn that operators must have instruction before running the machine. Ask questions; get training; read the Operator's Manual; don't just tell the supervisor you know what to do. Untrained operators can cause injury, fatality, and property damage.

This task sheet will help you to understand how to safely start and stop a skid steer.

Interlock Control System

Skid steers are equipped with an interlock system meaning the skid steer cannot be started unless the operator is physically in position to operate the machine. The machine cannot be started and operated except from the seat and with the seat belt fastened.

The seat belt and the operator restraint allow the lift, tilt, and traction functions of the skid steer to be activated. All of these functions are electronically interlocked with the start function. A lighted display on the instrument panel will indicate if these systems are functional (Figure 6.1.1.a.). You should see the lighted display for :

- Seat occupied/seat belt fastened/operator restraint bar
- Valve for lift and tilt functions
- Traction function (forward/ reverse
- Power to input controller supplied (the input controller provides power to all output functions electronically).

The owner or the operator should not attempt to disable the interlock system.

Start Procedure

Use the bucket or attachment steps, grab handles and safety treads to get on and off the skid steer. Always maintain 3 points of contact as you climb on the skid steer and face the machine as you do this.

Follow these steps to start:

1. Adjust seat position.
2. Fasten seat belt snugly or lower seat bar if so equipped.
3. Check that foot pedals and hand controls are in a neutral position.
4. Set engine speed control to a 1/2 speed position.
5. Turn key to start. If equipped with a cold temperature start (pre-heat) follow manu-facturer's recommendations.

1. Allow the engine and trans-mission oil to warm for 5 minutes in cold weather.

Figure 6.1.1.a. When the operator is seated in the skid steer seat with seat belt fastened or operator restraint bar in place, the instrument panel lights will indicate that system functions are ready to use.

Become familiar with all parts of the instrument panel.

Learning Goals

- To be able to safely start and stop the skid steer

Related Task Sheets:

Starting and Stopping Diesel and Gasoline Engines	4.7
Skid Steers	6.1

Stopping the Skid Steer

Stopping the engine may not be as simple as turning off the ignition key. Some manufacturers may instruct the user to let the machine idle for a few minutes to cool the engine, hydraulics, and hydrostatic transmission fluid. Become familiar with what each machine requires for shut down. Your supervisor should have this information readily available.

To stop the skid steer:

1. Idle back the engine speed to 1/2 throttle .
2. Set the parking brake.
3. Return lift arms and attachments to ground level. See Figure 2
4. If attachments with hydraulic hoses are to be changed, relieve the pressure in the auxiliary hydraulic system either by turning the ignition key past stop for a few seconds until the engine is stopped, or by moving the hydraulic control lever back and forth several times after the engine is stopped. This will make the use of the Quick Couplers® easier to disconnect and connect.
5. Turn the ignition key to off if not already completed.
6. Remove key if required.
7. Raise seat bar or remove seat belt, and dismount the machine using the grab handles while facing the machine. Maintain 3-points of contact as you exit the machine.

Figure 6.1.1.b.. Return lift arm and skid steer attachments to ground level when stopping the machine. Exiting the cab is made safer since an incidental bumping against control levers will not lower the components inadvertently on the operator.

Ask the supervisor if there are additional requirements in shutting down that should be followed.

Safety Activities

1. If you have never operated a skid steer, visit an equipment dealership and ask to sit in the skid steer cab to observe what controls are available and where they are located. This may be done with your employer's guidance as well.
2. Use the operator's manual for the skid steer you will operate to study the controls and instrument gauges as you sit in the operator's position.
3. Practice starting and shutting off the skid steer and using the lift and traction controls while sitting in the machine with the parking brake set.
4. Learn where the lift arm locking pins are located on the skid steer.
5. Ask a classmate to describe the conditions that must be met for an operator to exit the skid steer cab.

References

1. Various Skid Steer Manufacturer's Operation and Maintenance Manuals
2. Website, www.agsafety.psu.edu. Scroll to Publications, E47 Skid-Steer Safety for Farm and Landscape, Dennis J. Murphy and William Harshman
3. Safety Management for Landscapers, Grounds-Care Businesses and Golf Courses, 2001, First Edition, John Deere Publishing, Moline, Illinois

Contact Information

National Safe Tractor and Machinery Operation Program
The Pennsylvania State University
Agricultural and Biological Engineering Department
246 Agricultural Engineering Building
University Park, PA 16802
Phone: 814-865-7685
Fax: 814-863-1031
Email: NSTMOP@psu.edu

Credits

Developed by WC Harshman, AM Yoder, JW Hilton and D J Murphy, The Pennsylvania State University.

Version 3/2013

This material is based upon work supported by the National Institute of Food and Agriculture, U.S. Department of Agriculture, under Agreement No. 2010-41521-20839. Any opinions, findings, conclusions, or recommendations expressed in this publication are those of the author(s) and do not necessarily reflect the view of the U.S. Department of Agriculture.

Cooperation provided by The Ohio State University and National Safety Council.

SKID STEER– GROUND MOVEMENT

HOSTA Task Sheet 6.1.2

NATIONAL SAFE TRACTOR AND MACHINERY OPERATION PROGRAM

Introduction

Just as the name implies the skid steer (regardless of manufacturer) is steered by skidding the inside tires or rubber track while the outside drive wheels or track moves the machine in the direction of the skid. On soft soil or a manure packed barn area this happens easily. On a hard surface like a roadway the machine may grab the hard surface and bounce roughly.

This task sheet discusses safe and efficient ground movement of a skid steer. This includes moving it without damage to the machine, bystanders or property.

Forward, back, turn

Control levers are the steering "wheel" and ground movement control of a skid steer. Some skid steers use two levers (Figure 6.1.2.a.), while others may use a "joystick" type of control. Use the Operator's Manual and become familiar with the controls that you will be using.

To use control levers to move the skid steer forward or reverse:

- Push forward on both levers to go forward
- Pull both levers toward you to go in reverse
- Push on the left lever and pull back on the right lever to turn to the right
- Push on the right lever and pull back on the left lever to turn to the left.

Note: Maintain full load engine speed above 2900 rpm for efficient operation. Attempting to move the skid steer with low engine speed will often stall the engine.

Newer models of skid steers are equipped with "joystick" controls. Joysticks can control movement, steering, and the hydraulic functions of raising and lowering the bucket or tilting the bucket forward and back. Joysticks have internationally accepted symbols to indicate their function (Figure 6.1.2.b). In some cases there may be dual functions for the joy stick depending upon the mode of use selected.

Study the operation symbols or ask your supervisor to explain how the joystick or any other component you do not understand is used.

Figure 6.1.2.a. Control levers for moving or steering a skid steer are pushed forward, pulled back, or some combination to "skid" the machine in the direction you wish to go. Notice the lap bar which restrains the operator in the skid steer cab.

Figure 6.1.2.b Joystick controls reduce operator fatigue. The joystick can be rotated to many positions and has finger-tip button controls to move the skid steer, raise/lower lift arms, tip the bucket, and operate hydraulic powered accessories.

By letting go of the skid steer control levers they will return to a neutral position and you will stop moving.

Learning Goals

- To safely steer the skid steer in the direction you must travel

Related Task Sheets:

Skid Steers	**6.1**

Cooperation provided by The Ohio State University and National Safety Council.

Safety considerations

Once in the cab of the skid steer operator vision is reduced to the side and to the rear of the machine. While no bystanders, children, pets or livestock should be in the work zone, the operator must be aware of what is happening in the work area. Barn walls, supporting posts and beams and other machinery can be damaged by careless skid steer use.

Be especially careful when backing the skid steer. Not all skid steers have mirrors that let you see behind you.

Skid steers are not made for rough terrain work. When operated on sloping ground:

- Drive slowly.
- Keep the bucket as low to the ground as possible.
- Load the bucket evenly.

If you must move the skid steer over a sloped area follow these safe practices:

- Avoid crossing steep slopes.
- Keep the heaviest weight up the hill whether traveling up or down the hill. For example: with an empty bucket back up the hill. With a full bucket, drive forward up the hill (see Figure 6.1.2.c).
- Stay away from ditches, stream banks and silage pile edges to prevent an overturn.

Figure 6.1.2.c. Do not cross steep slopes, but travel up or down the slope with the heaviest part of the load carried toward the top of the hill.

Does your skid steer have a reverse alarm to warn others that you are backing the skid steer?

Safety Activities

1. Set up a skid steer loader course to practice moving the skid steer around and through obstacles. Do this in forward and in reverse. Include using the loaded bucket as part of the course.
2. Inspect the work area you are assigned to for: hidden obstacles, building parts that are close to the work area, overhead utility lines, ditches, and any other potential problem that might interfere with your moving the skid steer as you work.
3. Have an operator sit in the skid steer facing forward; approach the skid steer from different angles asking the operator to signal when they can see you coming. Mark these positions and discuss the restricted field of vision around the machine. Repeat the exercise, but use a caution (traffic) cone and see how the field of vision changes. What could be the results if a small child, pet, or by-stander entered that area as you operated the skid steer?

References

1. Safety Management for Landscapers, Grounds-Care Businesses and Golf Courses, 2001, First Edition, John Deere Publishing, Moline, Illinois
2. www.cdc.gov/niosh/nasd/Click op search by topic/ Scroll to Skid Steer.

Contact Information

National Safe Tractor and Machinery Operation Program
The Pennsylvania State University
Agricultural and Biological Engineering Department
246 Agricultural Engineering Building
University Park, PA 16802
Phone: 814-865-7685
Fax: 814-863-1031
Email: NSTMOP@psu.edu

Credits

Developed by WC Harshman, AM Yoder, JW Hilton and D J Murphy, The Pennsylvania State University.

Version 3/2013

This material is based upon work supported by the National Institute of Food and Agriculture, U.S. Department of Agriculture, under Agreement No. 2010-41521-20839. Any opinions, findings, conclusions, or recommendations expressed in this publication are those of the author(s) and do not necessarily reflect the view of the U.S. Department of Agriculture.

SKID STEER– ATTACHING ACCESSORIES

HOSTA Task Sheet 6.1.3

NATIONAL SAFE TRACTOR AND MACHINERY OPERATION PROGRAM

Introduction

Skid steers can be used for a variety of tasks. Commonly the skid steer is equipped with a scoop bucket to move soil, gravel, feed, and more. Attachments like pallet forks, post hole augers, soil preparation tools, and powered brooms will mean that you will change one of these accessories for another.

This task sheet will discuss the skid steer quick attachment procedures and how to do this important job safely.

Know the parts

Boom mounted attachments can be changed quickly. The parts of the system include:

A. Pivoting mounting plate attached to the boom lift arms

B. Latch handles to lock the attachment to the pivoting mounting plate

C. Attachment saddle (part of the attachment)

How to do it

To mount an attachment, the latching handles must be in the fully "up" position. If not the lock pins will not be retracted.

Align the skid steer mounting plate with the attachment's saddle by moving the skid steer while hydraulically raising or lowering

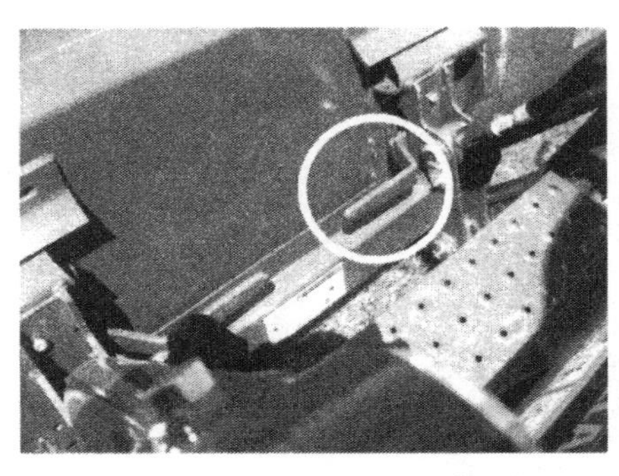

Figure 6.1.3.a. The standard latch handle lock for securing the attachment to the skid steer mounting plate is circled. The latch handle must be placed completely down to lock it in place.

Figure 6.1.3.b. If the skid steer loader mounting plate does not have latch handles to lock the attachment, look for the hydraulically operated pins that provide the locking mechanism and for the hydraulic control that activates the lock pin.

the top of the mounting plate under the attachment saddle. Raise the mounting plate using the foot or hand lever controls until the back surface of the attachment rests against the mounting plate. The attachment can then be lowered with the bucket rolled forward (bucket does not touch the ground). It is ready to be locked into place.

To be safe, turn off the engine, set the parking brake, and exit the skid steer. Push the locking levers down firmly to engage the lock pins into the retaining tabs.

Note: Some Skid Steers may be equipped with a push-button attachment locking system electrically activating hydraulic pins from the operator's seat.

Reverse the process to remove the attachment. When attachment is free, lower the boom slightly and slowly back away from the attachment. Be sure the attachment is setting in a stable position.

Do not attempt to lock the manual lock pins from inside the operator's cab. Keep all body parts inside the cab.

Learning Goals

- To safely attach skid steer attachments.

Related Task Sheets:

Implements With Hydraulic Components	**5.5**
Skid Steers	**6.1**

Cooperation provided by The Ohio State University and National Safety Council.

Removing a hydraulic powered attachment

To remove an attachment that is hydraulically powered involves not only the mechanical connection, but the hydraulic hoses as well.

Follow these steps to disconnect the hydraulic hoses.

- Make sure the attachment is in a stable position before disconnecting the mechanical linkage and hydraulic connectors.
- With the lift boom arms lowered move the hydraulic control levers back and forth a few times to release the static (load) pressure.
- Push back on the lock ring.
- Remove the hydraulic hoses from the couplings.
- Replace the dust caps on each connector.
- Hang the hoses on the equipment.
- Keep the hoses off the ground.

If you are not sure of these steps, seek assistance to prevent damage to the machine or injury to yourself.

Figure 6.1.3.c. A hydraulically operated tiller attachment is shown.

Figure 6.1.3.d. Hydraulic hoses are attached to the skid steer at conveniently located quick connection points.

Don't forget to remove the hydraulic lines before pulling away.

Safety Activities

1. Ask your supervisor to demonstrate how the bucket attachment is removed and replaced on the skid steer.
2. Practice removing and re-attaching the skid steer load bucket or other attachment being used. Pay special attention to the lower bucket tabs where the lock pins hold the bucket/attachment in place. The lock pins must fit into the lower bucket or attachment tabs to be secure. Do not raise the attachment if these lock pins are not engaged in the attachments lower tabs.

References

1. Skid Steer manufacturer's Operator's Manuals.
2. Safety Management for Landscapers, Grounds-Care Businesses and Golf Courses, 2001, First Edition, John Deere Publishing, Moline, Illinois

Contact Information

National Safe Tractor and Machinery Operation Program
The Pennsylvania State University
Agricultural and Biological Engineering Department
246 Agricultural Engineering Building
University Park, PA 16802
Phone: 814-865-7685
Fax: 814-863-1031
Email: NSTMOP@psu.edu

Credits

Developed by WC Harshman, AM Yoder, JW Hilton and D J Murphy, The Pennsylvania State University.

Version 3/2013

This material is based upon work supported by the National Institute of Food and Agriculture, U.S. Department of Agriculture, under Agreement No. 2010-41521-20839. Any opinions, findings, conclusions, or recommendations expressed in this publication are those of the author(s) and do not necessarily reflect the view of the U.S. Department of Agriculture.

Cooperation provided by The Ohio State University and National Safety Council.

SKID STEER– USING HYDRAULIC SYSTEM ATTACHMENTS

HOSTA Task Sheet 6.1.4

NATIONAL SAFE TRACTOR AND MACHINERY OPERATION PROGRAM

Introduction

The skid steer is a hydraulic machine powered by an engine. Everything that happens when you start the engine is a hydraulic action for ground movement, steering, lift arm control, bucket position, or attachment operation.

This task sheet will help you to understand and properly care for and use the skid steer and the hydraulically operated attachments you may encounter.

Figure 6.1.4.a. Missing hydraulic connection covers signify a problem in this picture. What will happen to the hydraulic system if the dust covers are not kept in place?

Hydraulic Power

The term "hydraulic" refers to fluids under pressure. Any liquid can be placed under pressure, but not all liquids are used for hydraulic work. An un-drained garden hose left lying in the sun serves as an example. When we turn the nozzle on, solar-heated water erupts from the hose with great force. Water, however, becomes steam at 212 degrees Fahrenheit and could not be used as a working hydraulic fluid.

Oil is the common hydraulic fluid used with farm equipment. Hydraulic oil system components are briefly shown on page 2, Figure 6.1.4.b. Turn to Figure 6.1.4.b. before reading further.

Hydraulic fluids work through systems with very small openings and are under great pressure.

Precautions When Using Hydraulics

To safely and correctly operate hydraulic systems, understand these three points:

- Oil needs to be clean
- Heat is generated by use
- Oil leaks under pressure

Be sure you understand each point. If necessary, discuss these points with a knowledgeable supervisor or hydraulic technician.

Clean Oil Needs:

Hydraulic pumps and control valves operate with minute clearances and close tolerances. Grit, grime, and dirt pushed through these openings can eventually wear the surfaces and damage the system. Clean hydraulic oil must be used. The fill area and connections must be kept clean as well. Dirt is the greatest source of hydraulic system damage.

Hydraulic connector covers should be in place to keep out dust, dirt, grease, and moisture. Clean connections keep systems working longer.

Learning Goals

- To safely and correctly connect and use the skid steer hydraulically operated machine attachments.

Related Task Sheets:

Using Implements with Hydraulic Components	**5.5**
Skid Steers	**6.1**
Skid Steer-Attaching Accessories	**6.1.3**

Cooperation provided by The Ohio State University and National Safety Council.

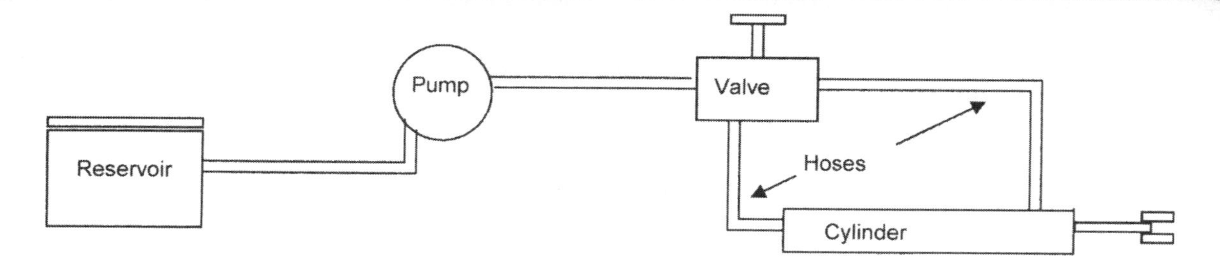

Figure 6.1.4.b. Hydraulic systems are closed systems which move and control fluid (hydraulic oil) for the purpose of operating cylinders and/or motors. This drawing gives a general look at hydraulic components. Much more detail is involved in these systems than this drawing shows. Filters, pressure relief valves, accumulators, etc. are included as well. Consult a machinery owner's manual to see drawings of more complex hydraulic systems.

Hydraulic system pressure may exceed 2000 psi. Pin hole leaks can develop.

Hydraulic Use Precautions (continued)

Heat Generated by Use Hazards:

As hydraulic fluid moves through the closed system, the fluid meets resistance from the load to be lifted or moved. Pressure increases and heat in the lies and hydraulic controls builds. Under extreme load conditions, the reinforced hoses can become hot, however, metal connections, fittings, and piping can become super-heated. *Place the back of your hand near the connection to sense for heat before touching the connection.* If hot, allow the hydraulic system to cool down before touching the heated connections.

High Pressure Oil Leaks:

Pressure within the hydraulic system can exceed 2000 pounds per square inch (psi). Reinforced hoses can develop pin hole leaks and hydraulic connections can vibrate loose allowing oil to leak from the system.

Hydraulic leaks may be hard to see. Never check for these leaks with your hand. The high pressure can inject oil droplets under your skin. Oil injected under your skin is a medical emergency and will require immediate medical care. Gangrene can occur, and limb amputation may be necessary.

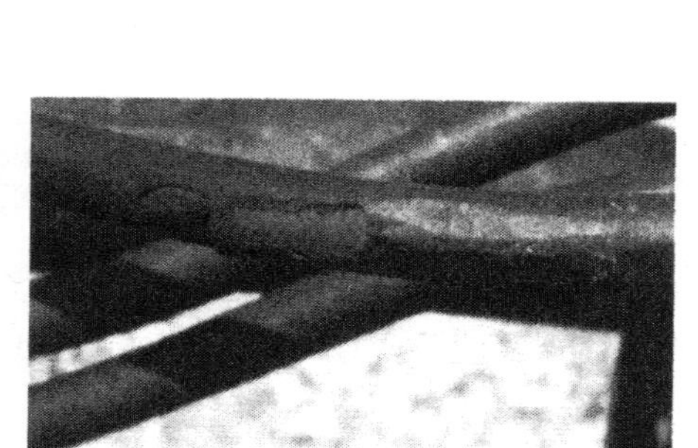

Figure 6.1.4.c. Hydraulic hoses may be reinforced, but damage to the outer covering plus pin holes from high pressure can cause serious injury (e.g. amputation) and machinery down time.

Figure 6.1.4.d. Hydraulic hoses and fittings can become hot during use. Place your hand near them to check for heating. Do not just grab them!

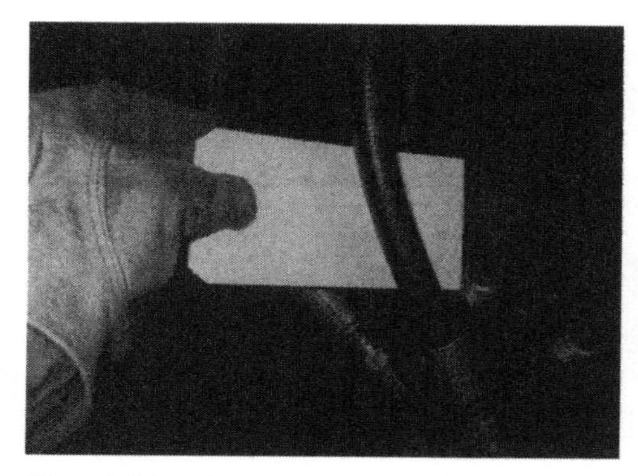

Figure 6.1.4.e. Use a mirror or piece of cardboard to check for high pressure hydraulic leaks. Do not use your hand! Pin hole leaks are often invisible.

Cooperation provided by The Ohio State University and National Safety Council.

Fittings and Connections

Figure 6.1.4.e. The female half of the hydraulic coupler is considered to be part of the skid steer. Dust covers protect the quick release fitting which includes the lock ring.

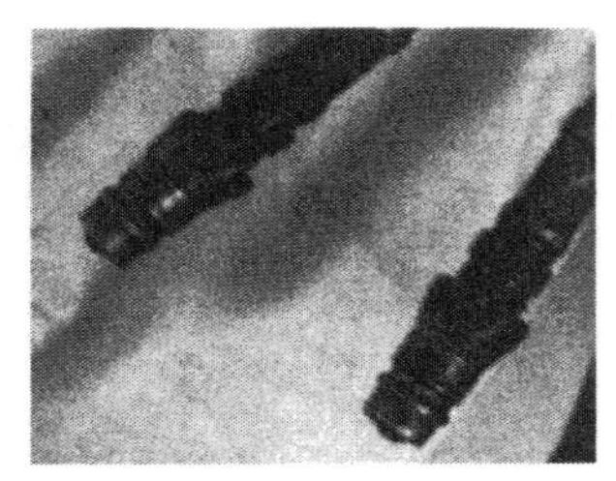

Figure 6.1.4.f. The male half of the coupler is part of the cylinder or hydraulic device system. Be sure to wipe dirt and grime from the hose end fitting.

Connecting Hydraulic Hoses to Couplers

Hydraulic couplers make the connections quick and simple. Follow these steps.

1. Use gloves or a wipe cloth to remove dirt and grit from the couplers.
2. Remove the dust covers from the couplers.
3. Push the couplers together until the lock ring snaps the two parts securely. Older style lock levers and manual pull lock rings may be also found. Ask for a demonstration of these connectors.

If you cannot easily make the connection, try the following:

a. While seated on the tractor where no hydraulic lift arms or other moving parts can crush you, move the hydraulic control levers back and forth to release any static pressure. The previous operator may have failed to do this.

b. Move the locking ring of the female coupler back and forth to be sure that dirt has not blocked its movement.

In some circumstances, the hoses leading to the hydraulic cylinders may have become reversed. The system will still operate. However, using the system with hoses reversed will result in the control valves/levers causing the opposite action of what is expected. This can lead to hazardous situations where operators must react quickly and adjust their knowledge and skills to the new condition.

To correct the reversal problem, disconnect the hydraulic hoses and switch them to the opposite female coupler.

If hydraulic repairs have changed the standard coupling set-up, you may find that you must ask for help in determining which hose goes to which coupler.

Hydraulic systems can produce pressure in excess of 2000 psi. Oil trapped in a hydraulic component may still be under enough pressure to cause mechanical problems or hazardous situations to develop. Someone's faulty repairs may have created several problems that the beginning operator cannot solve.

Figure 6.1.4.g. The lock ring of the female end of the coupling will secure the two fittings together.

Disconnecting Hydraulic Hoses

To disconnect hydraulic hoses:

- relieve the static pressure by moving the control lever
- push back on the lock ring
- remove the hydraulic hose
- replace the dust caps on each connector
- hang the hoses on the implement
- keep hoses off the ground

Figure 6.1.4.h. A firm grip will be needed to insert the hydraulic hoses from the implement (male end) into the coupling on the tractor (female end).

Safety Activities

1. Identify all the hydraulic system components that are external to the skid steer. You may wish to name the parts and their purpose to a friend or mentor or supervisor.
2. Check the hydraulic fluid level of a skid steer. Could you find where to check the fluid level? If not use the Operator's Manual to find the location of the hydraulic fluid fill and/or check point.
3. Practice connecting the hydraulic hoses to the skid steer coupler until you can do this easily..
4. Use the skid steer hydraulic system for practice:
   a. raising and lowering the lift arms
   b. tipping the bucket forward and rolling it back

5. Answer these questions:

   A. What is the greatest source of damage to a hydraulic system?

   1. Water 2. Dirt 3. Air 4. None of these

   B. The term hydraulic refers to:

   1. Fluid under pressure 2. Air under pressure 3. Gas under pressure

   C. Hydraulic pressures on skid steers and attachments may exceed ______________psi

   1. 2000 psi 2. 4000 psi 3. 10,000 psi

   D. The safe way to check for pin hole leaks in the hydraulic system is to:

   1. Rub your hand over the hose.
   2. Hold a match near where you suspect the leak.
   3. Hold a piece of metal or cardboard near where you suspect the leak.

References

1. www.asae.org/Click on Technical Library/Scroll to Standards/Type hydraulic couplers/Open PDF file S366, December 2001.
2. Safety Management for Landscapers, Grounds-Care Businesses and Golf Courses, 2001, 1st Edition, John Deere Publishing, Moline, Illinois.
3. Operator's Manuals for various skid steer models.

Contact Information

National Safe Tractor and Machinery Operation Program
The Pennsylvania State University
Agricultural and Biological Engineering Department
246 Agricultural Engineering Building
University Park, PA 16802
Phone: 814-865-7685
Fax: 814-863-1031
Email: NSTMOP@psu.edu

Credits

Developed by WC Harshman, AM Yoder, JW Hilton and D J Murphy, The Pennsylvania State University.
Version 3/2013

This material is based upon work supported by the National Institute of Food and Agriculture, U.S. Department of Agriculture, under Agreement No. 2010-41521-20839. Any opinions, findings, conclusions, or recommendations expressed in this publication are those of the author(s) and do not necessarily reflect the view of the U.S. Department of Agriculture.

ATVS AND UTILITY VEHICLES

HOSTA Task Sheet 6.2

NATIONAL SAFE TRACTOR AND MACHINERY OPERATION PROGRAM

Introduction

They look like fun. They can go fast. They can travel in the woods. They can kill and injure. What are they? They are ATVs and utility vehicles.

In a recent year, 90,000 injuries and 120 deaths were reported due to use of these fun vehicles. The U.S. Consumer Product Safety Commission reports that 4 of every 10 people treated in hospital emergency rooms are younger than age 16. Why would this be the case?

This task sheet discusses safe use of ATVs and utility vehicles as they are used for work and recreational purposes.

All-Terrain Vehicles

As the name implies, all-terrain vehicles (ATVs) can travel almost anywhere. Rough terrain, steep slopes, rutted mountain roads, and muddy conditions make ATV use appealing. Sportsmen, leisure time enthusiasts, and workers use ATVs. ATVs have become a valuable tool for farm and ranch tasks.

ATVs are designed for work. Other task sheets discuss tractor and skid steer stability. Review Task Sheets 4.12, 4.13, and 7.1. Then consider these ATV design features.

- stability
- suspension
- drive lines
- power and speed

Stability: A four-wheel ATV is more stable than a three-wheel ATV. Heavy loads, steep slopes, and "popping the clutch" can cause the ATV to roll or flip backward. Overturns occur with operator actions that change the center of gravity.

Note: Three-wheeler sales have been banned for several years.

Suspension: ATV suspension systems vary with the machine. Less expensive models may use only balloon tires for suspension. These ATVs can bounce and pitch sideways at high speeds. More expensive models use coil springs and shock absorbers to improve traction and steering control.

Drive lines: ATV drive mechanisms vary greatly. Several combinations of clutches, driveshafts, and differential locks are used. Higher speeds and sharp turns can increase the risk of side overturns if the drive wheels are locked together for traction.

Power and Speed: ATV engines vary in size from 100 cc to 700 cc or greater. Transmission gear ratios vary also. Some ATVs can travel over 50 mph. High-speed operation of the ATV increases the risk of loss of control and rollovers.

Remember, ATVs are not toys. They are powerful machines.

Figure 6.2. a. A four-wheeler, or ATV, can be used for many purposes. Respect the ATV for the powerful machine it is.

It is good advice to dispose of a three-wheeler ATV. They turn over easily.

Learning Goals

- To safely use ATVs and utility vehicles for work and recreational purposes

Related Task Sheets:

Injuries Involving Youth	2.1
Age-Appropriate Tasks	2.4
Mechanical Hazards	3.1
Tractor Hazards	4.2
Tractor Stability	4.12
Using the Tractor Safely	4.13
Skid Steers	7.1

Cooperation provided by The Ohio State University and National Safety Council.

Injury from ATV use most often occurs because of:

A) extra passengers
B) excessive speed
C) road travel

Figure 6.2.b. ATV use as a farm tool calls for strength to control the machine, skill to move and direct the machine, and maturity to understand the consequences of unsafe ATV use. An adult supervisor should work with you to help you learn how to work safely with an ATV.

ATV Operation and Safety

Safety training for ATV use is the first step in being a qualified ATV operator. Local ATV dealers, ATV clubs, and safety professionals from Cooperative Extension, state Departmenrts of Conservation and Natural Resources and farm organizations may offer safe ATV operation programs. The Specialty Vehicle Institute of America (SVIA) provides training as well. Visit them on the Internet at www.svia.org. At a minimum, use the operator's manual and the safety signs on the ATV to help educate yourself before using the machine.

Here are some guidelines for safe ATV use:

- Manufacturers recommend that ATVs with engine sizes greater than 70cc be sold only for children 12 and older and that ATVs with engines greater than 90cc be sold only for individuals 16 and older. The child's strength, skills, and maturity determine readiness to operate an ATV.
- Carrying passengers increases the risk of overturn injury and death. A second person changes the center of gravity of the machine and the machine's steering ability.
- Know the machine's limitations. Operating on steep terrain, pulling heavy loads, excessive speed, and "wheelie" type starts can result in ATV turnover.
- Wear a full-face shield helmet. The helmet should fit snugly and securely. It should be labeled with the American National Standards Institute (ANSI) Z90.1 label.
- If a face shield is not part of the helmet, wear goggles or a separate face shield, especially at high speeds or in wooded terrain. The protective lens should carry the ANSI Z78.1 label.
- Over-the-ankle shoes with sturdy heels and soles are necessary.
- Gloves and long sleeves are needed for specific jobs.
- Use lights, reflectors, and highly visible flags to increase the ATV's visibility.
- Avoid public roads. Paved and unpaved roads are designed for truck and automotive traffic. ATVs are designed for off-road use. Increased risk for rollovers of ATVs on road surfaces has been shown.
- Check your state's vehicle code for use of the ATV as an agricultural machine. Use of the ATV for agricultural purposes and only incidental road travel may be permitted in your state.

Utility Vehicles

Utility vehicles are similar to golf carts except they are fitted with cargo boxes to carry work material. The utility vehicle can have four, five, or six wheels depending upon its use. The UV weighs about 1,000 pounds and can carry several hundred pounds of cargo. The machine can be diesel, gasoline, electric, or hydrogen fuel cell powered.

Like other farm machines, the utility vehicle is made for work purposes. Hauling feed, mulch materials, and supplies makes it a convenient transport for small jobs. Like an ATV, the utility vehicle is a tool and not a toy.

Safe operation of the utility vehicle requires the same safe work habits as used with tractors, skid steer loaders, and ATVs.

Safe Utility Vehicle Use

Use the operator's manual and safety signs/decals found on the machine to learn how the utility vehicle operates and what safety practices to observe. A successful operator becomes familiar with a machine before attempting to use it. Ask a qualified operator to show you what to do if no training materials can be found.

The following safety practices should be followed in operating a utility vehicle:

- Some manufacturer's specifications suggest that no operator younger than age 16 should be permitted to operate a utility vehicle.
- With increased amounts of cargo, the utility vehicle's center of gravity is raised. Risk of an overturn increases. Drive slowly and turn smoothly.
- To prevent overturns, secure the load from shifting sideways.
- Avoid driving on steep slopes. It is safer to drive uphill or downhill rather than across a slope. Avoid sharp turns to prevent overturns. Drive to the top or bottom of a slope to make a turn. When approaching a downhill slope, reduce speed before you reach the slope. This will help reduce wear on the brakes.
- Reduce speed over rough terrain to prevent the utility vehicle from bouncing. Operator and riders have been thrown from utility vehicles.
- A second rider should occupy the passenger seat. Do not permit extra riders to ride in the cargo box. Use the handholds. If the utility vehicle has a roll-bar, buckle the seat belt.
- Do not drive near ditches or embankments. Remember if the ditch is 6 feet deep, stay back from the edge by at least 6 feet.
- Use your tractor, skid steer loader, and ATV knowledge to safely operate a utility vehicle.

As with all machinery, use the device as it was designed. Utility vehicles are tools, not toys.

Figure 6.2.c. A utility vehicle is versatile. It can do the smaller jobs that a pick-up truck may be unsuited to do. Remember that the utility vehicle has limitations. Overloading, shifting loads, and sharp braking can cause turnovers.

Utility vehicles can overturn at high speeds and while making sharp turns.

Figure 6.2.d. Avoid steep banks. Utility vehicles can easily overturn. The driver must know the machine and the work area to reduce potential risk of injury. *Safety Management for Landscapers, Grounds-Care Businesses, and Golf Courses, John Deere Publishing, 2001. Illustrations reproduced by permission. All rights reserved.*

Safety Activities

1. Use the Internet website www.atvsafety.org to solve crossword puzzles or to play word search games related to all-terrain vehicle (ATV) safety.

2. Visit the John Deere website, www.JohnDeere.com, or the Bobcat website, www.bobcat.com, to learn about all-terrain vehicle (ATV) specifications for weight, payload, and engine size.

3. Collect newspaper, magazine, or Internet news articles about ATV and utility vehicle injuries and deaths. Create a poster presentation to display at a local ATV or utility vehicle dealership.

4. What does the designation "100cc engine" represent? Using the math formula for volume of a cylinder (ask your teacher), calculate the diameter and height of the cylinder that would represent a 100cc engine cylinder. Use a sheet of paper to construct the cylinder. Answer the same question for a 500cc engine cylinder.

References

1. Safety Management for Landscapers, Grounds-Care Businesses, and Golf Courses, John Deere Publishing, 2001. Illustrations reproduced by permission. All rights reserved.
2. www.cdc.gov/nasd/ Search the National Ag Safety Database site by topic for ATV information.
3. www.atvsafety.org/Search site for interactive quizzes, word searches, and puzzles.
4. www.svia.org/Search the Specialty Vehicle Institute of America site for ATV information.

Contact Information

National Safe Tractor and Machinery Operation Program
The Pennsylvania State University
Agricultural and Biological Engineering Department
246 Agricultural Engineering Building
University Park, PA 16802
Phone: 814-865-7685
Fax: 814-863-1031
Email: NSTMOP@psu.edu

Credits

Developed by WC Harshman, AM Yoder, JW Hilton and D J Murphy, The Pennsylvania State University. Reviewed by TL Bean and D Jepsen, The Ohio State University and S Steel, National Safety Council. Revised 3/2013

This material is based upon work supported by the National Institute of Food and Agriculture, U.S. Department of Agriculture, under Agreement Nos. 2001-41521-01263 and 2010-41521-20839. Any opinions, findings, conclusions, or recommendations expressed in this publication are those of the author(s) and do not necessarily reflect the view of the U.S. Department of Agriculture.

TELEHANDLERS

HOSTA Task Sheet 6.3

NATIONAL SAFE TRACTOR AND MACHINERY OPERATION PROGRAM

Introduction

Large volumes of agricultural crops and inputs stored in large facilities have created the need for equipment that can reach higher and farther. Telehandlers can lift up to 10,000 lbs. and their booms can extend outwards 30-40 feet. Since they could be mistakenly overloaded or operated on sloped ground a thorough understanding of telehandler use is a must. Understanding the safe operation of the telehandler is the focus of this task sheet.

Figure 6.3.a. Large package hay or straw bales can be stored higher and deeper into the storage area with the telehandler.

What is a telehandler?

Telehandlers are becoming more common on farms. These powerful units go by many names such as material handlers, telehandlers, and many localized versions. Technically, telehandlers are **rough terrain variable reach forklifts** and are considered a class 7 powered industrial truck. They operate entirely different however than a forklift.

The Boom- The telescopic (variable reach) boom can extend/ retract to 30-40 feet and elevate to an angle of 70 degrees from the horizontal. Capacity may reach 10,000 lbs. Booms are marked in increments to alert the operator how far the boom is extended, (Figure 6.3.c. and 6.3.d.).

Frame Tilt- When operated on sloping ground the telehandler frame can be altered relative to the ground by 10 to 15 degrees in either direction to keep the boom vertical in position. A frame tilt/ level indicator is mounted in the cab to assist the operator in keeping the frame level relative to the slope.

Steering- Most telehandlers have three steering options for various work locations. These include:

a. Front wheel steering where the front wheels turn as with our automobiles

b. Circle steering allows the front and rear wheels to react in opposite directions to permit a tight turning radius

c. Crab steering which creates the ability to move diagonally over the ground as all four wheels react in the same direction as the steering wheel is turned.

Continued on page 2.

The telehandler requires as much understanding of stability as operating a tractor does.

Learning Goals

- To understand how a telehandler operation
- To understand the concepts of machine and load stability in operating a telehandler

Related Task Sheets:

Hand Signals	2.9
Mechanical Hazards	3.1
Tractor Hazards	4.1
Tractor Stability	4.12
Skid Steers	6.1

Cooperation provided by The Ohio State University and National Safety Council.

What is a telehandler? (continued)

The Carriage- This is the lifting attachment or attachments which can tilt further forward and back than a forklift and can also rotate slightly to tilt a load into an unusual space (Figure 6.3.a.).

Outriggers– Larger telehandlers may be equipped with outriggers for heavy load stability when loaders are operated stationary (Figure 6.3.e).

Figure 6.3.b. The telehandler load capacity chart must be used to determine if the weight of the round bales and the distance the boom must be extended to store them remains within the capacity range of the machine. Exceeding the weight and reach capacity of the telehandler could cause it to tip forward.

Know how to interpret the load capacity charts found in the operator's cab.

Load Capacity

Telehandler load capacity depends upon many variables. Lifting large loads to high storage areas may exceed the capacity of a machine. The operator must understand that each load and each position to which the load is lifted or moved has the opportunity to change the center of gravity and stability of the telehandler.

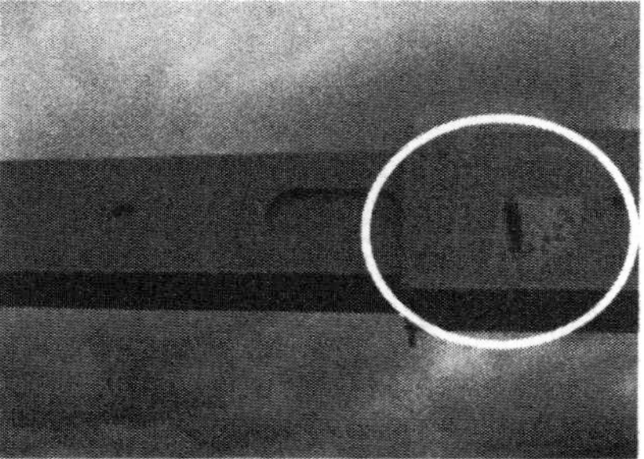

Figure 6.3.d. A close-up of the boom angle indicator.

Figure 6.3.c. As the boom is extended a letter (circled) or number markings are revealed that are referenced in the operator's cab to give restrictions on how much weight can be lifted safely. A boom angle indicator is included. *Insert photo courtesy of JLG, Inc.*

Figure 6.3.e. If the telehandler is equipped with outriggers use them to stabilize the machine during use.

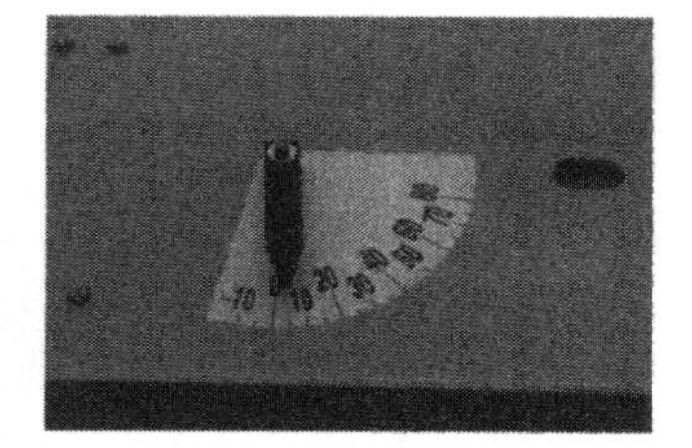

Figure 6.3.f. A typical boom angle indicator is visible to the operator from the cab seat. A similar indicator is found on the instrument panel to show frame angle (frame relative to ground slope). *Photo courtesy of JLG, Inc.*

Cooperation provided by The Ohio State University and National Safety Council.

(Continued from page 2)

Each of the following items, if handled incorrectly, can lead to a mishap.

- Boom angle and extension. See Figure 6.3.c and 6.3.d. for boom markings and Figure 6.3.e for the boom angle indicator)
- Load weight
- Use of outriggers, if equipped
- Rubber tires vs. rigid tires
- Grade, or slope
- Wind
- Lifting attachment

The operator's station has several charts to reference load capacity versus boom extension, boom angle, and frame angle. Use these charts to determine maximum machine angles and settings before lifting a load.

Start-up procedure

Following your training period on the telehandler, use this reminder on how to start the machine.

1. Complete a pre-inspection of the machine.
2. Fasten your safety belt.
3. Observe that all controls are in the neutral position.
4. Turn ignition switch to pre-heat position if so equipped; start the engine when signaled.
5. Warm up the engine at 1/2 throttle.
6. Close the cab door.
7. Check lights, back-up alarm, and horn.

Figure 6.3.g. When finished with the work, park on level ground with the boom retracted and lowered. Set the park brake. Remove the keys so that an untrained person cannot move the machine.

Moving/Using the Telehandler

Before moving the telehandler:

- Check the steering and braking controls.
- Be sure the boom extension and leveling controls are operational, but test these on level ground.
- Lower the outriggers before lifting.
- Practice using the lift and leveling controls before moving a load.
- Check that other personnel and machines are not in the area.
- Plan your travel for best visibility.
- Keep the boom retracted and as close to the ground as possible.

If the telehandler has outriggers, be sure they are lowered during lifting (for stability) and then raised for travel.

- Start, stop, turn and brake smoothly.
- Slow down for turns and uneven surfaces before reaching those hazards.
- Avoid overhead utility lines to prevent electrocution.
- Raise the outriggers before moving the machine.

Cooperation provided by The Ohio State University and National Safety Council.

Shut-down procedure

When the work is completed park in a safe location on level ground away from other equipment and traffic.

Follow these steps:

1. Apply the park brake.
2. Shift transmission to neutral
3. Retract boom and lower boom and attachments (see Figure 4).
4. Let engine idle for 3-5 minutes to cool.
5. Shut off engine and remove key as directed by the employer/supervisor.
6. Remove seat belt.
7. Use grab handles and exit the machine safely.
8. Block wheels if parking on a slope is unavoidable.
9. Some models may have a master electrical switch to disconnect the battery from service. Disconnect if so.

Safety Activities

1. If you have never operated a telehandler, visit an equipment dealership and ask to sit in telehandler's cab to observe what controls are available and where they are located. This may be done with your employer's guidance as well.
2. Use the operator's manual for the telehandler you will operate to study the controls and instrument gauges as you sit in the operator's position.
3. Practice starting and stopping the telehandler, raising and extending/retracting the boom, leveling the frame of the telehandler while sitting with the parking brake set, and/or lowering and raising the outriggers.
4. Practice driving the telehandler with no load. Use the 2-wheel, 4-wheel, and crab drive functions.
5. Practice picking up and lifting a load, extending the loaded boom, and lowering the load.

References

1. Websites for various telehandler manufacturers Operator's Manual
2. Website, www.agsafety.psu.edu. Scroll to Publications, E47 Skid-Steer Safety for Farm and Landscape

Contact Information

National Safe Tractor and Machinery Operation Program
The Pennsylvania State University
Agricultural and Biological Engineering Department
246 Agricultural Engineering Building
University Park, PA 16802
Phone: 814-865-7685
Fax: 814-863-1031
Email: NSTMOP@psu.edu

Credits

Developed by WC Harshman, AM Yoder, JW Hilton and D J Murphy, The Pennsylvania State University.

Version 3/2013

This material is based upon work supported by the National Institute of Food and Agriculture, U.S. Department of Agriculture, under Agreement No. 2010-41521-20839. Any opinions, findings, conclusions, or recommendations expressed in this publication are those of the author(s) and do not necessarily reflect the view of the U.S. Department of Agriculture.

Cooperation provided by The Ohio State University and National Safety Council.

USING A TRACTOR FRONT-END LOADER

HOSTA Task Sheet 6.4

NATIONAL SAFE TRACTOR AND MACHINERY OPERATION PROGRAM

Introduction

A front-end loader (high-lift with bucket or other accessories) mounted on a tractor is a valuable tool for lifting, moving, dragging, and pushing items such as soil, gravel, large round bales, equipment parts, and road repair materials. Using the front-end loader requires an understanding of machine capacity limitations, center of gravity, and an awareness of work surroundings.

This task sheet discusses safely using a front-end loader mounted on a farm tractor.(Similar task sheets dealing with skid steer and material handlers also discuss these safety ideas.)

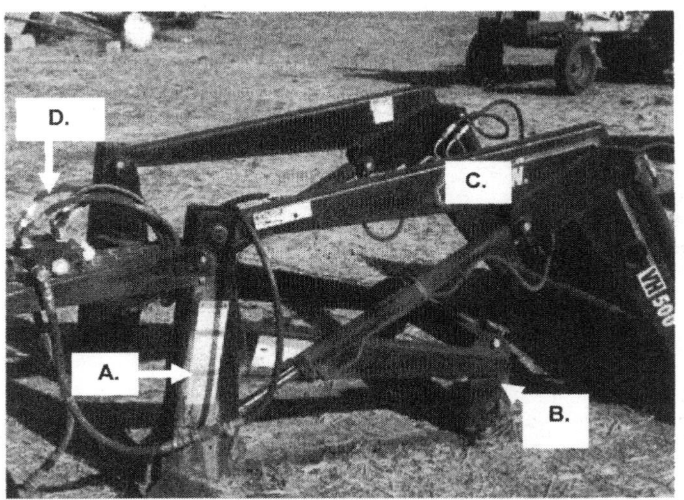

Figure 6.4.a. Front end loaders may be parked and have to be mounted to the tractor you will operate. Components include: **A.** The tower columns which are hooked to the tractor's mid-frame and often serve as the support legs. Some smaller units may have a light duty rod that is inserted as the support leg. **B.** Bar and saddle which hooks to the tractor's front frame. **C.** The lift arms and frame. **D.** The hydraulic control system. Study the components before attempting to connect to the tractor. Tighten all connections securely

Front–end loader components

Oftentimes the front-end loader is used on a tractor dedicated to that attachment. In other cases the front-end loader is parked on its support legs and must be attached to the tractor's frame and hydraulic system for use. The tractor will have mounting points from which to attach the front-end loader.

Loader components include (Figure 6.4.a.):

A. Tower columns, that are the back of the loader and attach to the tractor's mid-frame and may serve as the support legs. Some loaders will have support legs which support the front end loader frame when it has been removed from the tractor if different than the tower columns.

B. Bar and Saddle, found behind the bucket and which mounts the loader to the front of the tractor's frame

C. Lift arms and frame

D. Hydraulic system components (hoses and control valves)

Manufacturer's have made the attachment of the front-end loader to the tractor relatively simple. Use the Operator's Manual to identify the loader components before attempting to attach it to the tractor.

Check that all components are secure before operating the front-end loader.

Learning Goals

- To safely use a tractor-mounted, front-end loader.
- To understand how the center of gravity of a farm tractor changes as the front-end loader is used.

Related Task Sheets:

Cooperation provided by The Ohio State University and National Safety Council.

Center of gravity

Review the following task sheet(s), Tractor Hazards- **4.2**, Tractor Stability- **4.12** and Using the Tractor Safely-**4.13**.

The tractor's center of gravity is engineered to maintain a stable "footprint" with the ground. Anything that moves the center of gravity outside of the stability baseline can lead to a tractor overturn. Raising the front-end loader bucket with or without a load raises the center of gravity. See Figure 6.4. With the center of gravity raised, while operating on uneven surfaces, or on rough roadways, an overturn can occur more easily. Always travel with the bucket as low to the ground as possible. If working on a hill with a loaded bucket, travel uphill in forward and downhill in reverse.

Figure 6.4.b. When operating the front-end loader keep the loaded or empty bucket as low to the ground as possible while in transport. Sideway overturns can result if you try to travel with the bucket in the up position. *Farm and Ranch Safety Management, John Deere Publishing, 2009. Illustrations reproduced by permission. All rights reserved.*

Load capacity

Agricultural inputs of feed, fertilizer, chemical bulk packs, and large package hay bales have all become heavier and bulkier. All loads, whether heavy or light, when lifted with a front-end loader changes the center of gravity of any tractor. The tractor can be more easily tipped as a result. Loads can also roll back onto the operator. To reduce risk of overturn or load rollback:

- Use a wide front-end tractor for loader work rather than a narrow front wheel tractor to improve stability.
- Understand that some loads are going to be too heavy for the tractor to lift. Know your machines capacity.
- Make sure the load must not be bigger in size than the bucket. Load rollback can result.
- Use several trips to complete the work rather than trying to move too much material in one load.
- Keep loads close to the ground during transport (maintain a low center of gravity).
- Avoid slopes or rough terrain when transporting a load with a front-end loader.
- Move slowly if loads must be carried high and avoid jerky movements.
- Consider if the tractor needs more ballast (weight) added to the rear before you continue to use it. If the tractor feels like it is tipping during loading, reduce the load or add ballast to the rear end of the tractor.

Heavy loads lifted high can tip the tractor. Materials can roll back off the bucket if too large.

Pinch and crush points

Front end loader lift arms move closely to the tractor frame and mounting points. Other close fitting parts include the connections between the bucket or attachments and the lift frame as these parts are rolled back during loading. **These are places where a person can be pinched or crushed.** To avoid pinching and crushing injuries:

- Be sure that bystanders or helpers move away from the front end loader during use.
- Before making needed repairs, lower the front-end loader, shut off the engine, and relieve the hydraulic pressure by moving the hydraulic control lever back and forth a few times.

Figure 6.4.c. Extending the front-end loader with bucket or fork attachment moves bulky loads easily, but also raises the tractor's center of gravity. This can allow the tractor to tip sideways more easily. Use the Operator's Manual to determine safe loading capacity.

Work surroundings

Farm equipment operators must be aware of their surroundings as they go about the work to be done. Before operating the front-end loader check these points:

- Location of fellow workers
- Location of children and pets
- Location of livestock and livestock equipment
- Location of building corners and overhangs
- Location of utility lines

Equipment operators can become so focused on their work they can overlook where other persons or animals have moved in the work zone. Farms have children, bystanders, and pets that may not understand what you are doing or anticipate your movements. Be alert to these situations.

Buildings have received damage from equipment operation. Be sure that you understand the width and height of the front-end loader and any cargo you are carrying in it. Avoid working too near buildings if possible.

Most importantly know the location of overhead power utility lines. Maintaining a safe distance from the utility lines. Contacting power lines with front-end loaders or the cargo can result in electrocution.

To avoid electrocution do not use the front-end loader bucket to dig into the ground unless you know where underground utilities are buried.

Number 1 rule: Keep it low; drive slow.

Figure 6.4.d. To prevent electrocution, lower the front end loader bucket to avoid power lines crossing the work area. *Farm and Ranch Safety Management, John Deere Publishing, 2009. Illustrations reproduced by permission. All rights reserved.*

Cooperation provided by The Ohio State University and National Safety Council.

Safety Activities

1. Answer these questions.

   A. What happens to a tractor's center of gravity when you raise the front-end loader?

   B. Why should the front -end loader be lowered to the ground before you leave the operator's seat?

   C. Describe a situation where load rollback can occur.

2. Become familiar with front-end loader controls by practicing the raising and lowering of the front-end loader and then rolling the bucket or attachment forward and back. Do this until you are not confused what each control movement does as you use it. Request that a supervisor observe your initial efforts.

3. Using a pile of sand or sawdust or mulch, practice scooping up the while observing how you approach the pile of material. Pay attention to how much you scoop into the bucket, and how heavy it is to lift. Slowly move the loaded material to another nearby location and dump it. Request that a supervisor observe your initial efforts.

4. When you have mastered safety activity 2 and 3 above, repeat the activity, but place the load into a dump truck or other container by approaching the dump truck or container slowly, raising the bucket fully, adjusting the final approach and then dumping the load carefully without damage to the truck or container.

References

1. Safety Management for Landscapers, Grounds-Care Businesses and Golf Courses, 2001, First Edition, John Deere Publishing, Moline, Illinois
2. Operator's Manuals from various manufacturers.

Contact Information

National Safe Tractor and Machinery Operation Program
The Pennsylvania State University
Agricultural and Biological Engineering Department
246 Agricultural Engineering Building
University Park, PA 16802
Phone: 814-865-7685
Fax: 814-863-1031
Email: NSTMOP@psu.edu

Credits

Developed by WC Harshman, AM Yoder, JW Hilton and D J Murphy, The Pennsylvania State University. Version 3/2013

This material is based upon work supported by the National Institute of Food and Agriculture, U.S. Department of Agriculture, under Agreement No. 2010-41521-20839. Any opinions, findings, conclusions, or recommendations expressed in this publication are those of the author(s) and do not necessarily reflect the view of the U.S. Department of Agriculture.

Cooperation provided by The Ohio State University and National Safety Council.

DUMP TRUCKS AND TRAILERS–FARM USE ONLY

HOSTA TASK SHEET 6.5

NATIONAL SAFE TRACTOR AND MACHINERY OPERATION PROGRAM

Introduction

Farm producers have increasingly turned to the use of dump trucks and trailers to quickly and efficiently move the inputs and products of the farm.

Serious property damage, crippling injury, and even death can result when these trucks overturn or the dump bed falls onto a worker during repairs or use.

This task sheet identifies risks associated with using dump trucks and trailers and how to avoid common hazards. This task sheet refers only to farm use of dump trucks and trailers and does not discuss road use of these vehicles

In-experienced or untrained operators should not be assigned to operate dump truck and trailers. Age restrictions in CDL licensing requirements may further remove younger employees from being considered for this duty.

Figure 6.5.a. A truck with dump trailer. May be called a semi-trailer, end dump truck.

Figure 6.5.b. A pup trailer.

Figure 6.5.c. A regular dump truck.

Only properly licensed, experienced operators should be assigned to road use of farm dump trucks and trailers.

Types and Uses

Dump trucks and dump trailers are used for many farm chores including hauling grain, silage, firewood, sawdust, wood chips, soil, gravel, sand, debris, and other pulverized or loose items.

Among the types of large trucks and trailers found on farms are semi-trailer end-dump trucks (Figure 6.5.a), pup trailers (Figure 6.5.b), and regular and long-bed straight dump trucks (Figure 6.5.c). Hereafter, these units will be referred to as farm dump trucks and trailers.

Farm dump trucks and trailers use hydraulic hoists to raise the bed during unloading (Figure 6.5.d). Raising the bed allows the contents of the load to slide out of the bed for unloading by gravity.

The remainder of this task sheet discusses the variable conditions, types of load and unloading surfaces and mechanical issues that can create hazards in using farm dump trucks and trailers.

Learning Goals

- To understand the concept of center of gravity of farm dump trucks and its potential to cause overturns.
- To safely use farm dump trucks and trailers during on-farm use.

Related Task Sheets:

Hazard Warning Signs	**2.8**
Mechanical Hazards	**3.1**
Electrical Hazards	**3.6**
Tractor Hazards	**4.2**

Cooperation provided by The Ohio State University and National Safety Council.

Figure 6.5.d. A hydraulic lift cylinder raises the dump bed for the unloading of materials. As the dump truck or trailer bed is raised the center of gravity is raised as well. When unloading, park the truck on a level, firm surface.

Dump trucks have a center of gravity just like a farm tractor.

Dump truck hazards

Three situations present the most potential for a fatal injury involving on-farm use of large farm dump trucks and trailers.

One situation involves trucks and/ or trailers tipping over while emptying loads. These rollovers happen for a variety of reasons and will be discussed in the Instability Hazard section.

The second situation occurs when a person works under a raised dump truck or trailer bed. Usually the person has not physically blocked the dump bed from coming down unexpectedly. The crushing injury from the dump bed or loaded bed is usually fatal.

The third scenario occurs when the dump truck or trailer bed comes into contact with overhead electrical lines where the driver or a ground located helper is electrocuted. This usually happens when the driver pulls away after emptying the load without lowering the bed.

In all three cases, a person can be crushed or electrocuted within seconds. There have been at least eight incidents in Pennsylvania involving farm dump trucks in the past few years. Many incidents go unreported if a fatality is not involved therefore the number of incidents may be much higher.

Instability Hazard

A dump truck or trailer becomes less stable as its bed is raised, especially when the ground is not level or not firm (Figure 6.5.e.). The greater the length and height of the bed, and the greater the degree of slope, the greater is the hazard for tipping over. As the bed is raised, it is important the load center of gravity stay between the frame rails of the bed, preferably right in the center. Even when the ground is relatively flat, a slight slope can be created by one set of tires settling into a hole or deep set of ruts, low tire pressure on one side, or a depression created on one side as an off-center load is unloaded on soft ground. Driving or unloading to close to a ditch, road bank, or material pile edge can cause the rig to tip over as well. It is often a combination of these conditions that result in instability and a tip over.

Off-center and shifting loads often contribute to overturns. These can occur for a variety of reasons including the load not distributed properly when loaded, (e.g., top-heavy or too much on one side) and material not flowing evenly out of the bed (e.g., wet or frozen material stuck to the sides or floor). Materials such as silage, high moisture corn, and damp sand can hang up causing uneven loads.

To reduce risk of overturns:

- unload on a firm level surface
- keep the rig in a straight line; not jack-knifed
- avoid unloading in high winds
- consider a dump bed equipped with two hydraulic cylinders to reduce the chance of overturns

Figure 6.5.e. Raising the dump bed on sloped ground increases the risk of rollover. Longer trailers, windy conditions, and uneven loads add to the risk. Never attempt to unload a rig that is parked in a jack-knifed position.

Use the manufacturer's mechanical block device to prevent the bed from coming down during repairs and maintenance.

Crushing Hazard

Workers have been crushed to death when dump beds come down unexpectedly. Checking an unfamiliar noise or malfunction, performing routine maintenance (e.g. greasing) or repair may mean the operator enters the space between the dump bed and truck frame. In this work position the risk of a crushing fatality can occur if the bed is lowered inadvertently by a worker or co-worker tripping a hydraulic control lever, or by a mechanical or hydraulic component failure.

To reduce the risk of dump bed crushing injury or fatality:

1. Completely understand the dump bed controls and how the system works.
2. Use the manufacturers locking device to secure the dump bed when working between the dump bed and truck frame (Figure 6.5.f).
3. If a problem exists for which you are not trained, immediately call upon a knowledgeable operator to help correct the malfunction.

Electrocution Hazard

Overhead power lines are a hazard. The possibility of electrocution exists with raised dump trucks or trailer beds contacting power lines. This may happen because of forgetfulness, haste or impatience when the driver doesn't want to wait for the bed to completely lower before pulling away (Figure 6.5.g). If contact with power lines occurs, the driver is normally protected from electrocution because they are insulated from the electrical charge by the truck tires. If the driver leaves the cab and is in contact with the ground and the truck they could be electrocuted. Some drivers, either from panic or lack of knowledge, attempt to leave the cab and are electrocuted. A person touching any part of the truck or trailer in contact with an overhead power line may also be electrocuted.

To avoid electrocution risks:

- Know the location of overhead power lines in relation to dump locations
- Have a helper to observe for potential power line contact and to signal you for safe clearances.
- Always move the unit with the bed down.

Figure 6.5.f. A wooden plank is not a suitable blocking device to secure a dump bed.

Figure 6.5.g. Avoid contact with overhead power lines.

Cooperation provided by The Ohio State University and National Safety Council.

Safety Activities

1. Conduct a farm community survey to determine the extent of dump trucks, semi-trailer, end-dump truck and pup trailer use in your community.

2. Conduct a survey of local farmers to find out how they received training in safe dump truck and trailer use.

3. Copy and mail this Task Sheet to local farmers as a community service of your 4-H and/or FFA club.

4. Visit your local Driver Test Station to locate information on weight restrictions and CDL licensing requirements for off-farm or road use of the dump truck and trailer.

5. Build a table top demonstration to show how dump trucks and trailers can tip over on a sloped surface. Use the following plan. Locate a model dump truck or semi-truck end dump trailer to place on the tilting deck. Raise the tilting deck, use a protractor to measure the angle where the truck overturns with the bed down, with the bed raised and with the bed raised with some load involved. The load may be any items you may find to place in a small cloth bag. Damp sand in a plastic baggy could be used for an example. Material that shifts in the raised bed as the angle increases can be demonstrated. For comparison secure a model dump truck with a round bed to compare the results.

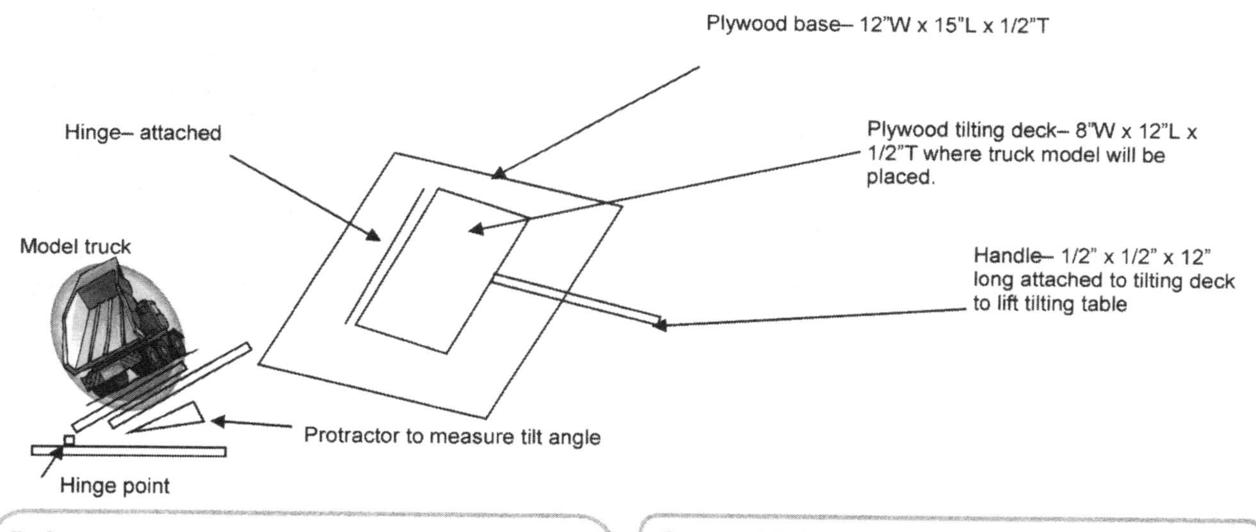

References

1. Website, www.agsafety.psu.edu . Scroll to Publications, E44 Farm Dump Truck and Trailer Safety, Dennis J. Murphy and William C. Harshman

Contact Information

National Safe Tractor and Machinery Operation Program
The Pennsylvania State University
Agricultural and Biological Engineering Department
246 Agricultural Engineering Building
University Park, PA 16802
Phone: 814-865-7685
Fax: 814-863-1031
Email: NSTMOP@psu.edu

Credits

Developed by WC Harshman, AM Yoder, JW Hilton and D J Murphy, The Pennsylvania State University. Version 3/2013

This material is based upon work supported by the National Institute of Food and Agriculture, U.S. Department of Agriculture, under Agreement No. 2010-41521-20839. Any opinions, findings, conclusions, or recommendations expressed in this publication are those of the author(s) and do not necessarily reflect the view of the U.S. Department of Agriculture.